Security in Wireless Mesh Networks

WIRELESS NETWORKS AND MOBILE COMMUNICATIONS

Series Editor: Yan Zhang

AUERBACH PUBLICATIONS

www.auerbach-publications.com
To Order Call: 1-800-272-7737 • Fax: 1-800-374-3401
E-mail: orders@crcpress.com

Security in Wireless Mesh Networks

Edited by
Yan Zhang
Jun Zheng
Honglin Hu

CRC Press
Taylor & Francis Group
Boca Raton   London   New York

CRC Press is an imprint of the
Taylor & Francis Group, an **informa** business

AN AUERBACH BOOK

Auerbach Publications
Taylor & Francis Group
6000 Broken Sound Parkway NW, Suite 300
Boca Raton, FL 33487-2742

© 2009 by Taylor & Francis Group, LLC
Auerbach is an imprint of Taylor & Francis Group, an Informa business

No claim to original U.S. Government works
Printed in the United States of America on acid-free paper
10 9 8 7 6 5 4 3 2 1

International Standard Book Number-13: 978-0-8493-8250-5 (Hardcover)

Library of Congress Cataloging-in-Publication Data

Zhang, Yan, 1977-
    Security in wireless mesh networks / Yan Zhang, Jun Zheng, and Honglin Hu.
        p. cm.
    Includes bibliographical references and index.
    ISBN 978-0-8493-8250-5 (alk. paper)
    1. Wireless communication systems--Security measures. 2. Computer
networks--Security measures. 3. Routers (Computer networks) I. Zheng, Jun,
Ph.D. II. Hu, Honglin, 1975- III. Title.

TK5103.2.Z53 2007
005.8--dc22   2007011243

Visit the Taylor & Francis Web site at
http://www.taylorandfrancis.com

and the Auerbach Web site at
http://www.auerbach-publications.com

Contents

PART III: SECURITY STANDARDS, APPLICATIONS, AND ENABLING TECHNOLOGIES

List of Contributors

Garhan Attebury
University of Nebraska-Lincoln
Lincoln, Nebraska

Noureddine Boudriga
CNAS Research Lab
University of Carthage
Carthage, Tunisia

Thomas M. Chen
Southern Methodist University
Dallas, Texas

Yi Cui
Department of Electrical Engineering
    and Computer Science
Vanderbilt University
Nashville, Tennessee

Falko Dressler
Autonomic Networking Group
Department of Computer Sciences
University of Erlangen
Nuremberg, Germany

A. Antony Franklin
Indian Institute of
    Technology Madras
Chennai, Tamilnadu, India

J.J. Garcia-Luna-Aceves
Computer Engineering
University of California
Santa Cruz, California

Chin-Tser Huang
University of South Carolina
Columbia, South Carolina

Sanjay K. Jha
School of Computer Science
    and Engineering
University of New South Wales
Sydney, Australia

Salil S. Kanhere
School of Computer Science
    and Engineering
University of New South Wales
Sydney, Australia

Moazzam Khan
Manitoba University
Manitoba, Winnipeg, Canada

Neila Krichene
CNAS Research Lab
University of Carthage
Carthage, Tunisia

Geng-Sheng Kuo
Beijing University of Posts
    and Telecommunications
Beijing, China

Zhenjiang Li
Computer Engineering,
University of California, Santa Cruz
Santa Cruz, California

Zheng-Ping Li
Beijing University of Posts
    and Telecommunications
Beijing, China

Jelena Misic
Manitoba University
Manitoba, Winnipeg, Canada

Hassnaa Moustafa
France Telecom R&D
Paris, France

C. Siva Ram Murthy
Indian Institute of
    Technology Madras
Chennai, Tamilnadu, India

Anjum Naveed
School of Computer Science
    and Engineering
University of New South Wales
Sydney, Australia

Bart Preneel
Department of Electrical
    Engineering
Katholieke Universiteit
Leuven, Belgium

Shah Rahman
Cisco Systems
San Jose, California

Byrav Ramamurthy
University of Nebraska-Lincoln
Lincoln, Nebraska

Stefaan Seys
Department of Electrical Engineering
Katholieke Universiteit
Leuven, Belgium

Dave Singelée
Department of Electrical
    Engineering
Katholieke Universiteit
Leuven, Belgium

Yong Wang
University of Nebraska-Lincoln
Lincoln, Nebraska

Nancy-Cam Winget
Cisco Systems
San Jose, California

Taojun Wu
Department of Electrical Engineering
    and Computer Science
Vanderbilt University
Nashville, Tennessee

Yuan Xue
Department of Electrical Engineering
    and Computer Science
Vanderbilt University
Nashville, Tennessee

Manel Guerrero Zapata
Technical University
    of Catalonia
Barcelona, Spain

Guo-Mei Zhu
Beijing University of Posts
    and Telecommunications
Beijing, China

INTRODUCTION

I

Chapter 1

An Introduction to Wireless Mesh Networks

A. Antony Franklin and C. Siva Ram Murthy

Contents

Wireless mesh networking has emerged as a promising concept to meet the challenges in next-generation wireless networks such as providing flexible, adaptive, and reconfigurable architecture while offering cost-effective solutions to service providers. Several architectures for wireless mesh networks (WMNs) have been proposed based on their applications [1]. One of the most general forms of WMNs interconnects the stationary and mobile clients to the Internet efficiently by the core nodes in multi-hop fashion. The core nodes are the mesh routers which form a wireless mesh backbone among them. The mesh routers provide a rich radio mesh connectivity which significantly reduces the up-front deployment cost and subsequent maintenance cost. They have limited mobility and forward the packets received from the clients to the gateway router which is connected to the backhaul network/Internet. The mesh backbone formed by mesh routers provides a high level of reliability. WMNs are being considered for a wide variety of applications such as backhaul connectivity for cellular radio access networks, high-speed metropolitan area mobile networks, community networking, building automation, intelligent transport system networks, defense systems, and citywide surveillance systems. Prior efforts on wireless networks, especially multi-hop ad hoc networks, have led to significant research contributions that range from fundamental results on theoretical capacity bounds to development of efficient routing and transport layer protocols. However, the recent work is on deploying sizable WMNs and other important aspects such as network radio range, network capacity, scalability, manageability, and security. There are a number of research issues in different layers of the protocol stack and a number of standards are coming up for the implementation of WMNs for WANs, MANs, LANs, and PANs. The mesh networking testbeds by industries and academia further enhanced the research in WMNs. The mesh networking products by different vendors are making WMNs a reality.

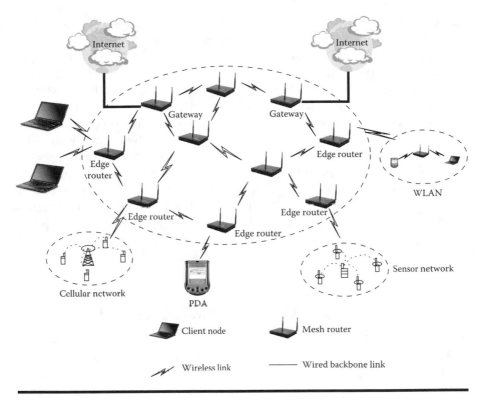

Figure 1.1 Architecture of a wireless mesh network.

1.1 Introduction

WMNs are multi-hop wireless networks formed by mesh routers and mesh clients. These networks typically have a high data rate and low deployment and maintenance overhead. Mesh routers are typically stationary and do not have energy constraints, but the clients are mobile and energy constrained. Some mesh routers are designated as gateway routers which are connected to the Internet through a wired backbone. A gateway router provides access to conventional clients and interconnects ad hoc, sensor, cellular, and other networks to the Internet, as shown in Figure 1.1. A mesh network can provide multi-hop communication paths between wireless clients, thereby serving as a community network, or can provide multi-hop paths between the client and the gateway router, thereby providing broadband Internet access to clients. As there is no wired infrastructure to deploy in the case of WMNs, they are considered cost-effective alternatives to WLANs (wireless local area networks) and backbone networks to mobile clients. The existing wireless networking technologies such as IEEE 802.11, IEEE 802.15, IEEE 802.16, and IEEE 802.20 are used in the implementation of WMNs. The IEEE 802.11 is a set of WLAN standards that define many aspects of wireless networking. One such aspect is mesh networking, which is currently under development by the IEEE 802.11 Task Group. Recently, there has been growing research and practical interest in WMNs. There are numerous ongoing projects on wireless mesh networks in academia, research labs, and companies. Many academic institutions developed their own testbed for research purposes. These efforts are toward developing various applications of WMNs such as home, enterprise, and community networking. As the WMNs use multi-hop paths between client nodes or between a client and a gateway router, the existing protocols for multi-hop ad hoc wireless networks are well suited for WMNs. The ongoing work in WMNs is on increasing the throughput and developing efficient protocols by utilizing the static nature of the mesh routers and topology.

1.1.1 Single-Hop and Multi-Hop Wireless Networks

Generally, wireless networks are classified as single-hop and multi-hop networks. In a single-hop network, the client connects to the fixed base station or access point directly in one hop. The well-known examples of single-hop wireless networks are WLANs and cellular networks. WLANs contain special nodes called access points (APs), which are connected to existing wired networks such as Ethernet LANs. The mobile devices are connected to the AP through a one-hop wireless link. Any communication between mobile devices happens via AP. In the case of cellular networks, the geographical area to be covered is divided into cells which are usually considered to be hexagonal. A base station (BS) is located in the center of the cell and the mobile terminals in that cell communicate with it in a single-hop fashion. Communication between any two mobile terminals happens through one or more BSs. These networks are called infrastructure wireless networks because they are infrastructure (BS) dependent. The path setup between two clients (mobile nodes), say node A and node B, is completed through the BS, as shown in Figure 1.2.

In a multi-hop wireless network, the source and destination nodes communicate in a multi-hop fashion. The packets from the source node traverse through one or more intermediate/relaying nodes to reach the destination. Because all nodes in the network also act as routers, there is no need for a BS or any other dedicated infrastructure. Hence, such networks are also called infrastructure-less networks. The well-known forms of multi-hop networks are ad hoc networks, sensor networks, and WMNs. Communication between two nodes, say node C and node F, takes place through the relaying nodes D and E, as shown in Figure 1.3.

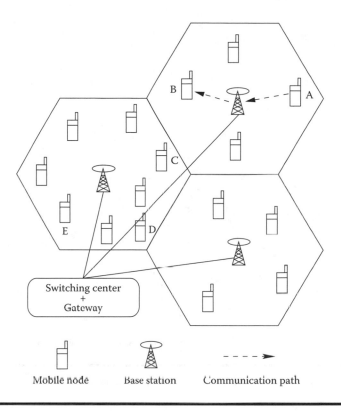

Figure 1.2   Single-hop network scenario (cellular network).

In the case of single-hop networks, complete information about the clients is available at the BS and the routing decisions are made in a centralized fashion, thus making routing and resource management simple. But it is not the case in multi-hop networks. All the mobile nodes have to coordinate among themselves for communication between any two nodes. Hence, routing and resource management are done in a distributed way.

1.1.2   Ad hoc Networks and WMNs

In ad hoc networks, all the nodes are assumed to be mobile and there is no fixed infrastructure for the network. These networks find applications where fixed infrastructure is not possible, such as military operations in the battlefield, emergency operations, and networks of handheld devices. Because of lack of infrastructure the nodes have to cooperate among themselves to form a network. Due to mobility of the nodes in the network, the network topology changes frequently. So the protocols for ad hoc networks have to handle frequent changes in the topology. In most of the applications of ad hoc networks, the mobile devices are energy constrained as

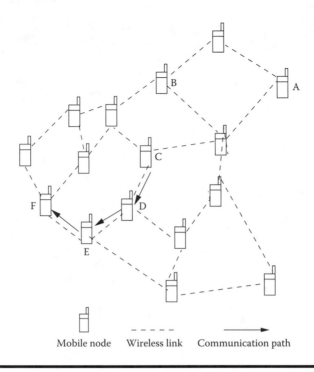

Mobile node   Wireless link   Communication path

Figure 1.3   Multi-hop network scenario (ad hoc network).

they are operating on battery. This requires energy-efficient networking solutions for ad hoc networks. But in the case of WMNs, mesh routers are assumed to be fixed (or have limited mobility) and form a fixed mesh infrastructure. The clients are mobile or fixed and utilize the mesh routers to communicate to the backhaul network through the gateway routers and to other clients by using mesh routers as relaying nodes. These networks find applications where networks of fixed wireless nodes are necessary. There are several architectures for mesh networks, depending on their applications. In the case of infrastructure backbone networking, the edge routers are used to connect different networks to the mesh backbone and the intermediate routers are used as multi-hop relaying nodes to the gateway router, as shown in Figure 1.1. But in the case of community networking, every router provides access to clients and also acts as a relaying node between mesh routers.

1.2 Architecture of WMNs

There are two types of nodes in a WMN called mesh routers and mesh clients. Compared to conventional wireless routers that perform only routing, mesh routers have additional functionalities to enable mesh networking. The mesh routers have multiple interfaces of the same or different communications technologies based on the requirement. They achieve more coverage with the same transmission power by using multi-hop communication through other mesh routers. They can be built on general-purpose computer systems such as PCs and laptops, or can be built on dedicated hardware platforms (embedded systems). There are a variety of mesh clients such as laptop, desktop, pocket PCs, IP phones, RFID readers, and PDAs. The mesh clients have mesh networking capabilities to interact with mesh routers, but they are simpler in hardware and software compared to mesh routers. Normally they have a single communication interface built on them. The architecture of WMNs (shown in Figure 1.1) is the most common architecture used in many mesh networking applications such as community networking and home networking. The mesh routers shown have multiple interfaces with different networking technologies which provide connectivity to mesh clients and other networks such as cellular and sensor networks. Normally, long-range communication techniques such as directional antennas are provided for communication between mesh routers. Mesh routers form a wireless mesh topology that has self-configuration and self-healing functions built into them. Some mesh routers are designated as gateways which have wired connectivity to the Internet. The integration of other networking technologies is provided by connecting the BS of the network that connects to WMNs to the mesh routers. Here, the clients communicate to the BS of its own network and the BS in turn communicates to the mesh router to access the WMN.

1.3 Applications of WMNs

WMNs introduce the concept of a peer-to-peer mesh topology with wireless communication between mesh routers. This concept helps to overcome many of today's deployment challenges, such as the installation of extensive Ethernet cabling, and enables new deployment models. Deployment scenarios that are particularly well suited for WMNs include the following:

- Campus environments (enterprises and universities), manufacturing, shopping centers, airports, sporting venues, and special events
- Military operations, disaster recovery, temporary installations, and public safety
- Municipalities, including downtown cores, residential areas, and parks
- Carrier-managed service in public areas or residential communities

Due to the recent research advances in WMNs, they have been used in numerous applications. The mesh topology of the WMNs provides many alternative paths for any pair of source and destination nodes, resulting in quick reconfiguration of the path when there is a path failure. WMNs provide the most economical data transfer coupled with freedom of mobility. Mesh routers can be placed anywhere such as on the rooftop of a home or on a lamppost to provide connectivity to mobile/static clients. Mesh routers can be added incrementally to improve the coverage area. These features of WMNs attract the research community to use WMNs in different applications:

■ Home Networking: Broadband home networking is a network of home appliances (personal computer, television, video recorder, video camera, washing machine, refrigerator) realized by WLAN technology. The obvious problem here is the location of the access point in the home, which may lead to dead zones without service coverage. More coverage can be achieved by multiple access points connected using Ethernet cabling, which leads to an increase in deployment cost and overhead. These problems can be solved by replacing all the access points by the mesh routers and establishing mesh connectivity between them. This provides broadband connectivity between the home networking devices and only a single connection to the Internet is needed through the gateway router. By changing the location and number of mesh routers, the dead zones can be eliminated. Figure 1.4 shows a typical home network using mesh routers.

■ Community and Neighborhood Networking: The usual way of establishing community networking is connecting the home network/PC to the Internet with a cable or DSL modem. All the traffic in community networking goes through the Internet, which leads to inefficient utilization of the network resources. The last mile of wireless connectivity might not provide coverage outside the home. Community networking by WMNs solves all these problems and provides a cost-effective way to share Internet access and other network resources among different homes. Figure 1.5 shows wireless mesh networking by placing the mesh routers on the rooftop of houses. There are many advantages to enabling such mesh connectivity to form a community mesh network. For example, when enough neighbors cooperate and forward each others' packets, they do not need individual Internet connectivity; instead they can get faster, cost-effective Internet access via gateways distributed in their neighborhood. Packets dynamically find a route, hopping from one neighbor's node to another to reach the Internet through one of these gateways. Another advantage is that neighbors can cooperatively deploy backup technology and never have to worry about losing information due to a

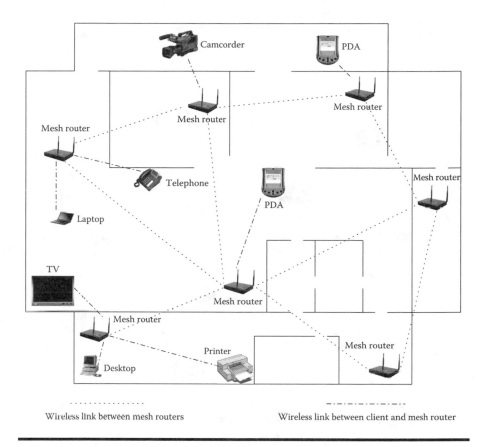

Figure 1.4 Wireless mesh network-based home networking.

catastrophic disk failure. Another advantage is that this technology alleviates the need for routing traffic belonging to community networking through the Internet. For example, distributed file storage, distributed file access, and video streaming applications in the community share network resources in the WMNs without using the Internet, which improves the performance of these applications. Neighborhood community networks allow faster and easier dissemination of cached information that is relevant to the local community. Mesh routers can be easily mounted on rooftops or windows and the client devices get connected to them in a single hop.

■ Security Surveillance System: As security is turning out to be of very high concern, security surveillance systems are becoming a necessity for enterprise buildings and shopping malls. The security surveillance system needs high bandwidth and a reliable backbone network to communicate surveillance information, such as images, audio, and

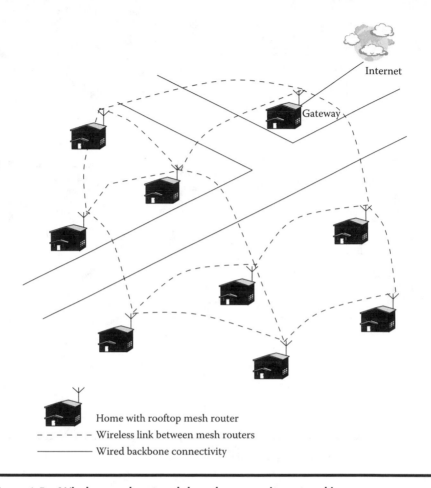

Figure 1.5   Wireless mesh network-based community networking.

video, and low-cost connectivity between the surveillance devices. The recent advances of WMNs provide high bandwidth and reliable backbone connectivity and an easy way of connecting surveillance devices located in different places with low cost.

■ Disaster Management and Rescue Operations: WMNs can be used in places where spontaneous network connectivity is required, such as disaster management and emergency operations. During disasters like fire, flood, and earthquake, all the existing communication infrastructures might be collapsed. So during the rescue operation, mesh routers can be placed at the rescue team vehicle and different locations which form the high-bandwidth mesh backbone network, as shown in Figure 1.6. This helps rescue team members to communicate with each other. By providing different communication

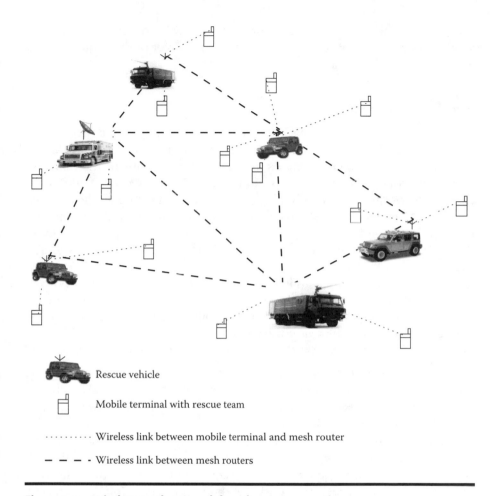

Figure 1.6 Wireless mesh network-based rescue operation.

interfaces at the mesh routers, different mobile devices get access to the network. This helps people to communicate with others when they are in critical situations. These networks can be established in less time, which makes the rescue operation more effective.

1.4   Issues in WMNs

Various research issues in WMNs are described in this section. As WMNs are also multi-hop wireless networks like ad hoc networks, the protocols developed for ad hoc networks work well for WMNs. Many challenging issues in ad hoc networks have been addressed in recent years. WMNs have inherent characteristics such as a fixed mesh backbone formed by mesh routers, resource-rich mesh routers, and resource-constrained clients compared to ad hoc networks. Due to this, WMNs require considerable work to address the problems that arise in each layer of the protocol stack and system implementation.

1.4.1 Capacity

The primary concern of WMNs is to provide high-bandwidth connectivity to community and enterprise users. In a single-channel wireless network, the capacity of the network degrades as the number of hops or the diameter of the network increases due to interference. The capacity of the WMN is affected by many factors such as network architecture, node density, number of channels used, node mobility, traffic pattern, and transmission range. A clear understanding of the effect of these factors on capacity of the WMNs provides insight to protocol design, architecture design, and deployment of WMNs.

In [2] Gupta and Kumar analytically studied the upper and lower bounds of the capacity of wireless ad hoc networks. They showed that the throughput capacity of the nodes reduces significantly when node density increases. The maximum achievable throughput of randomly placed n identical nodes each with a capacity of W bits/second is $\Theta(\frac{W}{\sqrt{n*log(n)}})$ bits/second under a non-interference protocol. Even under optimal circumstances the maximum achievable throughput is only $\Theta(\frac{W}{\sqrt{n}})$ bits/second. The capacity of the network can be increased by deploying relaying nodes and using a multi-hop path for transmission.

The IEEE 802.11 standard [4] provides a number of channels in the available radio spectrum, but some of them may be interfering with each other. If the interfering channels are used simultaneously, then the data gets corrupted at the receiving end. But the non-overlapping channels can be used simultaneously by different nodes in the same transmission range without any collision of the data. IEEE 802.11b [6] provides 3 such non-overlapping channels at 2.4 GHz band and IEEE 802.11a [5] provides 13 non-overlapping channels at 5 GHz band. These orthogonal channels can be used simultaneously at different nodes in the network to improve the capacity of the network. In multi-channel multi-radio communication each node is provided with more than one radio interface (say m) and each interface is assigned one of the orthogonal channels available (say n). If each node has n number of radio interfaces ($m = n$) and each orthogonal channel is assigned to one interface, then the network can achieve n-fold increase in capacity because the n interfaces can transmit simultaneously without any interference with each other. But normally the number of interfaces is less than the number of available channels ($m < n$) due to the cost of the interfaces and the complexity of the nodes. In this case an m-fold increase in capacity can be achieved by assigning m interfaces with m

different orthogonal channels. Moreover, when $m < n$ the capacity bound of a multi-channel multi-radio wireless mesh network depends on the ratio of n and m [7].

1.4.2   Physical Layer

The network capacity mainly depends on the physical layer technique used. There are a number of physical layer techniques available with different operating frequencies and they provide different transport capacity in wireless communications. Some existing wireless radios even provide multiple transmission rates by different combinations of modulation and coding techniques [6]. In such networks, the transmission rate is chosen by link adaptation techniques. Normally, link signal-to-noise ratio (SNR) or carrier-to-noise ratio (CNR) from the physical layer is considered for link adaptation, but this alone does not describe the signal quality in the environment like frequency-selective fading channel. To overcome the problems with RF transmission, other physical layer techniques have been used for wireless communications. Some high-speed physical layer techniques are available which improve the capacity of the wireless networks significantly. Some of the techniques for improving the capacity of WMNs are described in this section.

■ Orthogonal Frequency Division Multiplexing (OFDM): The OFDM technique is based on the principle of Frequency Division Multiplexing (FDM) with digital modulation schemes. The bit stream to be transmitted is split into a number of parallel low bit rate streams. The available frequency spectrum is divided into many sub-channels and each low bit rate stream is transmitted by modulating over a sub-channel using a standard modulation scheme such as Phase Shift Keying (PSK) and Quadrature Amplitude Modulation (QAM). The primary advantage of OFDM is its ability to work under severe channel conditions, such as multi-path and narrow-band interference, without complex equalization filters at the transmitter and receiver. The OFDM technique has increased the transmission rate of IEEE 802.11 networks from 11 to 54 Mbps.

■ Ultra Wide Band (UWB): UWB technology provides much higher data rate (ranges from 3 to 10 GHz) compared to other RF transmission technologies. A significant difference between traditional radio transmission and UWB radio transmission is that traditional radio transmission transmits information by varying the power, frequency, or phase in distinct and controlled frequencies while UWB transmission transmits information by generating radio energy at specific times with a broad frequency range. Due to this, UWB transmission is immune to multi-path fading and interference,[1] which are common in any radio transmission technique. UWB wireless links have the characteristic that the bandwidth decreases rapidly as the distance increases. On the other hand UWB provides hundreds of non-interfering channels within radio range of each other. Hence, UWB is applicable for only short-range communications such as WPAN. Mesh architecture combined with UWB wireless technology allows a very easy installation of communications infrastructure in offices or homes by deploying many repeater modules every 10 meters. As these repeater modules require power to operate on, they have to be placed with ceiling lights or floor power boxes. The IEEE 802.15 TG4a standard for WPAN uses a UWB physical layer technique consisting of a UWB impulse radio (operating in unlicensed UWB spectrum) and a chirp spread spectrum (operating in unlicensed 2.4 GHz spectrum).

■ Multiple-Input Multiple-Output (MIMO): The use of multiple antennas at the transmitter and receiver, popularly known as MIMO wireless, is an emerging, cost-effective technology that makes high bandwidth wireless links a reality. MIMO significantly increases the throughput and range with the same bandwidth and overall transmission power expenditure. This increase in throughput and range is by exploiting the multi-path propagation phenomena in wireless communications. In general, the MIMO technique increases the spectral efficiency of a wireless communications system. It has been shown by Telatar that the channel capacity (a theoretical upper bound on system throughput) for a MIMO system increases as the number of antennas increases, proportional to the minimum of transmitter and receiver antennas [8]. MIMO can also be used in conjunction with OFDM and is part of the IEEE 802.16 standard.

■ Smart Antenna: The smart antenna technique improves the capacity of wireless networks by adding the directionality for transmission and reception of signals at the transmitter and receiver antenna. This also helps in increasing energy efficiency. In cellular networks, due to complexity and cost of smart antennas, it is implemented in BS alone. The directional antenna system is actively researched in ad hoc networks also. There are some directional antenna systems available that can be tuned to certain directions by electronic beam forming. This technique improves the performance of wireless

---

[1] In RF transmission, when the transmitted signal is reflected by mountains or buildings the radio signal reaches the receiving antenna along two or more paths. The effect of this multi-path reception includes constructive and destructive interference and phase shifting of the signal.

networks by reducing interference between the transmissions of different nodes in the network. But the use of a directional antenna necessitates special MAC (Medium Access Control) protocols to support directionality in transmission and reception.

1.4.3 Medium Access Scheme

The MAC (Medium Access Control) protocols for wireless networks are limited to single-hop communication while the routing protocols use multi-hop communication. The MAC protocols for WMNs are classified into single-channel and multi-channel MAC. They are discussed in this section.

- Single-Channel MAC: There are several MAC schemes which use single-channel for communication in the network. They are further classified as (1) contention-based protocols, (2) contention-based protocols with a reservation mechanism, and (3) contention-based protocols with a scheduling mechanism.
  - Contention-based protocols: These protocols have a contention-based channel access policy among the nodes contending for the channel. All the ready nodes in the network start contending for the channel simultaneously and the winning node gains access to the channel. As the nodes cannot provide guaranteed bandwidth, these protocols cannot be used in carrying real-time traffic, which requires QoS (quality of service) guarantees from the system. Some of the contention-based protocols are MACAW (a media access protocol for Wireless LANs) [9], FAMA (Floor Acquisition Multiple Access protocol) [10], BTMA (Busy Tone Multiple Access protocol) [11], and MACA-BI (Multiple Access Collision Avoidance By Invitation) [12].
  - Contention-based protocols with a reservation mechanism: Because the contention-based protocols cannot provide guaranteed access to the channel, they cannot be used in networks where real-time traffic has to be supported. To support real-time traffic, some protocols reserve the bandwidth *a priori*. Such protocols can provide QoS support for time-sensitive traffic. In this type of protocol, the contention occurs during the resource (bandwidth) reservation phase. Once the bandwidth is reserved, the nodes get exclusive access to the reserved bandwidth. Hence, these protocols can provide QoS support for time-sensitive traffic. Some of the examples for these type of protocols are D-PRMA (Distributed Packet Reservation Multiple Access protocol) [13], CATA (Collision Avoidance Time Allocation

protocol) [14], HRMA (Hop Reservation Multiple Access protocol) [15], and RTMAC (Real-Time Medium Access protocol) [16].

■ Contention-based protocols with scheduling mechanism: These protocols focus on packet scheduling at nodes and also scheduling nodes for access to the channel. The scheduling is done in such a way that all nodes are treated fairly and no node is starved of bandwidth. These protocols can provide priorities among flows whose packets are queued at nodes. Some of the existing scheduling-based protocols are DWOP (Distributed Wireless Ordering Protocol) [17], DLPS (Distributed Laxity-based Priority Scheduling) [18], and DPS (Distributed Priority Scheduling) [19].

Contention-based protocols that use single-channel for communication cannot completely eliminate contention for the channel. In the case of WMNs the end-to-end throughput significantly reduces due to the accumulating effect of the contention in the multi-hop path. Further, an ongoing transmission between a pair of nodes refrains all the nodes which are in a two-hop neighborhood of nodes participating in the transmission from transmitting on the channel during the transmission period. To overcome these problems multi-channel MAC and multi-channel multi-radio MAC protocols are proposed.

■ Multi-Channel MAC (MMAC): Multi-channel MAC [20] is a link layer protocol where each node is provided with only one interface, but to utilize the advantage of multi-channel communication, the interface switches among different channels automatically. In MMAC the communication time is split into a number of beacon intervals. In the beginning of each beacon interval, during an ATIM (Ad hoc Traffic Indication Message) window period all the nodes in the network tune their radio to a common control channel and negotiate for the channel to be used for the remaining period of the beacon interval. Each node maintains a data structure called PCL (Preferred Channel List — usage of the channels within the transmission range of the node). When a source node S1 wants to send data to receiver node R1, during the ATIM window node S1 sends an ATIM packet with its PCL. Upon receiving the ATIM packet from node S1, node R1 compares the PCL of node S1 with its PCL and decides which channel is to be used during the beacon interval. Then node R1 sends an ATIM-ACK carrying the ID of the preferred channel. Node S1, on receiving the ATIM-ACK, confirms the reservation by sending an ATIM-RES packet to node R1. When other nodes in the vicinity of node R1 hear the ATIM-ACK, they choose a different channel for their communication. The throughput of MMAC is higher than that of IEEE 802.11 when the network load is high. This increase in throughput is due to the fact that each node uses an orthogonal channel, thereby increasing the number of simultaneous transmissions in the network. Though MMAC increases the throughput, there are some drawbacks with it. When a node has to send a packet to multiple destinations, it can send only to one destination in a beacon interval, because the nodes have to negotiate during the ATIM window in the control channel. Due to this restriction the per-packet delay increases significantly. MMAC does not have any scheme for broadcasting.

Slotted Seeded Channel Hopping protocol (SSCH) is another multi-channel link layer protocol using a single transceiver [21]. SSCH is implemented in software over an IEEE 802.11-compliant wireless Network Interface Card (NIC). SSCH uses a distributed mechanism for coordinating the channel switching decision. By this channel hopping at each node, packets of multiple flows in the interfering range of each other are transmitted simultaneously in an orthogonal channel. This improves the overall capacity of the multi-hop wireless network if the network traffic pattern has multiple flows in the interfering range of each other. Each node in the network finds the channel hopping schedule for it and schedules the packets within each channel. Each node transmits its channel hopping schedule to all its neighboring nodes and updates its channel hopping schedule based on traffic pattern. SSCH yields significant capacity improvement in both single-hop and multi-hop network scenarios.

■ Multi-Radio Multi-Channel MAC: In the application scenarios where the cost of the node and power consumption are not big issues, nodes can be provided with multiple wireless interfaces which are tuned to non-overlapping channels and can communicate simultaneously with multiple neighboring nodes. If nodes have multiple interfaces, then the MAC protocol has to handle the orthogonal channel assignment to each interface and schedule the packets to the appropriate interface. The Multi-radio Unification Protocol (MUP) [22] is one such protocol to coordinate the operation of the multiple wireless NICs tuned to non-overlapping channels. MUP works as a virtual MAC which requires no changes to the higher layer protocols and works with other nodes which do not have MUP. So these type of nodes can be added incrementally even after deployment. For the higher layer protocols the MUP looks like a single MAC running. It monitors the channel quality on each of the NICs to each of its neighbors. When the higher layer protocol sends packets to the MUP, it selects the right interface to forward the packets.

Kyasanur and Vaidya [23,24] proposed a link layer protocol for the scenario of nodes having more than one interface. The interfaces of a node are grouped into two fixed interfaces where interfaces are assigned a channel for long intervals of time and switchable interfaces where interfaces are assigned dynamically for short spans of time. The channel assigned to fixed interfaces is called a fixed channel and that assigned to switchable interfaces is called a switchable channel. Each node has both a fixed channel and a switchable channel. During a flow initiation, each node finds the channel in the switchable interface based on the fixed channel of the next-hop neighbor to transmit the data to it. Once the switchable interfaces are switched to a channel there is no need for switching the channel for the subsequent packets for that flow unless another flow requires channel switching on the switchable interface.

1.4.4 Routing

There are numerous routing protocols proposed for ad hoc networks in the literature. Because WMNs are multi-hop networks, the protocols designed for ad hoc networks also work well for WMNs. The main objective of those protocols is quick adaptation to the change in a path when there is path break due to mobility of the nodes. Current deployments of WMNs make use of routing protocols proposed for ad hoc networks such as AODV (Ad hoc On-Demand Distance Vector) [25], DSR (Dynamic Source Routing) [26], and TBRPF (Topology Broadcast based on Reverse Path Forwarding) [27]. However, in WMNs the mesh routers have minimal mobility and there is no power constraint, whereas the clients are mobile with limited power. Such difference needs to be considered in developing efficient routing protocols for WMNs. As the links in the WMNs are long lived, finding a reliable and high throughput path is the main concern rather than quick adaptation to link failure as in the case of ad hoc networks.

1.4.4.1 Routing Metrics for WMNs

Many ad hoc routing protocols such as AODV and DSR use hop count as a routing metric. This is not well suited for WMNs for the following reasons. The basic idea in minimizing the hop count for a path is that it reduces the packet delay and maximizes the throughput. But the assumption here is that links in the path either work perfectly or do not work at all and all links are of equal bandwidth. A routing scheme that uses the hop count metric does not take the link quality into consideration. A minimum hop count path has higher average distance between nodes present in that path compared to a higher hop count path. This reduces the strength of the signal received by the nodes in that path and thereby increases the loss ratio at each link [28]. Hence, it is always possible that a two-hop path with good link quality provides higher throughput than a one-hop path with a poor/lossy link. A routing scheme that uses the hop count metric always chooses a single-hop path rather than a two-hop path with good link quality. The wireless links usually have asymmetric loss rate as reported in [29]. Hence, new routing metrics based on the link quality are proposed in the literature. They are ETX (Expected Transmission Count), per-hop RTT (Round-Trip Time), and per-hop packet pair. Couto et al. proposed ETX to find a high throughput path in WMNs [28]. The metric ETX is defined as the expected number of transmissions (including retransmissions) needed to successfully deliver a packet over a link. As per IEEE 802.11 standard, a successful transmission requires acknowledgment back to the sender. ETX considers transmission loss probability in both directions, which may not be equal as stated earlier. All nodes in the network compute the loss probability to and from its neighbors by sending probe packets. If p_f and p_r are respectively the loss probability in forward and reverse direction in a link, then the probability that a packet transmission is not successful in a link is given by $p = 1 - (1 - p_f)(1 - p_r)$. The expected number of transmissions on that link is computed as $ETX = \frac{1}{1-p}$. In [30] the routing metrics based on link quality are compared with the hop count metric. The routing metric based on link quality performs better than hop count if nodes are stationary. The hop count metric outperforms the link quality metric if nodes are mobile. The main reason for this is that the ETX metric cannot quickly track the changes in the value of the metric. If the nodes are mobile, the ETX value changes frequently as the distance between the nodes changes.

As stated earlier, to improve the throughput the multi-radio multi-channel architecture is used in WMNs. In this case the routing metric based on link quality alone is not sufficient. It should also consider the channel diversity on the path. A new routing metric WCETT (Weighted Cumulative Expected Transmission Time) is proposed in [31], which takes both link quality and channel diversity into account. The link quality is measured by a per-link metric called ETT (Expected Transmission Time; expected time to transmit a packet of a certain size over a link). If the size of the packet is S and the bandwidth of the link is B, then $ETT = ETX * \frac{S}{B}$. The channel diversity in the path is measured as follows. If X_j is the sum of ETTs of the links using the channel j in the path, then channel diversity is measured as $max_{1 \le j \le k} X_j$, where k is the number of orthogonal channels used. The path metric for path p with n links and k orthogonal channels is calculated as

$$WCETT(p) = (1 - \beta) * \sum_{i=1}^{n} ETT_i + \beta * max_{1 \le j \le k} X_j,$$

where β is a tunable parameter subject to $0 \le \beta \le 1$. WCETT can achieve a good trade-off between delay and throughput as it considers both link quality and channel diversity in a single routing metric.

The WCETT metric considers the quality of links and the intra flow interference along the path. But it fails to take into account inter flow interference on the path. In [32], a new routing metric MIC (Metric of Interference and Channel switching) is proposed for multi-channel multi-radio WMNs. This new metric considers the quality of links, inter flow interference, and intra flow interference altogether. This metric is based on Interference-Aware Resource Usage (IRU) and Channel Switching Cost (CSC) metrics to find the MIC for a given path. IRU captures the differences in the transmission rate and the loss ratios of the wireless link and the inter flow interference. The IRU metric for a link k which uses channel c is calculated as $IRU_k(c) = ETT_k(c) * N_k(c)$, where $ETT_k(c)$ is the expected transmission time of the link k on the channel c, and $N_k(c)$ is the number of nodes interfering with the transmission of the link k on channel c. The CSC metric captures the intra flow interference along the path. CSC for a node i is assigned a weight w_1 if links in the path connected to it have different channels assigned, and w_2 if they are the same, $0 \leq w_1 < w_2$. The path metric for a given path p, MIC(p), is calculated as follows:

$$MIC(p) = \alpha * \sum_{(link\ l\ \varepsilon\ p)} IRU_l + \sum_{(node\ i\ \varepsilon\ p)} CSC_i.$$

Here α is a positive factor which gives a trade-off between benefits of IRU and CSC.

1.4.4.2 Routing Protocols for WMNs

In [30], the authors proposed an LQSR (Link Quality Source Routing) protocol. It is based on DSR and uses ETX as the routing metric. The main difference between LQSR and DSR is getting the ETX metric of each link to find out the path. During the route discovery phase, the source node sends a Route Request (RREQ) packet to neighboring nodes. When a node receives the RREQ packet, it appends its own address to the source route and the ETX value of the link in which the packet was received. The destination sends the Route Reply (RREP) packet with a complete list of links along with the ETX value of those links. Because the link quality varies with time, LQSR also propagates the ETX value of the links during the data transmission phase. On receiving a data packet, an intermediate node in the path updates the source route with the ETX value of the outgoing link. Upon receiving the packet, the destination node sends an explicit RREP packet back to the source to update the ETX value of links in the path. LQSR also uses a proactive mechanism to update the ETX metric of all links by piggybacking Link-Info messages to RREQ messages occasionally. This Link-Info message contains the ETX value of all the links incident on the originating node.

A new routing protocol for multi-radio multi-channel WMNs called Multi-Radio Link Quality Source Routing (MR-LQSR) is proposed in [31], which uses WCETT as a routing metric. The neighbor node discovery and propagating the link metric to other nodes in the network in MR-LQSR are the same as that in the DSR protocol. But assigning the link weight and finding the path weight using the link weight are different from DSR. DSR uses equal weight to all links in the network and implements the shortest path routing. But MR-LQSR uses a WCETT path metric to find the best path to the destination.

In [32], the authors showed that, if a WCETT routing metric is used in a link state routing protocol, it is not satisfying the isotonicity property of the routing protocol and leads to formation of routing loops. To avoid the formation of routing loops by the routing metrics, they proposed Load and Interference Balanced Routing Algorithm (LIBRA) [32], which uses MIC as the routing metric. In LIBRA a virtual network is formed from the real network and decomposed the MIC metric into isotonicity link weight assignment on the virtual network. The objective of MIC decomposition is to ensure that LIBRA can use efficient algorithms such as Bellman–Ford or Dijkstra's algorithm to find the minimum weight path on the real network without any forwarding loops.

1.4.5 Transport Layer

There are several reliable transport protocols proposed for ad hoc networks. Some of them are modified versions of TCP (Transmission Control Protocol) that work well in ad hoc networks and others are designed specifically for an ad hoc network scenario from scratch.

TCP is the de facto standard for end-to-end reliable transmission of data on the Internet. TCP was designed to run efficiently on wireline networks. Using the TCP protocol on a wireless network degrades the performance of the network in terms of reduction in throughput and unfairness to the connections. This degradation in performance is due to the following reasons. The Bit Error Rate (BER) in wireless networks is very high compared to wireline networks. Frequency of path break in wireless networks is high due to mobility of nodes in ad hoc networks. If the packets get dropped in the network due to these reasons, the TCP sender misinterprets this event as congestion and triggers the congestion control mechanism to reduce the congestion window size. This reduces the effective throughput of the network.

■ TCP Variants for Wireless Networks: To solve the problem of degradation of throughput of TCP over wireless networks, various modifications to TCP protocols have been proposed. These modifications are mainly based on differentiating the congestion loss and non-congestion loss at the TCP sender when there is a packet loss in the network. The proposed protocols [33] and [34] rely on cooperation from the network, i.e., the intermediate nodes inform the source regarding the status of a path. In ELFN (Explicit Link Failure Notification) [33], the intermediate node informs the sender about the link failure explicitly. When the sender is informed that the link has failed, it disables its retransmission timer and enters into standby mode. In the standby mode the sender probes the network to check if the network connection is re-established by sending a packet from the congestion window periodically. Upon receiving an ACK from the receiver, i.e., after the connection is established, the sender resumes its normal operation. In TCPF (TCP-Feedback) [34], when an intermediate node detects path break, it sends an RFN (Route Failure Notification) message to the TCP sender. On receiving an RFN message, the TCP sender goes to snooze state. In this state the TCP sender stops sending packets and freezes all its variables such as retransmission buffer, congestion window, and packet buffer. Once the route is established again, the intermediate node sends an RRN (Route Re-established Notification) message to the sender. Upon receiving an RRN message from an intermediate node, the sender resumes its transmission using the same variable values that were being used prior to interruption. To avoid an infinite wait for an RRN message, TCPF uses a route failure timer, which is the worst-case route re-establishment time.

■ Other Transport Protocols for Wireless Networks: In [35], a transport protocol for wireless networks was proposed by not modifying the existing TCP protocol. This is done by introducing a thin layer called ATCP between the network layer and transport layer and it is invisible to transport layer. This makes nodes with ATCP and without ATCP interoperable with each other. ATCP gets information about congestion in the network from the intermediate nodes through ECN (Explicit Congestion Notification) and ICMP messages. Based on this, the source node distinguishes congestion and non-congestion losses and takes the appropriate action.

  ■ When the TCP sender identifies any network partitioning, it goes into persist state and stops all the outgoing transmissions.

  ■ When the TCP sender notices any loss of packets in the network due to channel error, it retransmits the packet without invoking any congestion control.

  ■ When the network is truly congested, it invokes the TCP congestion control mechanism.

1.4.6 Gateway Load Balancing

In WMNs the gateway nodes are connected to the backhaul network, i.e., to the Internet, which provides Internet connectivity to all nodes in the network. So the gateway may become a bottleneck for the connections to the

Internet. As many clients in the network generate traffic to the gateway, the available bandwidth should be utilized effectively. The traffic generated by client nodes aggregates at gateway nodes in the WMN. If some of the gateway nodes are highly loaded and other gateway nodes are lightly loaded, it creates load imbalance between gateway nodes, which leads to packet loss and results in a degradation in network performance. Hence, load balancing across gateway nodes in WMNs improves bandwidth utilization and also increases network throughput.

Load balancing across gateway nodes is obtained by distributing the traffic generated by the network to the backhaul network through all gateway nodes in the WMNs. The load balancing across multiple gateway nodes can be measured quantitatively by a metric called Index of Load Balance (ILB) [36] which is calculated as follows.

Load index (LI) of a gateway i is defined as the fraction of the gateway's backhaul link utilized by a given node k, $LI(i) = \frac{\sum_{k \in N} \beta_k(i) * T_k}{C(i)}$, where $\beta_k(i)$ is the fraction of node k's traffic that is sent through gateway i, T_k is the total traffic generated by node k, and $C(i)$ is the capacity of the backhaul link connected to the gateway node i. The LI value ranges from 0 to 1, with 1 representing 100 percent loaded gateway. The ILB of the network is calculated as

$$ILB = \frac{max\{LI(i)\} - min\{LI(i)\}}{max\{LI(i)\}}$$

Therefore a perfectly balanced network has ILB equal to zero and a highly imbalanced network has ILB equal to one. The objective of all load balancing techniques is to obtain ILB values as small as possible. Several techniques for load balancing across gateways were proposed in the literature. Some of them are discussed in this section.

- Moving Boundary-Based Load Balancing: A flexible boundary is defined for each gateway and the nodes which fall in the boundary are directed to communicate through that gateway. To adopt to variations in the traffic, the region of boundary is periodically redefined. The boundary can be defined in two different ways: (1) in a shortest path-based moving boundary approach, the boundary region for a gateway node is defined by distance of the node from the gateway, and (2) in a load index-based moving boundary approach, the gateways announce their load Index and the nodes join lightly loaded gateways. In this scheme the lightly loaded gateway serves more nodes and the heavily loaded gateway serves fewer nodes.
- Partitioned Host-Based Load Balancing: Here, the nodes in the network are grouped, and each group is assigned to a particular gateway. The main difference compared to the moving boundary-based

load balancing is that no clear boundary is defined. This can be done in both a centralized and distributed way. In the centralized method, a central server assumes the responsibility of assigning the gateway to the nodes. The central server collects the complete information about the gateway nodes and traffic requirements of all the nodes and then allocates nodes to the gateways. In the distributed method, a logical network is formed by the gateway nodes. Each node is associated with a gateway node known as a dominating gateway through which traffic generated by this node reaches the Internet. The nodes in the network periodically update their dominating gateway about their traffic demand. The gateway nodes exchange information about their load and capacity information through the logical network. When a gateway is highly loaded, hand-over takes place, i.e., the gateway delegates some nodes to other gateways which are lightly loaded.

■ Probabilistic Stripping-Based Load Balancing: In the techniques discussed above, each node in the network utilizes only one gateway, which may not lead to perfect load balancing among the gateways. In a probabilistic stripping-based load balancing scheme, each node utilizes multiple gateways simultaneously, which gives perfect load balancing theoretically. In this technique each node identifies all the gateway nodes in the network and attempts to send a fraction of its traffic through every gateway. Hence, the total traffic is split among multiple gateways. This technique is applicable in the case where the load can be split for sending through multiple gateways.

1.4.7 Security

As mentioned earlier, due to the unique characteristics of WMNs, they are highly vulnerable to security attacks compared to wired networks. Designing a foolproof security mechanism for WMNs is a challenging task. The security can be provided in various layers of the protocol stack. Current security approaches may be effective against a particular attack in a specific protocol layer, but they lack a comprehensive mechanism to prevent or counter attacks in different protocol layers. The following issues pose difficulty in providing security in WMNs:

■ Shared Broadcast Radio Channel: In a wired network, a dedicated transmission line is provided between the nodes. But the wireless links between the nodes in WMNs are broadcast in nature, i.e., when a node transmits, all the nodes within its direct transmission range receive the data. Hence, a malicious node could easily obtain data being transmitted in the network if it is placed in the transmission range of mesh routers or a mesh client. For example, if you have a WMN and so does your neighbor, then there is a scope for either snooping into private data or simply hogging the available bandwidth of a neighboring, but alien node.

■ Lack of Association: In WMNs, the mesh routers form a fixed mesh topology which forms a backbone network for the mobile clients. Hence, the clients can join or leave the network at any time through the mesh routers. If no proper authentication mechanism is provided for association of nodes with WMNs, an intruder would be able to join the network quite easily and carry out attacks.

■ Physical Vulnerability: Depending on the application of WMNs, the mesh routers are placed on lampposts and rooftops, which are vulnerable to theft and physical damage.

■ Limited Resource Availability: Normally, the mesh clients are limited in resources such as bandwidth, battery power, and computational power. Hence, it is difficult to implement complex cryptography-based mechanisms at the client nodes. As mesh routers are resource rich in terms of battery power and computational power, security mechanisms can be implemented at mesh routers. Due to wireless connectivity between mesh routers, they also have bandwidth constraints. Hence, the communication overhead incurred by the security mechanism should be minimal.

1.4.8 Power Management

The energy efficiency of a node in the network is defined as the ratio of the amount of data delivered by the node to the total energy expended. Higher energy efficiency implies that a greater number of packets can be transmitted by the node with a given amount of energy resource. The main reasons for power management in WMNs are listed below.

■ Power Limited Clients: In WMNs, though the mesh routers do not have limitations on power, clients such as PDAs and IP phones have limited power as they are operated on batteries. In the case of Hybrid WMNs, clients of the other networks that are connected to them, such as sensor networks, can be power limited. Hence, power efficiency is of major concern in WMNs.

■ Selection of Optimal Transmission Power: In multi-hop wireless networks, the transmission power level of wireless nodes affects connectivity, interference, spectrum spatial reuse, and topology of the network. Reducing the transmission power level decreases the interference and increases the spectrum spatial reuse efficiency and the number of hidden terminals. An optimal value for transmission power decreases the interference among nodes, which in turn increases the number of simultaneous transmissions in the network.

■ Channel Utilization: In multi-channel WMNs, the reduction in transmission power increases the channel reuse, which increases the number of simultaneous transmissions that improves the overall capacity of the network. Power control becomes very important for CDMA-based systems in which the available bandwidth is shared by all the users. Hence, power control is essential to maintain the required signal-to-interference ratio (SIR) at the receiver and to increase the channel reusability.

Several power efficient MAC protocols and power-aware routing protocols are proposed for ad hoc networks to efficiently utilize limited energy resource available in mobile nodes. These protocols consider all the nodes in the network power limited. In WMNs, some nodes are power limited and others have no limitation on power. So, when a power-efficient protocol is used in WMNs, it would not utilize the resource-rich mesh routers to reduce power consumption on power-limited mesh clients. Hence, new protocols are required which consider both types of nodes and efficiently utilize the power of the client nodes.

1.4.9 Mobility Management

In WMNs the mobile clients get network access by connecting to one of the mesh routers in the network. When a mobile client moves around the network, it switches its connectivity from one mesh router to another. This is called hand-off or hand-over. In WMNs the clients should have capability to transfer connectivity from one mesh router to another to implement hand-off technique efficiently. Some of the issues in handling hand-offs in WMNs are discussed below.

■ Optimal Mesh Router Selection: Each mesh client connects to one of the mesh routers in the WMN. Normally, each mesh client chooses the mesh router based on the signal strength it receives from the mesh routers. When a mobile client is in the transmission range of multiple mesh routers, it is very difficult to clearly decide to which mesh router the mobile client must be assigned.

■ Detection of Hand-off: Hand-off may be client initiated or network initiated. In the case of client initiated, the client monitors the signal strength received from the current mesh router and requests a hand-off when the signal strength drops below a threshold. In the case of network initiated, the mesh router forces a hand-off if the signal from the client weakens. Here the mesh router requires information from other mesh routers about the signal strength they receive from the particular client and deduces to which mesh router the connection should be handed over.

■ Hand-Off Delay: During hand-off, the existing connections between clients and network get interrupted. Though the hand-off gives continuous connectivity to the roaming clients, the period of interruption may be several seconds. All ongoing transmissions of the client are transferred from the current mesh router to a new mesh router. The time taken for this transfer is called hand-off delay. The delay of a few seconds may be acceptable for applications like file transfer, but for applications that require real-time transport such as interactive VoIP (Voice-over-IP) or videoconferencing, it is unacceptable.

■ Quality of Hand-Off: During hand-off some number of packets may be dropped due to hand-off delay or interruption on the ongoing transmission. The quality can be measured by the number of packets lost per hand-off. A good quality hand-off provides a low packet loss per hand-off. The acceptable amount of packet loss per hand-off differs between applications.

The hand-off mechanisms in cellular networks are studied in [37] and [38]. When a user moves from the coverage area of one BS to the adjacent one, it finds an uplink–downlink channel pair from the new cell and drops the link from the current BS. In WLANs, whenever a client moves from one AP to another, the link has to be reconfigured manually. In this case, all ongoing connections are terminated abruptly. It may be applicable in LAN environments as the clients have limited mobility around a limited area. But in the case of WMNs, the mesh clients may constantly roam around different mesh routers. Here, manual reconfiguration of mesh clients, whenever the client moves from one mesh router to another, is a difficult task. So the hand-off has to be done automatically and transparently. The users should not feel that the existing connections are transferred from one mesh router to another. For applications such as VoIP and IPTV in WMNs, sophisticated and transparent hand-off techniques are required.

1.4.10   Adaptive Support for Mesh Routers and Mesh Clients

Compared to other networking technologies where all the nodes in the network are considered to have similar characteristics, WMNs have different characteristics between mesh routers and mesh clients. The main differences between them which make the need for new networking protocols for WMNs are

■ Mobility: In many applications of WMNs, the mesh routers form a fixed backbone network by placing the mesh routers at fixed locations such as rooftops and lampposts. So the mesh routers are considered immobile, but the clients in the mesh network are highly mobile and can be connected to any mesh router based on signal strength received from different mesh routers.

■ Resource Availability: Normally, mesh routers are operated with electric power rather than battery power. They are placed in locations where the powerline is available, so the mesh routers do not have energy constraints. But the clients are operating with battery power and are considered energy constrained.

The existing protocols for ad hoc networks consider the characteristics of all nodes in the same way. The energy-aware protocols consider all nodes in the network battery operated. The protocols that take into account the mobility of nodes in the network consider all nodes in the network mobile. For example, a routing protocol designed for networks with high mobility and limited power when used in WMNs does not utilize the limited mobility and rich energy resource nature of mesh routers. Hence, it fails to improve the performance of WMNs. But due to the characteristics of mesh routers, the routing protocols become simple and efficient. So WMNs need efficient protocols that consider the differences between the mesh routers and mesh clients to improve the performance of WMNs.

1.4.11 Integration with Other Network Technologies

The integration of WMNs with other existing network technologies such as cellular, WiFi, WiMAX, WiMedia, and sensor networks can be achieved by bridging functions at the mesh routers. These bridging functions can be provided by adding network interfaces corresponding to the networking technology that the mesh router has to support. There are several issues to be addressed in integrating multiple networking technologies with WMNs:

■ Complexity of Mesh Router: The integration of multiple networking technologies with the mesh network increases the complexity of the mesh routers. For each networking technology to be supported by a mesh router, a network interface should be provided. This increases the hardware and software complexity of the mesh routers.

■ Cost of Mesh Router: The networking hardware or network interface for different networking technologies are not the same. Each networking technology needs specially designed hardware to operate on. Mesh routers have to be provided with the same number of interfaces as the number of networking technologies supported by them. This increases the cost of mesh routers.

- Services Provided by Integrated WMNs: The services provided by different networking technologies are different. Services not provided by IEEE 802.11 can be provided by cellular networks. Similarly the services provided by sensor networks cannot be provided by cellular networks. The integration of other networking technologies with WMNs provides many services to the users that are not provided by WMNs alone. Depending on the service requirement, the required networking technologies can be integrated with WMNs.
- Inter-Operability of Network Technologies: The protocols for different network technologies are independent and operating them together is a difficult task. For example, the routing protocols used by a cellular network and an IEEE 802.11 network are not the same. Further, the MAC protocols used by different networking technologies are not inter-operable. So the inter-operability of different networking technologies necessitates new software architectures or middleware implementations over the mesh networking platform.

Though the integration of multiple networking technologies with WMNs is a difficult job, the services rendered by this necessitate the researchers to come up with a feasible solution. The development of new network architectures and middleware solutions may solve some of these problems. The problem of implementation of many network interfaces in a single mesh router can be solved by using software-defined radios. The software-defined radio system is a software based communication system for modulation and demodulation of radio signal. This is done by advanced signal processing techniques implemented in a digital computer or in a reconfigurable digital electronic system. This technique produces different radios that can receive and transmit a new form of radio protocol just by running different software rather than designing new hardware. This helps in reducing the number of networking interfaces in mesh routers.

1.4.12 Deployment Considerations

- Scenario of Deployment: The capability required for deployments of different WMNs is not the same. For example, WMN deployment for community networking to share network resources among people is not the same as for rescue operations. Some of the deployment scenarios in which the deployment issues vary are
  - Emergency Operation Deployment: This kind of application scenario demands a quick deployment of a communication backbone network through which the mobile devices can communicate. For example, during disasters like flood, fire, and earthquake all the existing communication network infrastructure might be destroyed. Hence, a quick deployment of

a backbone communication network is essential. Most importantly, the network should provide support for time-sensitive traffic such as voice and video. The network should also provide support for different networking technologies to communicate using this network. Hence, the mesh routers should provide interfaces for other existing technologies which allow people to communicate using any communication equipment they have.

■ Commercial Broadband Access Deployment: The aim of this deployment is to provide an alternate network infrastructure for wireless communications in urban areas and areas where a traditional cellular BS cannot handle the traffic volume. This scenario assumes significance as it provides very low cost per bit transferred compared to the cellular network infrastructure. Another major advantage of this application is the resilience to failure of a certain number of nodes. Addressing, configuration, positioning of relaying nodes, redundancy of nodes, and power sources are the major issues in deployment. Billing, provisioning of QoS, security, and handling mobility are major issues that the service provider needs to address.

■ Home Network Deployment: The deployment of a home area network needs to consider the limited range of the devices that are to be connected by the network. Given the short transmission ranges of a few meters, it is essential to avoid network partitions. Positioning of mesh routers at certain key locations of a home area network can solve this problem; also network topology should be decided so that every mesh router is connected through multiple neighbors for availability.

■ Cost of Deployment: The commercial deployment of a communications infrastructure using a WMN essentially eliminates the requirement of laying cables and maintaining them. Hence, the cost of deployment is much less than that of the wired infrastructure. Only the mesh routers have to be placed in appropriate locations for efficient coverage. The mesh router manufacturers are providing mesh routers for outdoor placements. Mesh routers can be placed on poles on the street, which reduces the cost of deployment of mesh networks.

■ Incremental Deployment: In any WMN deployment, the coverage of a geographical area can be extended by adding mesh routers incrementally. With minimum configuration, the network starts functioning and mesh routers can be added incrementally for expanding the size of the network. For example, during the community networking deployment process whenever a mesh router is installed, it can be commissioned.

- Short Deployment Time: Compared to any wired communication infrastructure, WMNs have less deployment time due to absence of laying cables. Wiring the dense urban region is extremely difficult and time consuming, in addition to the inconvenience caused. Mesh routers can be placed based on the area of coverage and number of active users in the area. They can be deployed even on rooftops, provided that electrical power is available.
- Auto-Configurability: The incremental deployment of mesh networks to increase the coverage area or number of users leads to changes in topology of the network at later stages. The lossy nature of the wireless medium changes due to environmental changes, which leads the routing protocols to change the path very often. Due to this, the network needs re-configuration very often.
- Operational Integration with Other Infrastructure: Operational integration with other networking technologies such as satellite, cellular, and sensor networks can be considered to improve the performance or provide additional services to the end users. In the commercial world, the WMNs that service a given urban region can interoperate with the cellular infrastructure to provide better QoS and smooth hand-offs across the networks. Hand-offs to a different network can be done to avoid call drops when a mobile node with an active call moves into a region where service is not provided by the current network.
- Area of Coverage: In most of the cases, the area of coverage of WMNs is determined by the nature of application for which the network is set up. For example, for home networks the coverage of the mesh routers is within the home or within the room in which the router is placed. But in the case of wireless service providers, mesh routers should be covering a number of homes on a street. Long-range communication by fixed mesh routers can be achieved by means of directional antennas. The mesh routers' and mobile clients' capabilities such as transmission range and associated hardware, software, and power source should match the area of coverage required.
- Service Availability: Service availability is defined as the ability of a network to provide service even with failure of certain nodes. In WMNs the mesh routers form a fixed mesh backbone to provide multiple services to the mobile clients. These mesh routers may be placed in outdoor areas such as lampposts and rooftops. They are subject to failure due to power failure, environmental damage, physical damage, or theft. Due to this, the services provided by a WMN to mobile clients may not be available in certain areas. Hence, the mesh routers need to be placed in such a way that failure of some

of them does not lead to lack of service in that area. In such cases, redundant inactive mesh routers can be placed in such a way that, in the event of failure of active mesh routers, the redundant mesh routers can take over their responsibilities.

■ Choice of Protocols: The choice of protocols at different layers of the protocol stack is to be done by taking into consideration the deployment scenario. The MAC protocol should ensure provisioning of security at link level for military applications. The routing protocol also should be selected with care. In the case of integration of different networking technologies, end-to-end paths may have different types of nodes with different capabilities. It requires routing protocols that consider the resource limitations of the nodes. At the transport layer, depending upon the environment in which the WMN is deployed, the connection-oriented or connectionless protocols should be chosen. If the clients connected to the WMN are highly mobile, a frequent hand-off of the clients with the mesh routers takes place. This causes the higher-layer protocols to take necessary action appropriately; also, packet loss arising due to congestion, channel error, link break, and network partition is to be handled differently in different applications. The timer values at different layers of the protocol stack should be adapted to the deployment scenario.

1.5 WMN Deployments/Testbeds

For the deployment of WMNs to be viable, they must be easy to install. This is particularly important for home applications where people are unwilling to install highly technical networks. A number of IEEE standards such as 802.11, 802.15, 802.16, and 802.20 have emerged recently for wireless networks. Many task groups have been working on standardization of the protocols for WMNs, which leads to the development and interoperability of mesh networking products from different vendors. Many testbeds have been established to carry out research and development work in WMNs.

1.5.1 IEEE 802.11 WMNs

IEEE 802.11 [4] is the most popular WLAN standard that defines the specifications for the physical and MAC layer and has been adopted by many vendors of WLAN products. A later version of this standard is the IEEE 802.11b [6], commercially known as WiFi. The original standards for IEEE 802.11 promised a data rate of 1 to 2 Mbps in the license-free 2.4 GHz ISM (Industrial, Scientific, Medical) band. IEEE 802.11b defines operation in the

2.4 GHz ISM band at data rates of 5.5 and 11 Mbps. IEEE 802.11a [5] operates in the 5 GHz band (unlicensed national information infrastructure band). It supports data rates up to 54 Mbps. IEEE 802.11e deals with the requirements of time-sensitive applications such as voice and video. IEEE 802.11g aims at providing the high data rate of IEEE 802.11a in the ISM band. Under the 802.11 standard, mobile clients can operate in infrastructure mode and ad hoc mode. In infrastructure mode a mobile client communicates with others through one or more APs. In ad hoc mode mobile clients can communicate directly with each other without using an AP. The set of mobile clients associated with a given AP is called a Basic Service Set (BSS). A BSS is the basic building block of the network. BSSs are connected by means of a Distribution System (DS) to form an extended network. Any logical point through which non-IEEE 802.11 packets enter the system is called a portal. Portals are also used for integrating wireless networks with the existing wired network. The BSS, DS, and portals together with the mobile clients they connect constitute the Extended Service Set (ESS). Another working group in IEEE 802.11 [3], called 802.11s, has been formed recently to standardize the ESS for mesh networking. It defines architecture and protocols based on IEEE 802.11 MAC to create an 802.11-based Wireless Distribution System (WDS). This WDS supports both broadcast, multicast, and unicast delivery using radio-aware metrics over self-configuring multi-hop topologies. There are two main proposals for 802.11s by SEEMesh and Wi-Mesh. The main features of these proposals are as follows:

■ Supports single and multiple radios.
■ With authentication and key management procedures, it provides secure key distribution and secure exchange of routing information, supporting centralized and distributed models.
■ Supports QoS and power efficiency-aware routing with a WDS four-addressing extension that supports dynamic auto-configuration of MAC-layer data delivery.
■ Enables multiple routing algorithms for MAC address-based forwarding with a simple Hello message for mesh discovery and association and supporting extended mesh discovery.

1.5.2   IEEE 802.15 WMNs

The 802.15 WPAN Task Group [39] focuses on the development of consensus standards for Personal Area Networks or short-distance wireless networks. These WPANs address wireless networking of portable and mobile computing devices such as PCs, PDAs, peripherals, cell phones, pagers, and consumer electronics and allow these devices to communicate and interoperate.

The IEEE 802.15 Task Group 5 is chartered to determine the mechanisms that must be present in the PHY and MAC layers of WPANs to enable mesh networking. A mesh network is a PAN that employs one of the two connection arrangements, full mesh topology or partial mesh topology. In the full mesh topology, all nodes are in the transmission range of one another, i.e., each node can communicate with other nodes in one hop. In partial mesh topology, nodes in the network have one-hop communication with a few nodes only. The 802.15 mesh networks have the following capabilities:

■ Extension of network coverage without increasing transmit power or receiver sensitivity
■ Enhanced reliability via route redundancy
■ Easier network configuration
■ Better battery life of device due to fewer retransmissions

1.5.3   IEEE 802.16 WMNs

The Worldwide Interoperability for Microwave Access (WiMAX) forum describes WiMAX as "a standards-based technology enabling the delivery of last mile wireless broadband access as an alternative to cable and DSL." The 802.16 [40] standard requires line-of-sight towers and operates in the 10 to 66 GHz frequency band. But the 802.16a [41] extension does not require line-of-sight and operates in the 2 to 11 GHz frequency band. To allow the consumers to connect to the Internet while moving at vehicular speeds, the 802.16e [42] extension was developed. The main advantage of 802.16-based mesh networks compared to 802.11 is higher coverage range and bandwidth. As 802.16 uses TDMA-based scheduling of channel access, it provides efficient resource utilization. These advantages make 802.16 best suited for WMNs. The recent draft on 802.16 [43] integrated the mesh mode specification into the standard. This mesh mode supports Time Division Duplex (TDD), which separates downlink and uplink in time. The MAC frame has two sub-frames called control sub-frame and data sub-frame. Every control sub-frame consists of 16 transmission opportunities and each transmission opportunity equals seven OFDM symbols. The data sub-frame consists of mini slots, which are basic units for resource allocation. The scheduling algorithm in 802.16 allocates the time slots in the data frame. This is done by control message exchange in the control sub-frame so that there is no contention in the data sub-frame. In a transmission opportunity each node contends for channel and runs an election algorithm to compute whether or not it can win a slot, because other nodes may also try to transmit in the selected time slot. If it wins in the election algorithm, the node broadcasts its schedule to all the neighbors and repeats the procedures in the next transmission time. If it fails, the node selects the next transmission slot and continues contention until it wins. For a connection setup, a request/grant/confirm three-way handshake procedure is used.

1.5.4 Academic Research Testbeds

Many academic research institutes established testbeds to study realistic behavior of WMNs. Some of them are discussed in this section.

- MIT Roofnet [44–46]: MIT Roofnet is an 802.11b multi-hop network designed to provide broadband Internet connectivity to users in apartments of Cambridge, MA. It has about 50 nodes connected through 802.11b interfaces in multi-hop fashion and connected to the Internet through an Ethernet interface available in the apartments. Research on Roofnet includes link-level measurements of 802.11 interfaces, finding high-throughput routes in the face of lossy links, adaptive bit-rate selection, and developing new protocols which take advantage of radio's unique properties. The main feature of Roofnet is that it is an unplanned network, i.e., no configuration or planning is required.

- CalRadio-I [47]: California Institute for Telecommunications and Information Technology developed CalRadio-I, which is a radio/networking test platform for wireless research and development. This is a single integrated, wireless networking test platform which provides a simple, low-cost platform development from the MAC layer to a higher layer. All the MAC functionalities are coded in C language that runs on the DSP processor. Any modification to the MAC protocol can be done and tested in it. CalRadio-I functions as a test instrument, an AP, and as a WiFi client.

- BWN-Mesh Testbed at Georgia Tech [48]: The WMN tested by the Broadband and Wireless Network (BWN) Lab at Georgia Institute of Technology consists of 15 IEEE 802.11b/g-based mesh routers. Using this mesh network testbed, various experiments to investigate the effects of inter-router distance, backhaul placement, and clustering are performed by varying the mobility of the nodes. Other testbeds in the lab such as next-generation Internet testbed as backhaul access to the Internet are connected to a mesh testbed. The measurements using this testbed reveal that existing protocols for wireless ad hoc networks such as TCP for transport layer, AODV for network layer, and IEEE 802.11g for MAC do not perform well in terms of end-to-end delay and throughput in WMNs. So the research at BWN is focused on adaptive protocols for transport, routing, and MAC layers and their cross-layer design. Integration of other network technology testbeds such as WSNs (Wireless Sensor Networks), WSANs

(Wireless Sensor and Actor Networks), next-generation Internet, and WiMAX with WMNs testbed leads to design and evaluation of protocols for heterogeneous wireless networks.

■ UCSB MeshNet [49]: The University of California, Santa Barbara, deployed an experimental testbed on their campus. It consists of 25 nodes equipped with multiple IEEE 802.11a/b/g wireless radios. The main objective of the testbed is to design protocols for the robust operation of multi-hop wireless networks. Specifically, the testbed is being used to conduct research on scalable routing protocols, efficient network management, multimedia streaming, and QoS for multi-hop wireless networks.

1.5.5 Industrial Research in WMNs

Many companies started research in WMNs on their own and in collaboration with academic research institutions. Some of them recently came up with mesh networking products for implementing mesh network-based applications. In this section some of the industries working toward research aspects of WMNs and some of the industries providing mesh networking products are discussed.

■ Microsoft Research [50]: Microsoft researchers at Redmond, Cambridge, and Silicon Valley are working to create wireless technologies that allow neighbors to connect their home networks together (community networking). They deployed their own mesh network testbed in their office building and local apartment complex. They developed a software module called the Mesh Connectivity Layer (MCL) which implements ad hoc routing and link quality measurement. Architecturally, MCL is a loadable Windows driver. It implements a virtual network adapter, so that the ad hoc network appears as an additional (virtual) network link to the rest of the system. The routing protocol used by MCL is LQSR, which improves network performance by supporting link-quality metrics for routing. The MCL driver implements an interposition layer between the link layer and the network layer. To higher-layer software, MCL appears to be just another Ethernet link, albeit a virtual link. To lower-layer software, MCL appears to be just another protocol running over the physical link. This design has several significant advantages. First, higher-layer software runs unmodified over the ad hoc network. The testbed runs both IPv4 and IPv6 over the ad hoc network without requiring any modifications to the network layer. All network layer functionalities such as ARP, DHCP, and Neighbor Discovery work well. Second, the ad hoc routing runs over heterogeneous link layers as well. This implementation supports Ethernet-like physical link layers (e.g., 802.11 and 802.3), but the architecture accommodates link layers with arbitrary addressing and framing conventions. The virtual MCL network adapter can multiplex several physical network adapters, so the ad hoc network can be extended across heterogeneous physical links. Third, the design can support other ad hoc routing protocols as well.

■ Intel [51]: A wide variety of research and development efforts at Intel are geared toward understanding and addressing the technical challenges for realizing multi-hop mesh networks. Intel's Network Architecture Lab is aimed at overcoming many of the challenges faced by WMNs. They developed low-cost and low-power AP prototypes or nodes to enable further research on security, traffic characterization, dynamic routing and configuration, and QoS problems. Intel is also working with other industries to develop standards and protocols that support WMNs and enable interoperability between products from multiple vendors. Intel is working to simplify the entire installation process, including network node placement and configuration so that end users and businesses can easily realize the full benefits of multi-hop mesh networking.

1.5.6 Mesh Networking Products

■ Strix Systems [52]: The mesh networking products from Strix Systems are RF-independent supporting existing wireless standards 802.11a/b/g and 802.16 (WiMAX), designed to easily add in any future wireless technologies. The Strix Access/One® family of products delivers high-performance WMN systems by employing modular future-proof architecture supporting multi-radio, multi-channel, and multi-RF mesh networking technologies. The Access/One architecture delivers the industry's most scalable and flexible wireless networking platform by which the largest citywide and countrywide communication services can be built. Unlike competing single and other multi-radio products, the Access/One design makes secure full-duplex transmission, instant path switching, and application classification a reality. Strix Access/One networks are deployed in many different environments and used for many different applications around the world, enabling users to access wireless broadband applications at any place, anywhere, any time even while moving at 200 miles per hour. Strix Access/One is a scalable self-configuring and self-healing system designed to meet the needs of service providers, government agencies, and outdoor mobile enterprises.

■ Nortel [53]: Nortel's WMN solution addresses the market requirements for networks that are highly scalable and cost-effective, offering end user security, seamless roaming beyond traditional WLAN boundaries, and provides easy deployment in areas that do not (or cannot) support a wired backhaul. Nortel's WMN solution is well-suited for providing broadband wireless access in areas that traditional WLAN systems are unable to cover. Nortel provides a number of products for WMN solutions, which include wireless AP, wireless bridge, WLAN security switches, and enterprise network management system. These products provide a number of applications for the mobile users such as secure mobile networking and voice connectivity featuring flexible seamless mobility across campus environments, IP telephony and converged multimedia applications, and low-cost, high-capacity point-to-point broadband transmission.

■ Kiyon Mesh Network [54]: Kiyon also provides mesh networking products for realizing WMNs. The KAN254B wireless BACNet router provides a WMN solution to industry and converts all standard field controllers or supervisory controllers using BACnet MSTP, BACnet IP, or Ethernet IP to a WMN. It can also be used for security systems, video cameras, lighting systems, fire, and Internet applications. People have applied them in offices and warehouses and even to connect buildings together when running wires was prohibitive.

■ FireTide® [55]: FireTide mesh networking provides solutions to education, health care, hospitality, municipal government, and warehousing. The mesh networking products from FireTide such as Hotspot indoor and outdoor mesh nodes provide a high-capacity wireless mesh backbone for outdoor and indoor networks. These products are designed for maximum performance, scalability, and ease of use. They can operate in 2.4- and 5-GHz frequency spectrum. The public safety mesh nodes are ideal for public safety agencies. This operates in 4.940- to 4.990-GHz spectrum, which has been allocated for public safety agencies in the United States.

1.6 Summary

WMNs have emerged as a promising technology for next-generation networking. In WMNs, no cabling is required to connect the mesh routers. All mesh routers self-configure wirelessly to form a rich radio mesh backbone network. The wireless connectivity between routers significantly reduces the deployment and maintenance cost when compared with wired networks. Due to these attractive features of WMNs, they are considered for a wide variety of applications such as community networking, emergency operations, home networking, and hybrid wireless architectures. In this chapter, the major issues and applications of WMNs were described.

The design issues and deployment scenarios were also discussed. Providing high throughput is the major design goal of WMNs, which has been addressed in multiple layers. To improve the performance of WMNs, the multi-channel, multi-radio architecture has been suggested. The related protocols for this architecture in MAC and routing layer were discussed. Some routing metrics were described to find high-throughput paths by taking into account the channel quality and inter flow and intra flow interference. Security and standardization are the main concerns for the wide deployment of WMNs. Some of the security issues and standards such as IEEE 802.11s and IEEE 802.16 mesh were also discussed. Finally, to provide insight on real implementations of WMNs, some WMN testbeds and mesh networking products were also discussed.

References

[1] I. F. Akyildiz, X. Wang, and W. Wang, Wireless mesh networks: A survey, *Computer Networks Journal*, vol. 47, no. 4, pp. 445–487, March 2005.

[2] P. Gupta and P. R. Kumar, The capacity of wireless networks, *IEEE Transactions on Information Theory*, vol. 46, no. 2, pp. 388–402, March 2000.

[3] IEEE 802.11 Standard Group Website, http://www.ieee802.org/11/

[4] IEEE Std 802.11-1997, Part 11: Wireless LAN medium access control (MAC) and physical layer (PHY) specifications, *The Institute of Electrical and Electronics Engineers*, 1997.

[5] IEEE Std 802.11a-1999, Part 11: Wireless LAN medium access control (MAC) and physical layer (PHY) specifications: High-speed physical layer in the 5 GHz band, *The Institute of Electrical and Electronics Engineers*, 1999.

[6] IEEE Std 802.11b-1999, Part 11: Wireless LAN medium access control (MAC) and physical layer (PHY) specifications: Higher-speed physical layer extension in the 2.4 GHz Band, *The Institute of Electrical and Electronics Engineers*, 1999.

[7] P. Kyasanur and N. H. Vaidya, Capacity of Multi-Channel Wireless Networks: Impact of Number of Channels and Interfaces, *Proceedings of ACM MOBICOM 2005*, pp. 43–57, August 2005.

[8] I. Emre Telatar, Capacity of multi-antenna Gaussian channels, *European Transactions on Telecommunications*, vol. 10, no. 6, pp. 585–595, November/December 1999.

[9] V. Bharghavan, A. Demers, S. Shenker, and L. Zhang, MACAW: A Media Access Protocol for Wireless LANs, *Proceedings of ACM SIGCOMM 1994*, pp. 212–225, August 1994.

[10] C. L. Fullmer and J. J. Garcia-Luna-Aceves, Floor Acquisition Multiple Access Protocol for Wireless LANs, *Proceedings of ACM SIGCOMM 1995*, pp. 262–273, August 1995.

[11] F. A. Tobagi and L. Kleinrock, Packet switching in radio channels: Part II: The hidden terminal problem in carrier sense multiple access and the busy tone solution, *IEEE Transactions on Communications*, vol. 23, no. 12, pp. 1417–1433, December 1975.

[12] F. Talucci and M. Gerla, MACA-BI (MACA by Invitation): A Wireless MAC Protocol for High Speed Ad Hoc Networking, *Proceedings of IEEE ICUPC 1997*, vol. 2, pp. 913–917, October 1997.

[13] S. Jiang, J. Rao, D. He, and C. C. Ko, A simple distributed PRMA for MANETs, *IEEE Transactions on Vehicular Technology*, vol. 51, no. 2, pp. 293–305, March 2002.

[14] Z. Tang, and J. J. Garcia-Luna-Aceves, A Protocol for Topology-Dependent Transmission Scheduling in Wireless Networks, *Proceedings of IEEE WCNC 1999*, vol. 3, no. 1, pp. 1333–1337, September 1999.

[15] Z. Tang and J. J. Garcia-Luna-Aceves, Hop-Reservation Multiple Access (HRMA) for Ad Hoc Networks, *Proceedings of IEEE INFOCOM 1999*, vol. 1, pp. 194–201, March 1999.

[16] B. S. Manoj and C. Siva Ram Murthy, Real-Time Traffic Support for Ad Hoc Wireless Networks, *Proceedings of IEEE ICON 2002*, pp. 335–340, August 2002.

[17] V. Kanodia, C. Li, A. Sabharwal, B. Sadeghi, and E. Knightly, Ordered Packet Scheduling in Wireless Ad Hoc Networks: Mechanisms and Performance Analysis, *Proceedings of ACM MOBIHOC 2002*, pp. 58–70, January 2002.

[18] I. Karthikeyan, B. S. Manoj, and C. Siva Ram Murthy, A distributed laxity-based priority scheduling scheme for time-sensitive traffic in mobile ad hoc networks, *Ad Hoc Networks Journal*, vol. 3, no. 1, pp. 27–50, January 2005.

[19] V. Kanodia, C. Li, A. Sabharwal, B. Sadeghi, and E. Knightly, Distributed priority scheduling and medium access in ad hoc networks, *ACM/Baltzer Journal of Wireless Networks*, vol. 8, no. 5, pp. 455–466, September 2002.

[20] J. So and N. H. Vaidya, Multi-Channel MAC for Ad Hoc Networks: Handling Multi-Channel Hidden Terminals Using a Single Transceiver, *Proceedings of ACM MOBIHOC 2004*, pp. 222–233, May 2004.

[21] P. Bahl, R. Chandra, and J. Dunagan, SSCH: Slotted Seeded Channel Hopping for Capacity Improvement in IEEE 802.11 Ad Hoc Wireless Networks, *Proceedings of ACM MOBICOM 2004*, pp. 216–230, September 2004.

[22] A. Adya, P. Bahl, J. Padhye, A. Wolman, and L. Zhou, A Multi-Radio Unification Protocol for IEEE 802.11 Wireless Networks, *Proceedings of IEEE BROADNETS 2004*, pp. 344–354, October 2004.

[23] P. Kyasanur and N. H. Vaidya, Routing and Interface Assignment in Multi-Channel Multi-Interface Wireless Networks, *Proceedings of IEEE WCNC 2005*, vol. 4, pp. 2051–2056, March 2005.

[24] P. Kyasanur and N. H. Vaidya, Routing and link-layer protocol for multi-channel multi-interface ad hoc wireless networks, *ACM Mobile Computing and Communications Review*, vol. 10, no. 1, pp. 31–43, January 2006.

[25] C. E. Perkins and E. M. Royer, Ad Hoc On-Demand Distance Vector Routing, *Proceedings of IEEE Workshop on Mobile Computing Systems and Applications*, pp. 90–100, February 1999.

[26] D. B. Johnson, D. A. Maltz, and J. Broch, DSR: The Dynamic Source Routing Protocol for multi-hop wireless ad hoc networks, *Ad Hoc Networking*, Chapter 5, pp. 139–172, Addison-Wesley, 2001.

[27] R. Ogier, F. Templin, and M. Lewis, Topology dissemination based on reverse-path forwarding (TBRPF), *IETF RFC 3684*, February 2004.

[28] D. S. J. D. Couto, D. Aguayo, J. Bricket, and R. Morris, A High-Throughput Path Metric for Multi-Hop Wireless Routing, *Proceedings of ACM MOBICOM 2003*, pp. 134–146, September 2003.

[29] D. Aguayo, J. Bicket, S. Biswas, G. Judd, and R. Morris, Link-Level Measurements from an 802.11b Mesh Network, *Proceedings of ACM SIGCOMM 2004*, pp. 121–132, August 2004.

[30] R. Draves, J. Padhye, and B. Zill, Comparison of Routing Metrics for Static Multi-Hop Wireless Networks, *Proceedings of ACM SIGCOMM 2004*, pp. 133–144, August 2004.

[31] R. Draves, J. Padhye, and B. Zill, Routing in Multi-Radio, Multi-Hop Wireless Mesh Networks, *Proceedings of ACM MOBICOM 2004*, pp. 114–128, September 2004.

[32] Y. Yang, J. Wang, and R. Kravets, Interference-Aware Load Balancing for Multi-Hop Wireless Networks, Technical Report UIUCDCS-R-2005-2526, Department of Computer Science, University of Illinois at Urbana-Champaign, 2005.

[33] G. Hallond and N. Vaidya, Analysis of TCP Performance over Mobile Ad Hoc Networks, *Proceedings of ACM MOBICOM 1999*, pp. 219–230, August 1999.

[34] K. Chandran, S. Raghunathan, S. Venkatesan, and R. Prakash, A feedback based scheme for improving TCP performance in ad hoc wireless networks, *IEEE Personal Communications Magazine*, vol. 8, no. 1, pp. 34–39, February 2001.

[35] J. Liu and S. Singh, ATCP: TCP for mobile ad hoc networks, *IEEE Journal on Selected Areas in Communications*, vol. 19, no. 7, pp. 1300–1315, July 2001.

[36] Chi-Fu Huang, Hung-Wei Lee, and Yu-Chee Tseng, A two-tier heterogeneous mobile ad hoc network architecture and its load-balance routing problem, *Mobile Networks and Applications*, vol. 9, no. 4, pp. 379–391, August 2004.

[37] I. F. Akyildiz, J. McNair, J. S. M. Ho, H. Uzunalioglu, and W. Wang, Mobility Management in Next Generation Wireless Systems, *Proceedings of the IEEE*, vol. 87, no. 8, pp. 1347–1385, August 1999.

[38] I. F. Akyildiz, J. Xie, and S. Mohanty, A survey of mobility management in next-generation all-IP-based wireless systems, *IEEE Wireless Communications*, vol. 11, no. 4, pp. 16–28, August 2004.

[39] IEEE 802.15 Standard Group Website, http://www.ieee802.org/15/

[40] IEEE 802.16 Standard Group Website, http://www.ieee802.org/16/

[41] IEEE Std 802.16a-2003 (amendment to IEEE Std 802.16-2001), Part 16: Air interface for fixed broadband wireless access systems — Amendment 2: Medium access control modifications and additional physical layer specifications for 2-11 GHz, *The Institute of Electrical and Electronics Engineers*, 2003.

[42] IEEE Std 802.16e-2005, Part 16: Air interface for fixed and mobile broadband wireless access system — Amendment 2: Physical and medium access control layers for combined fixed and mobile operation in licensed bands and corrigendum 1, *The Institute of Electrical and Electronics Engineers Inc.*, 2006.

[43] IEEE Std 802.16-2004 (Revision of IEEE Std 802.16-2001), Part 16: Air interface for fixed broadband wireless access systems, *The Institute of Electrical and Electronics Engineers Inc.*, 2004.

[44] Roofnet, http://pdos.csail.mit.edu/roofnet/doku.php?id=roofnet

[45] J. Bicket, D. Aguayo, S. Biswas, and R. Morris, Architecture and Evaluation of an Unplanned 802.11b Mesh Network, *Proceedings of ACM MOBICOM 2005*, pp. 31–42, August 2005.

[46] D. Aguayo, J. Bicket, S. Biswas, D. S. J. De Couto, and R. Morris, MIT Roofnet Implementation, http://pdos.csail.mit.edu/roofnet/design/

[47] UCSD Mesh Networks Testbed, http://www.calit2.net/

[48] Wireless Mesh Networks, http://www.ece.gatech.edu/research/labs/bwn/mesh/

[49] UCSB MeshNet, http://moment.cs.ucsb.edu/meshnet/

[50] Self-Organizing Neighborhood Wireless Mesh Networks, http://research.microsoft.com/mesh/

[51] Multi-Hop Mesh Networks, http://www.intel.com/technology/comms/cn02032.htm

[52] Strix Systems, http://www.strixsystems.com/

[53] Wireless Mesh Network Solution, http://www.nortel.com

[54] Kiyon, http://www.kiyon.com/

[55] Firetide Instant Mesh Network, http://www.firetide.com/

Chapter 2

Mesh Networking in Wireless PANs, LANs, MANs, and WANs

Neila Krichene and Noureddine Boudriga

Contents

Wireless mobile mesh networks are made up by several mobile nodes, fully wirelessly interconnected, which adopt multi-hop communication for data transmission. This chapter intends to argue why mesh networking technology represents a new issue to address for wireless networks by presenting the mesh networking fundamentals in wireless PANs, LANs, MANs, and WANs. For this purpose, we will first study the mesh networking characteristics while stressing the targeted applications, the network architecture, and the particularities of the routing, quality of service (QoS) provision, and management protocols. Then, details of the IEEE standardization efforts targeting the network coverage ranging from PANs to WANs are presented. We conclude by presenting some of the deployed solutions and discussing advanced design issues aiming at providing scalable, low-cost, and easily deployable Wireless Mobile Mesh Networks.

2.1   Introduction

The mobile ad hoc networks (or MANET) have gained researchers' attention for 30 years [1]. MANET nodes share wireless links and can play the role of client and router at the same time without relying on any infrastructure; thus accomplishing large deployment ease and investments cost decrease. Besides, the ephemeral nature of MANETs particularly copes with critical applications such as disaster recovery and battlefield communications. Many research works have addressed the multi-hop communication issue in wireless networks, but the practical impact was not very important because users rarely operate in ad hoc mode. For instance, the targeted applications were limited to specialized missions inducing an unreasonable cost, while users searched mostly for cheap information sharing and Internet access. Client satisfaction has created a new research topic that aims at revising the MANET concept by considering the MANET network as a flexible and low-cost extension of wired infrastructure networks that integrates them. As a result, the wireless mesh networking paradigm, which inherits some MANET characteristics and targets civilian applications, was born. It is worth noticing that both the wired Internet and the public switched telephone network may be classed as mesh networks [2]; however, future wireless mesh networks should rely on a wireless infrastructure to interconnect mobile devices in a multi-hop fashion. Wireless mesh networks (WMNs) support home and enterprise networking applications; they also provide ubiquitous Internet access and enable the implementation of intelligent transportation systems and public safety applications. Besides, their deployment does not require important investments comparable to the deployment of wired solutions. In fact, wireless mesh routers can rapidly and easily integrate the wireless infrastructure as soon as the coverage needs to be extended. As a result, a growing number of cities have adopted this paradigm to attract visitors and citizens and start a long-lasting development process. Users can temporarily join the mesh network and act as clients and routers for other nodes, thus enhancing the network capacity, throughput, and reliability. Currently, one can find off-the-shelf and proprietary mesh networks solutions while IEEE standardization efforts are targeting network coverage ranging from PANs to WANs. The goal of this chapter is to present the mesh networking fundamentals in wireless PANs, LANs, MANs, and WANs. To this end, a general overview of the mesh networks architecture and characteristics is given while addressing general concepts such as the supported applications, the routing and management protocols, the QoS provision, and the security considerations. Then, the detail of the IEEE standardization efforts targeting the network coverage ranging from PANs to WANs is presented. We particularly address the physical layer and the MAC layer design issues for the mesh communication mode support while presenting the challenges that are particular to each network (PAN, LAN, MAN or WAN). An overview of the available commercial systems and deployed solutions is also given. We conclude by discussing some of the research issues aiming at designing scalable, low-cost, and easily deployable wireless mobile mesh networks.

2.2   Wireless Mesh Networking Fundamentals

2.2.1   Network Architecture

A wireless mesh network is a hierarchical network formed by fully wirelessly interconnected nodes, as illustrated in Figure 2.1. A fully meshed network is a network where every node directly connects to every other node; a partial mesh network is a network where each node is connected to a set of other nodes [47]. We distinguish routers nodes that act as layer 3 gateways and support meshing functions. Such nodes are usually equipped with multiple network interfaces for different access technologies; they can guarantee wider coverage with less power consumption thanks to the support of multi-hop communications. The network resulting from the mesh routers interconnection is called a wireless backbone; it guarantees the connectivity between nomadic users and wired gateways. The wireless mesh network includes also Access Points (APs), which can be viewed as special mesh routers provided with a high-bandwidth wired connection to the Internet. The wireless network formed by the interconnection of the AP and the mesh routers is called a backhaul. The latter enables the access to external networks while providing high-bandwidth and seamless multi-hop communication at a low cost.

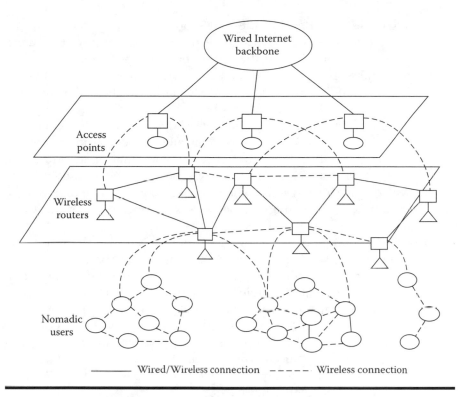

Figure 2.1   The wireless mesh network architecture.

Finally, mesh clients are generally equipped with a radio interface supporting mesh networking functions; that is why they can act as routers for other mesh nodes. However, they do not provide the bridge/gateway functionalities needed for Internet access and interoperability with other networking technologies. Mesh clients can be laptops, pocket PCs, PDAs, IP phones, etc.

2.2.2 Characteristics

Mesh networks are gaining a growing interest thanks to their special characteristics that enable the deployment of new applications at lower cost. The most important characteristics are as follows:

■ Multi-Hop Communication: The multi-hop communication scheme guarantees larger coverage zones and an enhancement of the network capacity. In fact, line-of-sight constraint no longer matters because the intermediate nodes relay the information to their neighbors on short wireless links using a reduced power transmission. As a result, the interferences are decreased and the throughput is augmented [3]. Besides, the multi-hop connectivity allows several devices to access the network at once by relying on other mesh nodes without affecting the overall network performance. Finally, mesh networks gain more capacity as the number of internal nodes increases and the data traffic can reach larger areas by crossing multiple hops until the final destination.

■ Wide Coverage and Cost Reduction: The wireless infrastructure supported by the mesh networks eliminates the deployment costs of a new wired backhaul through cities and rural areas. Moreover, the flexible infrastructure can easily be enforced by adding new wireless mesh routers anywhere, anytime the coverage needs to be enhanced. Only some APs need to be connected with the wired infrastructure to allow Internet access.

■ Self-Configuration and Self-Management: New mesh nodes that enter the network are transparently supported because meshing functions such as neighbors discovery and automatic topology learning are implemented. Wireless routers rapidly detect the presence of new paths, thus enhancing the overall performance and coverage.

■ Network Access and Interoperability: Backhaul devices are equipped with multiple network interfaces that support both Internet and peer-to-peer communications while guaranteeing access to existing wireless networks technologies such as traditional IEEE 802.11, WiMAX, ZigBee™, and cellular networks.

■ Mobility and Power Consumption: The mobility and power consumption vary with the nature of the mesh node. For example, mesh routers and APs have minimal mobility and reduced power constraints. However, mesh clients are mostly small mobile devices with reduced battery autonomy. Therefore, MAC and routing protocols supported by the backbone/backhaul do not need to be power efficient, but they cannot be implemented on simple mesh clients.

■ Reliability: Mesh networks rely on multi-hop communication and can use every internal node to route traffic to the destination. Therefore, multiple paths exist between two communicating endpoints and temporary path failures can be easily tolerated. Besides, mesh clients that need to communicate with external destinations (e.g., Internet) can choose between multiple egress points toward the wired network, thus tolerating router failures and reducing potential congestions.

2.2.3 Supported Applications

The mesh networks support a large number of applications dedicated to personal, local, metropolitan, and wide areas networks.

- Home Networking: Mesh networks can be deployed at home because they support bandwidth-greedy applications such as multimedia traffic transmission [5]. Mesh nodes can be desktop PCs, laptops, high-definition TV, and DVD players. Wireless APs or mesh routers can easily be added to cover dead zones without requiring wiring or complex configurations.

- Enterprise Networking: Traditional wireless LANs have been widely used in enterprises, but they have not succeeded in effectively reducing the deployment cost because the presence of a wired infrastructure is a must. Adopting mesh networks in enterprises enables the share of resources and an overall performance enhancement thanks to the multi-hop communication and the wireless infrastructure deployment. In fact, bottleneck congestion resulting from the one-hop access to the traditional APs is eliminated. Besides, the infrastructure can easily scale according to the network's needs without requiring complex configurations and wiring.

- Public Applications: Mesh networks support public applications at the metropolitan and wide area scale mainly because the line-of-sight constraint can be overcome. Wireless Internet access on the road, public safety, and implementation of intelligent transportation systems are highly appreciated by cities' inhabitants and visitors, and have already been deployed in many countries such as the United States, Taiwan, and Bangladesh.

The supported applications will be further detailed later in the chapter.

2.2.4 Routing Protocols

Wireless mesh networks are characterized by multi-hop communications and rely on a wireless backhauling system to access other external networks such as the Internet. Consequently, they need to address special constraints such as enhanced scalability, varying power constraints, and cross-layer design. These specificities require special routing capabilities that may be partially inherited from the ad hoc context, but that surely differ from those implemented in the wired and cellular networks. We believe that the specification of a wireless mesh routing protocol should provide new performance metrics that take into consideration the quality of the intermediate links while trying to minimize the path length. Meanwhile, the mesh routers and the mesh clients presenting different mobility and power constraints should implement an efficient hybrid routing protocol able to address those specificities. For instance, the Link Quality Source Routing (LQSR) based on the DSR protocol [49] selects the routes with respect to the expected transmission count (or ETX, [52]), the per-hop round-trip tune (RTT), and the per-hop packet pair. Results showed that adopting the ETX for stationary nodes guarantees a good performance although adopting the minimum hop count as route selection criteria for mobile nodes gives better results. New performance metrics that achieve good performances in the mesh context present a research issue that needs to be investigated.

In addition, fault-tolerance mechanisms that guarantee the rapid selection of a new path in case of link failure should be defined. Besides, the route selection should be based on the congestion status of the network to efficiently use the available resources. In fact, the mesh network presents multiple routes between communicating nodes so that alternative paths which offer the required QoS may be selected in case of mobility or link quality decrease. However, it is worth noticing that the route-establishment complexity increases as the network size grows. Meanwhile, the routing protocol should address the ephemeral nature of mesh nodes while guaranteeing the end-to-end QoS requirements, especially in the case of metropolitan and wide area mesh networks. When considering the ad hoc context, hierarchical routing protocols as presented in [53–55] adopt a self-organization scheme that groups the network nodes into clusters with a certain size. Each cluster is then managed by one or more clusterheads and nodes belonging to different clusters may communicate using other nodes as gateways. The routing mechanisms implemented inside a cluster may be proactive while intra-cluster routing may be on-demand. Such protocols achieve good performances especially when the node's density is high; however, they cannot be applied to the mesh context without adding some modifications. For instance, a mesh node selected as a clusterhead may not present sufficient power and processing capabilities, thus becoming a bottleneck. Geographic routing which is topology-based resists mobility better, but requires important processing resources. In addition, delivery is not always guaranteed even if a path exists between the communicating nodes. Open research issues need to be addressed if this routing principle is applied to the mesh networking context.

2.2.5 Network Management

Mesh networks management needs to address nodes' specificities in terms of mobility, location, and power to provide an up-to-date vision of the network status. The resulting accurate management data will serve especially for enhancing the overall performances and making the wise decisions to overcome the encountered problems.

- ■ **Mobility Management:** Mobility management addresses the location management and the hand-over. Location management addresses the location registration and the call delivery; it guarantees that active nodes remain always reachable despite their mobility. The hand-over process, also known as hand-off, consists in transferring

a communication; therefore, it requires a new connection generation and implements the control of the data flow. Advanced mobility
management mechanisms have been proposed for cellular and IP
networks; however, the adopted schemes are centralized because
they rely on the base stations. As mesh networks present an ad
hoc architecture, distributed or hierarchical location and hand-over
management functions should be adopted while taking into consideration the nodes' nature (routers or clients) and their different
mobility schemes. In fact, backbone nodes present reduced mobility
while mesh clients frequently roam across different mesh routers.
Proposing a multi-layer mobility management framework that
addresses mesh specificities is a hot research topic that needs to be
investigated. More specifically, location management functions may
be used at MAC and routing layers to provide better performances
and permit the development of new location-based applications for
the mesh scenarios.

■ **Power Management:** Mesh networks are made up of mesh routers
and mesh clients. While the routers present reduced mobility and
power constraints, the clients are tiny pieces of equipment, such
as IP phones and sensors, which are battery-dependent. Besides, it
is always preferable to reduce the transmission power to save the
resources and reduce the interferences while increasing the spectrum spatial-reuse efficiency. Consequently, power-efficient protocols need to be developed while paying particular attention to some
constraints as the hidden nodes scenario to avoid the performance
degradations at the MAC level.

■ **Network Monitoring:** Mesh routers need to calculate their own
statistics to report them for monitoring servers. Servers should then
analyze the data and process anomaly detection. They can then
trigger alarms or reactively respond, depending on the scenario. Few
networking management protocols have been proposed for the ad
hoc context [56]; however, they do not address the scalability issue
of the mesh networks. Besides, new data processing algorithms that
address the mesh network's specificity need to be developed.

2.2.6   QoS Provision

A service in a communication network is defined by the International
Telecommunication Union (ITU) as a service provided by the service plane
to an end user (e.g., a host [end system] or a network element) and which
utilizes the IP transfer capabilities and associated control and management
functions for delivery of the user information specified by the service level
agreement (SLA) [69]. In the telecommunications area, the quality of service

is intrinsic, perceived, or assessed. Intrinsic QoS is a technical measure considered by engineers and network service providers; it is always objectively compared to the expected performance not affected by customers' perceptions. Perceived QoS reflects the end user's view about a service while assessed QoS is a factor that the customer decides whether or not to continue using the service [69]. It is clear that the most challenging issue in providing QoS is to specify the requirements and then quantify them based on a set of measurable QoS parameters such as the delay, the jitter, and the bandwidth.

Today, most Internet protocols provide best-effort IP forwarding while QoS support is required to satisfy multimedia applications needs. To address this issue, two major QoS models have been proposed: the Integrated Service (IntServ) [73] and the Differentiated Service (DiffServ) [74]. IntServ is a QoS model which adopts virtual circuit connection mechanisms and offers per-flow end-to-end reservations. The Resource ReSerVation Protocol (RSVP) is used as a signaling protocol to set up and maintain virtual connections and reserve resources along a route. IntServ provides hard QoS guarantees; however, the adopted per-flow granularity leads to a scalability problem because the amount of state information increases with the number of flows and nodes. DiffServ was designed to overcome the difficulty of implementing and deploying IntServ and RSVP. In fact, the DiffServ scalable solution provides QoS on the wired Internet by defining a set of QoS classes and then classifying packets into them according to an SLA negotiated with the Internet Service Provider (ISP). Edge routers perform the complicated flows classification while the core routers do not keep per-flow information, but aggregate different packets that were assigned to different classes on a per-hop behavior (PHB). DiffServ aims to provide service differentiation among traffic aggregates over a long timescale, but it does not fit to a fast topology-changing context.

QoS routing algorithms deployed in the mesh networks adopt either an IntServ or a DiffServ approach according to the network size (coverage area and nodes numbers) and the mobility scheme. For instance, MeshDynamics proposes a technique for wireless mesh PANs called heartbeats [7], which relies on the information provided by each intermediate node to establish paths satisfying the QoS requirements from source to destination. Besides, [21] proposes a QoS routing protocol called WMR (Wireless Mesh Routing) [21] for a wireless mesh LAN infrastructure. WMR supports multimedia applications by guaranteeing minimum bandwidth and maximum end-to-end delay for all intra-BSS and inter-BSS communications; it also guarantees a per-flow granularity and processes a full, on-demand hop-by-hop routing with no route caching [21]. To fulfill the broadband wireless access QoS requirements in MAN networks and address the scalability issues, the IEEE 802.16 standard defines four classes of service while [68] presents a Wireless DiffServ architecture for the wireless mesh backbone.

2.2.7 Security Considerations

Mesh networks need to provide advanced security mechanisms to encourage client subscribing to reliable services. More specifically, the mesh traffic travels through multiple intermediate nodes on the particularly vulnerable wireless channels, thus increasing the hacking probability. Currently, mesh networks provide the same security services deployed in the WLANs and encrypt the backhaul communications which represent the important part of the whole traffic [4]. However, they have some characteristics that render them particularly vulnerable [6]. In fact, the adopted multi-hop communication which relies on the cooperation of the network nodes suffers from selfish behaviors. For instance, some selfish nodes may obtain free services while refusing to participate in routing and affecting the system availability. Besides, the lack of authentication provides attacking nodes with free-of-charge services. Consequently, hackers may cause denial of service by sending arbitrary traffic or advertise high rates, thus affecting network performance. Moreover, the routing service which adapts to the topology changes and the environment conditions can be attacked in several ways. In fact, malicious nodes can mislead targeted actors by pretending higher or reduced utility values to create an inaccurate representation of the network status, thus leading to serious denial of service attacks. To address this issue, each node should locally verify the consistency of the collected information and base its routing decision on the deduced conclusion.

2.2.8 Scheduling and Multimedia Support

Mesh networks adopt broadcast scheduling to coordinate transmissions between the communicating nodes. We mainly distinguish two types of scheduling which vary according to the scheduling-messages contention resolution procedure [30]. For instance, in the distributed scheduling adopted by the IEEE 802.16 standard, the nodes share their scheduling data within the two-hop range and cooperate to avoid contention while resources are granted, thanks to a connection establishment procedure. However, mesh BS collects resource requests from the nodes within a certain range and then allocates the resources in a centralized manner [38]. Such resource reservation procedures are implemented in the MAC layer to establish high-speed broadband mesh connections needed by multimedia applications. In fact, scheduling supplies guaranteed bandwidth and delay based on the flow priority requirements in both metropolitan and wide area networks [72]. In PAN context, beacons are used to allow isochronous transmission by reserving Channel Time Allocation (CTA) slots. We may state that the QoS provision mechanisms proposed for mesh networks differ from one network to another. In the following sections, we further detail the implementations of MAC and routing aware QoS that intend to support multimedia applications.

2.3 Wireless Mesh PANs

2.3.1 Background and Objectives

Wireless mesh PANs aim to provide short-range communications between small groups of fixed and mobile computing devices such as PCs, PDAs, peripherals, cell phones, pagers, and consumer electronics. As the network nodes have power constraints, the multi-hop communication is adopted to increase the coverage area while reducing transmission power and increasing the throughput. Besides, the nodes do not rely on an infrastructure as in wireless LANs; they have to play the role of clients and routers at the same time. Therefore, the network reliability and stability need to be guaranteed despite routers' mobility. In addition, wireless mesh PANs intend to provide multimedia applications that require the design of appropriate QoS routing protocols [7]. More specifically, multimedia home networking with high-speed streaming media and streaming content download, environmental monitoring, automatic meter reading, and plenty of commercial- and industrial-type applications monitoring need to be supported [9].

2.3.2 Challenges

The reliability of the QoS routing service is a major concern for wireless mesh PANs. In fact, in the ad hoc networks context, each node maintains a connectivity graph defining a path for every other node in the network. However, the node's mobility leads to a constant change in the routing tables and result in an important overhead as the number of the network members increases. To address these issues, mesh routings protocols select the next relay based on the local information stating which node has the strongest signal and is closest to the sender. Unfortunately, this local approach is efficient only in the case of small networks; besides, it is not able to guarantee QoS for mission-critical applications. A global approach based on the exchange of compact control messages for the routing tables updates needs to be found. On the other hand, the routing service needs to proactively adapt to the power constraints of the nodes to avoid paths breakage and QoS violations. The third wireless mesh PANs challenge is related to beacon alignment issues. In fact, traditional PANs use beacons to provide isochronous transmissions. A beacon is formed by CTA and Contention Access Period (CAP) time slots, as depicted in Figure 2.2.

CTA time slots are reserved slots for regular transmissions of traffic with hard QoS constraints such as video streaming over a multi-hop network.

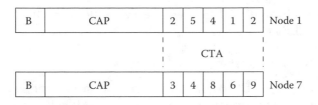

Figure 2.2   Two beacons experiencing interferences.

The Pico Net Controllers (PNCs) send the beacon synchronization pulses to coordinate the transmissions between the managed nodes. However, a node may not receive this pulse due to radio interference from other devices in other pico nets. Consequently, the PNCs should coordinate their transmissions with their managed nodes despite the fact that interference may occur at anytime (during the beaconing period [B], the CAP, or the CTA period).

2.3.3   Architecture

A mesh PAN can either be organized in a full mesh topology or a partial mesh topology. When each node is directly connected to all others, we obtain a fully meshed network [9]. In a partial mesh topology, only some nodes are directly connected to all others; the remaining ones are connected only to nodes with which they frequently communicate. A mesh PAN topology is made up of a PAN Coordinator (PAN-C) that is partially or fully connected with other Full Function Devices (FDDs). Each FDD is then interconnected with a set of Reduced Function Devices (RFDs). FDDs support enhanced functionalities such as routing and link coordination; RFDs are simple send/receive devices. This mesh topology allows better network coverage extension and provides enhanced reliability via route redundancy because nodes may act as routers and relay data in case of link breakage. In fact, data which has not reached its destination is forwarded to one or more neighbors by nodes that act as repeaters. Each node keeps a routing table that indicates which neighbor to contact when a packet with a particular address is forwarded. Moreover, an easier network configuration is fulfilled and the battery lives are extended due to short links usage.

2.3.4   The IEEE 802.15.5 Standard

The IEEE 802.15.5 Working Group was created in May 2004 to define a complete framework that provides a reliable and scalable wireless connectivity for mesh nodes based on the specification of the low-rate wireless

PANs specified in IEEE 802.15.4 standard and the high-rate wireless PANs specified in IEEE 802.1.5.3 [11,13].

2.3.4.1 Meshing and the Ultra Wide Band

The Ultra Wide Band (UWB) is a high-speed physical technique that particularly fits short-range communications. In fact, UWB enhances the meshing capabilities by having low power and cost constraints while guaranteeing precise location information and important throughput. This radio technology transmits signals with extremely wide spectrum (e.g., the bandwidth of the transmission can be several GHz wide [18]) at a very low transmission power so that the resulting Power Spectrum Density (PSD) is very low, thus allowing a massive frequency reuse [10]. For example, 1 W of total power spread across 1 GHz of frequency spectrum puts only 1n W of power into each hertz band of frequency. The resulting reduction of the consumed power allows tiny devices to save their battery life while resisting fading and interference. However, UWB applies only to short-range communications because the bandwidth decreases rapidly as distance increases [3,10]. Consequently, if the same throughput offered by the UWB needs to be provided for wireless mesh LANs or MANs, new physical layer transmission techniques need to be developed. UWB allows the coexistence of tens and even hundreds of simultaneous non-interfering channels within radio distance of each other. Using a mesh topology enables us to trade some channels to increase the overall performance, as illustrated by Figure 2.3.

In fact, nodes A and B are direct neighbors distant by 10 m and having 100 Mbps as available bandwidth. Besides, node C is a common neighbor distant by 5 m from A and B. This shorter distance implies 250 Mbps of available bandwidth between both A and C and B and C. If A wishes to communicate with B, it will be wise to choose the path A -> C -> B with an available bandwidth of 250 Mbps, which is two times faster than the direct one. Meshing also increases the coverage because nodes which are not in direct range can communicate by using other network members as relays. Using large UWB increases the available bandwidth as the number of nodes increases. To conclude, the combination of the UWB technology

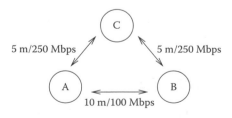

Figure 2.3   Meshing increases the throughput.

and a mesh topology guarantees a very easy and cheap deployment of communication networks for homes and offices.

2.3.4.2 Overview of the ZigBee IEEE 802.15.4 Standard

The ZigBee IEEE 802.15.4 standard specifies the PHY and MAC layers implementation which intend to support low-rate wireless communications in a PAN, which can be either a star, a mesh, or a cluster tree [70]. ZigBee also addresses the third layer functionalities and combines tree routing with on-demand non-tree routing while eliminating single point of failure.

Routes forming a tree branch are optimally traced based on the hop count, link quality, and power. Meanwhile, optimal on-demand paths are orthogonal and connect different tree branches. As a consequence, the tree routes and the on-demand ones interconnect all the nodes within the network and result in a mesh.

Besides, the network defines three logical devices depending on their functionalities. In fact, we distinguish the ZigBee coordinator, which is an FDD; the ZigBee router that can act as a coordinator within its operating area, and the ZigBee end device, which can be either an FDD or an RFD.

The mesh topology defined by ZigBee is also known as the peer-to-peer topology. It defines one PAN coordinator, allows any device to communicate with any other neighboring device, and enables multi-hop transmissions [70], thus forming an ad hoc self-healing and self-forming network.

2.3.4.3 IEEE 802.15.4 Physical Layer

The physical layer defines two services: the physical data and the physical management service. It manages the activation and deactivation of the radio transceiver, the energy detection (ED), the link quality indication (LQI), the channel selection, the clear channel assessment (CCA), and the transmitting and reception of packets across the physical medium [70]. The adopted modulation technique is the direct sequence spread spectrum (DSSS), which offers data rates of 250 kbps at 2.4 GHz, 40 kbps at 915 MHz and 20 kbps at 868 MHz. The low frequencies offer an extended range while the high frequency provides a high throughput. Besides, a single channel is defined between 868 and 868.6 MHz, ten channels are defined between 902.0 and 928.0 MHz, and 16 channels lie between 2.4 and 2.4835 GHz, thus enabling channel reallocation within the spectrum. Receiver sensitivities are −85 dBm for 2.4 GHz and −92 dBm for 868/915 MHz while the maximum transmit confirms with local regulations.

2.3.4.4 IEEE 802.15.4 MAC Layer

The ZigBee MAC layer provides two services: the MAC data service and the MAC management service interfacing to the MAC sub-layer management entity service access point (MLMESAP). The coordinator devises the super-frame into 16 equally sized slots and bounds it by network beacons. In fact, the beacon frame is sent in the first slot of each superframe to synchro-nize the attached devices, identify the PAN, and describe the superframe structure [70]. Besides, the superframe may have an inactive portion during which the coordinator enters in a low-power mode and an active portion consisting of the CAP and the contention free period (CFP). Devices that wish to communicate during the CAP period need to compete to gain access using a slotted (CSMA/CA) approach. On the other hand, the CFP presents guaranteed time slots, which may occupy more than one slot period [70]. The beacon is transmitted at the start of slot 0 without the use of CSMA while all frames except acknowledgment frames or any data frames that immedi-ately follow the acknowledgment of a data request command transmitted in the CAP shall use slotted CSMA-CA to access the channel. A transmission in the CAP shall be complete one IFS period before the end of the CAP, where an IFS (Inter Frame Space) period is the amount of time necessary to process the received packet by the physical layer. If the transmission is im-possible, it will be deferred until the CAP of the following superframe. The CFP starts on a slot boundary immediately following the CAP and extends to the end of the active portion of the superframe. Its length is determined by the length of the combined guaranteed time slots [70].

2.3.4.5 Overview of the IEEE 802.15.5 Standard

A wireless mesh PAN should guarantee isochronous and asynchronous data transmissions and provide high throughput and low latency while sup-porting a high spatial frequency reuse and a decentralized monitoring. To address these issues, [71] proposes the adoption of a superframe with a slot-ted structure at the MAC layer, as depicted in Figure 2.4. This superframe

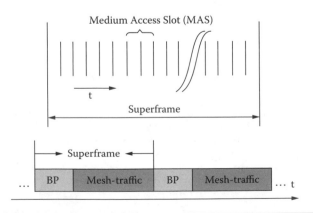

Figure 2.4 A 802.15 MAC superframe structure.

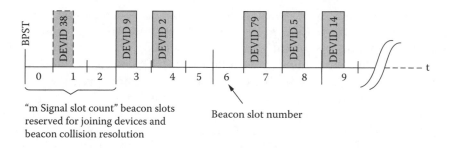

"m Signal slot count" beacon slots
reserved for joining devices and
beacon collision resolution

Beacon slot number

Figure 2.5 A new device joining the announcement period.

is made up of multiple Medium Access Slots (MASs) and divided into a
beacon period and a mesh traffic period, as shown in Figure 2.4.

The beacon period is used to exchange network and topology management information while data is transmitted during the mesh traffic period.
In fact, each device should transmit a beacon which provides the device
ID and the neighborhood and synchronization information along with the
neighbors and the medium access information. The beacon size may vary
and the number of transmitted beacon slots during MAS is determined using
the Adaptive Beacon protocol. Several empty beacon slots may be used by
the joining devices. In fact, a new device joining the beacon period should
indicate its presence within the announcement period, as in the case of the
device 38 in Figure 2.5 [71].

Thereafter, the joining node selects one of the available beacon slots,
as depicted in Figure 2.6. It is worth noticing that the neighbors provide
information about the empty slots and the beacon period duration.

During the beacon period, devices continually listen to the stated information to store the power indication for each beacon and then combine
the power and beacon device ID, thus deducing the neighborhood and
interference graph.

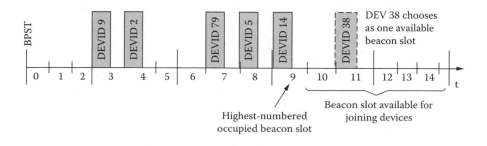

DEV 38 chooses
as one available
beacon slot

Highest-numbered
occupied beacon slot

Beacon slot available for
joining devices

Figure 2.6 Final beacon occupancy.

Data transmission is scheduled during the data transmission period in a distributed fashion. Data may be VoIP flows and multimedia streaming transmitted with QoS guarantees. The distributed QoS support is guaranteed by the Distributed Reservation Protocol (DRP), which acts as follows: communicating devices announce the desired transmissions, the receiver and transmitter may negotiate using the beacons, which carry information on the other reservations. In fact, the transmitter announces its desired transmission with its beacon and the receiver may accept or refuse to communicate. High-priority traffic may replace low-priority traffic and data is transmitted in a unidirectional fashion while the interference awareness allows parallel transmissions. Small frames are aggregated into larger Protocol Data Units (PDUs) and may be transmitted to multiple receivers.

2.3.4.6 Routing and QoS Support

This section presents two different proposals related to routing in wireless mesh PANs that have been submitted to the IEEE 802.15.5 Working Group.

- MeshDynamics Proposal: MeshDynamics has submitted a proposal for the IEEE 802.15.5 Work Group that addresses the QoS routing issue in wireless mesh PANs. In fact, wireless mesh PANs are characterized by the mobility of nodes which play the role of routers, thus affecting the routing performance and the QoS provision. To adapt to the changing topology and the environment conditions, a distributed control layer (Figure 2.7) has been proposed.

  Based on the application requirements in terms of latency and throughput and the nodes' status and setting in terms of mobility

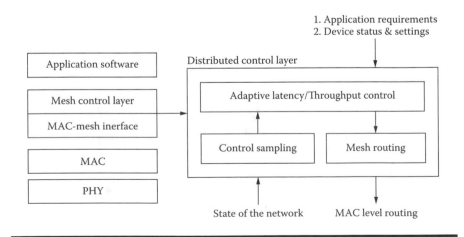

Figure 2.7 Proposal of a distributed control layer.

and power constraints, a QoS mesh routing is performed. For this purpose, the distributed control layer coordinates the mesh routing and adapts it to the power status of the nodes. For example, a relay node which needs to enter the sleep mode has just to change its mode to a low power setting and send a sleep mode message so that other entities will communicate with it only if it is their final destination. Another path that provides the same QoS but does not include the sleeping node is then proactively elected.

To guarantee a QoS routing service, MeshDynamics proposes a technique called Heartbeats [7]. For instance, each node within the network should send a heartbeat including toll-cost and hop-cost information, beacon alignment data, link state, and distance vector information. Each node that has to route a packet should enlist the intermediate entities that need to cooperate to guarantee delivery while providing the required QoS. As intermediate nodes need sometimes to reduce the traffic load when they need to provide better service for the traffic they are generating, they will raise their toll cost, which is the cost of using them as relay. Consequently, nodes with higher priorities will pay a higher hop cost for a shorter path (lower delay path) with increasing toll cost. Meanwhile, traffic with softer QoS constraints will be routed on longer routes and may experience congestion at popular nodes.

In addition, MeshDynamics proposes a software layer on the MAC layer that addresses isochronous transmission in Simultaneous Operating Piconets by managing the beacon alignment issues without modifying the MAC IEEE specifications. The principle consists of applying a theory to determine if there are common reachable nodes that may experience interference. That is, two PNCs that share a common reachable list of neighbors are not allowed to transmit beacons simultaneously; they should stagger their transmission. A PNC that cannot hear any of the other PNCs should hear neighboring intermediate devices that act as repeaters on behalf of their PNCs by sending the heartbeats periodically or as a request response (e.g., a node that hears a request asking for location and neighbors' identities sends the last beacon transmitted by its PNC). A more detailed description of the protocol can be found in [7].

■ Samsung Proposal: This sub-section intends to present the Samsung proposal for the 802.15.5 wireless mesh PAN targeting the low-rate mesh architecture based on the Meshed-Tree approach and addressing Meshed Tree routing, multicasting, and key pre-distribution. The proposal defines the Adaptive Robust Tree (ART) paradigm, which is based on an adaptive assignment of logical addresses reflecting the network topology during the tree definition. The ART defines three phases: the initialization (or configuration) phase, the operation phase, and the recovery phase. The initialization phase is triggered when new nodes join the network and reorganize themselves to form the ART. The tree formation requires the execution of two sub-phases: the association and the address assigning. Then, each node keeps track of the ART branches in the ART table (ARTT). Those branches are assigned one or more blocks of consecutive logical addresses. Communication between nodes starts during the operation phase. However, new nodes may integrate the network and lead to changes in the topology during this phase; hence many reconfigurations may take place to provide an up-to-date status. The recovery phase is triggered when nodes leave the network and cause link breakage. In this case, only the affected tree part is recovered without changing any assigned address; the other nodes still in the operation phase may continue their communications.

The ART formation begins by the association stage during which new nodes gradually join the network beginning from the tree root. After the bottom is reached, a reverse procedure is used to calculate the number of nodes along each branch. After the number of entities is calculated from the bottom to the tree root, each node may indicate its number of addresses. The end of the address assignment procedure is marked by the definition of the ARTT at each node. A meshed tree can then be built on the top of an ART. This can be done by adding additional links so that the network looks like a mesh while each individual link perceives a tree as depicted in Figure 2.8.

The Meshed Adaptive Robust Tree (MART) allows routing a packet through a shorter path; single points of failure can be avoided. For

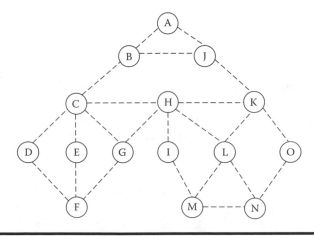

Figure 2.8   A Meshed Adaptive Robust Tree (MART).

instance, if the link between H and B is broken, packets from H to C or to E can still be routed. However, paths are still non-optimal in most cases.

Samsung also proposes a key distribution scheme called KEYDS to provide security services. The mesh nodes that form the backbone should provide security services to the rest of the network entities. Every pair of backbone members shares a secret key that is used to secure the communication between them. Besides, a group key is shared among all backbone members to allow backbone message broadcasts. All mesh points should participate in the key pre-distribution scheme and should be able to perform common pair-wise keys computations. The initial setup of the distribution key management begins when each node within the mesh network obtains its ID. Then, every mesh point obtains the key block from KEYDS with a corresponding column of the incidence matrix. A member of the backbone, as any other mesh point, also obtains the key block from KEYDS.

In addition, every member of the backbone obtains the corresponding key block from the trivial key pre-distribution scheme. Then, every mesh point (except members of the backbone) obtains the final hash-value of the hash chain and the lengths of the chain with respect to that final value. Finally, every member of the backbone obtains the start hash-value (the seed) of the hash-chain and the current length of the chain with respect to the final value given to the mesh points. Key refresh decisions are then taken by the backbone members when needed. When the network topology changes, the key pre-distribution scheme executes the mesh point exclusion, the mesh point association, and the lost mesh points' recovery to adapt to the new network needs.

2.4 Wireless Mesh LAN

Wireless mesh LANs have an extended coverage area compared to mesh PANs; they always adopt an infrastructure-based architecture and rely on reduced-mobility APs. Therefore, the PANs router mobility is no longer a challenging issue. Nevertheless, mesh LANs need to provide QoS guarantees and address hand-off and roaming issues.

2.4.1 Introduction and Advantages

A wireless mesh LAN may be seen as a wireless LAN where all the APs are wirelessly interconnected. Traditional mobility management functionalities such as hand-over and roaming are supported; however, inter-AP communication within the same Extended Service Set (ESS) is done in a hop-by-hop fashion. The transmission scenario in a wireless mesh LAN is done as follows: the AP managing the source forwards the traffic to its neighboring AP instead of sending it to all the APs in the ESS. Then, the neighboring APs sends the same packet to the next hop in the same way until the AP managing the destination is reached. At this time, the traffic is forwarded to the destination end node.

If we compare traditional wireless LANs to wireless mesh LANs, we notice that the latter offers particular advantages related with the deployment costs, offered services, and nature of the supported applications. For instance, deploying a mesh node needs no special wiring and configuration. With little investment and easy configuration process, the network is more reliable because we can simply add as many wireless nodes as needed to increase the performances and cover new zones. Mesh LANs also guarantee load-balancing and optimal resources utilization because wireless nodes may act as routers or APs when the nearest AP is congested and route data to the closest low-traffic node. Fault tolerance is also provided because the clients communicate in a multi-hop fashion, exploiting the redundancy of paths in case of failures. The traffic is automatically rerouted while the failed routers are rapidly detected and recovered or replaced. Furthermore, deploying wireless mesh LANs addresses line-of-sight constraints, especially in outdoor environments. The provided applications in the mesh context fit particularly to the multi-hop architecture as explained as follows [16, 17]:

- Warehousing: Warehousing or broadband home networking applications can be supported by traditional wireless LANs. However, the APs are mainly installed on the roofs to provide good coverage; besides, an expensive deployment of a wired backhaul is needed. Adopting wireless mesh LANs optimally addresses the pre-described deployment issues. In fact, APs are wirelessly interconnected and can be added anytime and anywhere to improve the scalability, the reliability, and the network performance. Moreover, fault-tolerant paths can be used to route the traffic between the mesh nodes until the final destination while congestion resulting from the traditional access to the hub is eliminated.
- Enterprise networking: An enterprise local area network aims at sharing the enterprise resources while guaranteeing high transmission rates and supporting advanced applications. It can be deployed in a small office, or it can interconnect multiple offices in the same building or multiple offices in separate sites. Traditional wireless LANs have been widely adopted to reduce the internetworking costs while improving the scalability. Nevertheless, the need of deploying a wired infrastructure has been always present. Moreover, adding new APs to the backhaul locally enhances the WLAN capacity, but

does not guarantee the fault tolerance and the congestion reduction. Adopting wireless mesh LAN architecture enables the share of resources and an overall performance enhancement, thanks to the multi-hop communication and the wireless infrastructure deployment. In fact, bottleneck congestion resulting from the one-hop access to the traditional APs is eliminated. Besides, the infrastructure can easily scale according to the network's needs without requiring complex configurations and wiring.

■ Healthcare: The hospitals are always built to prevent the propagation of electromagnetic waves because any disruption can have catastrophic consequences. However, exchanging voluminous monitoring and diagnosis data such as high-resolution radiographs at real-time and sharing information between the hospital crew is becoming a pressing need. The deployment of a wired network only interconnects some fixed medical devices while the adoption of a traditional wireless LAN induces high backhaul-wiring costs and many dead zones. The optimal solution consists of deploying a wireless mesh LAN where the mesh nodes and routers are placed according to propagation characteristics and capacity needs.

2.4.2 Architecture Technologies

The mesh wireless LAN has two possible architectures. The infrastructure architecture is formed by different APs interconnected wirelessly within an ad hoc network. The resulting wireless backhaul reacts to any topology changes by processing automatic topology learning and dynamic path configuration. The IEEE 802.11s standard defines the physical and MAC functions needed by the interconnected APs to manage the mesh clients such as the reliable unicast or multicast/broadcast delivery. The infrastructure architecture aims at reducing deployment costs while enhancing network coverage and reliability. More specifically, it becomes easy to add new APs to enforce the existing backhaul network and cover dead zones without any need of wire deployment and complex configurations. The infrastructure meshing is the most used because it allows good scalability and supports gateway functions such as bridging, thus enabling the connection to the Internet and the integration with other network technologies.

The client meshing architecture does not require the backhaul; in fact, mesh nodes can play the role of APs and be clients and routers at the same time forming a dynamic ad hoc network. To do so, the mesh nodes communicate in a peer-to-peer fashion and perform layer 3 routing while supporting auto-configuration and providing end user services. Packets are transmitted within flat network architecture from one hop to another until the final destination; however, congestion occurs more frequently and the network performance rapidly decreases when the number of mobile nodes grows. The hybrid architecture combines the infrastructure and the client meshing to achieve enhanced performances. Mesh clients can be managed by APs, but may also directly communicate with other peers. This mode is still not used very often in case of WiFi meshes.

2.4.3 Challenges

A wireless LAN implementing the IEEE 802.11 standards is formed by one or more APs responsible for central management and a set of mobile stations equipped with a 802.11-compliant interface. An AP and the stations situated in its coverage zone form a cell or Basic Service Set (BSS). The mobile stations may also form an Independent Basic Service Set (IBSS) when they directly communicate in an ad hoc fashion without requiring a central AP. A set of APs may be interconnected by a wired distribution system, thus forming an ESS which can be viewed as a single 802.11 network segment.

In the mesh context, the meshing APs have to form a wireless infrastructure; therefore, they need to implement auto-configuring mechanisms to automatically integrate the ad hoc network formed by the neighboring APs. Besides, the mesh traffic originated by a node is handled by the managing AP which is responsible for its delivery to the destination. This traffic may cross multiple intermediate nodes before reaching the recipient and each crossed node will introduce some latency, thus hardening the QoS provision in terms of minimum delay and jitter. Meanwhile, APs need to exchange data on wireless channels; therefore, mesh networks should guarantee the coexistence of intra-BSS and inter-BSS communication by eliminating possible interference while guaranteeing the required QoS [16]. Hidden and exposed terminals problems should also be addressed. Last but not least, APs forward the arriving packets to their MAC layer, which adopts a drop-tail queue management without taking into consideration the number of crossed hops. This management strategy may lead to a severe unfairness problem because neighboring or smaller hop length flows arrive more frequently at APs and fill up the link layer buffer. Consequently, packets coming from far away nodes face a full buffer and will systematically be dropped.

2.4.4 The IEEE 802.11s Standard

As described so far, the IEEE 802.11 standards define physical and MAC mechanisms for one-hop communications, rely on a wired infrastructure, and are subject to throughput degradation and unfairness when applied to multi-hop communication scenarios. Being aware of the tremendous advantages offered by mesh networks, industrial actors and researchers formed a separate task group in May 2004 under the 802.11 Working Group called IEEE 802.11s ESS Mesh that aims at specifying the physical and MAC extensions needed for the multi-channel support. Two main proposals, denoted by SEEMesh and Wi-Mesh, merged in January 2006 and were confirmed unanimously in March 2006. This fusion has resulted in the embryo of the 802.11s standard that will probably be approved in 2008.

The 802.11s standard aims at specifying the architecture and protocols required for the implementation of a Wireless Distribution System (WDS). The mesh mobile nodes will process an automatic self-configuration as soon as they enter the mesh network while the routing protocol will be integrated in the MAC layer to allow a dynamic path configuration for broadcast/multicast and unicast traffic. When the mesh traffic should reach a destination which is not associated with the AP of the sender, the AP will not send the packets to all APs within its ESS as in IEEE 802.11; it will rather send them to the next AP on the path. The mobile devices will support the multi-channel communications and can be equipped with multiple radios using the same mode while the targeted frequency band will be the unlicensed 2.4 to 5 GHz to guarantee the interoperability with other 802.11 standards.

2.4.4.1   IEEE 802.11s Device Classes

The 802.11s architecture is based on different classes of devices, as illustrated in Figure 2.9. A Mesh Point (MP) may be an AP or a mobile station which provides a partial or full mesh relaying function. An MP processes neighbor discovery and selects the channel to communicate and forward

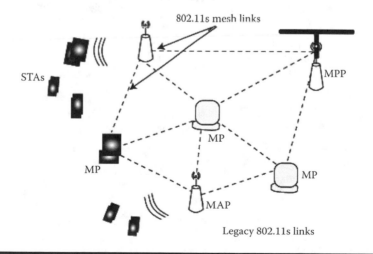

Figure 2.9   The proposed 802.11s architecture.

the traffic for other MPs using bidirectional channels. Mobile stations or end-user devices or stations are traditional stations with no mesh capabilities. Such devices will be wirelessly interconnected to a mesh AP (MAP) which is a particular MP able to operate in one of the legacy 802.11 modes. The 802.11s standard defines the mesh portal (MPP) that interconnects multiple WLAN meshes. The MPP can also play the role of an entry or exit to a wired network and support advanced functions such as transparent bridging, address learning, layer 3 routing, and bridge-to-bridge communications. Finally, an MPP may be configured for topology building and elected to become the root of the default forwarding tree, thus becoming a root portal. Each mesh network is identified by a mesh ID which is the equivalent of a service set identifier (SSID) representing an ESS in legacy 802.11 networks.

2.4.4.2 Medium Access Control: The Medium Access Coordination Function

The Medium-access Coordination Function (MCF) is a MAC sub-layer which is built on the top of the physical layer to provide the mesh services. As depicted in Figure 2.10, the MCF is responsible for guaranteeing the mesh configuration and management, the mesh security services based on the 802.11i standard, the topology discovery and association, the topology learning,

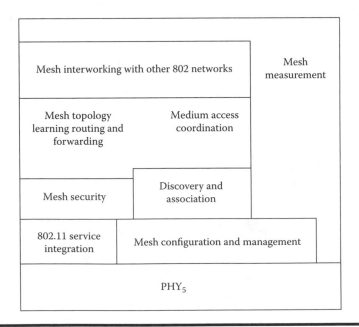

Figure 2.10 The 802.11s MCF function.

the routing and forwarding functions, the medium access coordination, the mesh measurement, and the mesh internetworking with other IEEE 802 networks.

■ Mesh Topology Learning, Routing, and Forwarding: The mesh topology learning and forwarding function is processed by the MP to discover its neighbors. It allows automatic topology learning and enables the link establishments and the dynamic paths discovery for data delivery purposes.

  When a new MP enters the mesh network, it begins by collecting information from neighboring MPs either by sending a probe request or passively listening to the periodic beacons. The candidate MP can then choose to associate with another peer to form the mesh topology. This association highly depends on the peer's capability, its power, its security information, and its link quality.

■ Path Selection Protocol: The MCF sub-layer implements the routing function at the MAC. In fact a hybrid routing protocol supporting both fixed and mobile MPs and including proactive and reactive schemes should be defined to handle unicast and multicast/broadcast traffic delivery. The 802.11s Standard Committee has chosen to mix the Ad-hoc On-demand Distance Vector (AODV, [51]) and the Optimized Link State Routing (OLSR) protocols while defining a set of radio-aware metrics reflecting the link status to enhance the routing reliability. For instance, an airtime metric reflecting of channel, path, and packet error rate has been proposed in [57] while the WRALA metric (Weighted Radio and Load Aware [19]) reflects the protocol overhead at the MAC and PHY layers, size of the frame, bit rate, link load, and error rate.

■ Forwarding Scheme: The wireless LAN mesh network uses four-address data frames with two extensions for QoS support and mesh control, as depicted in Figure 2.11. Each MP which receives a data frame begins by checking its authenticity and destination MAC and then forwards it if everything is OK. As STAs transmit three-address frames, the correspondent MPA needs to convert them to the four-address format before forwarding them toward the destination. Multicast and broadcast traffic is also forwarded if it uses the four-address format; moreover, the time to live (TTL) sub-field is decremented by each intermediate MP to monitor the broadcast data in the WLAN mesh.

■ Medium Access Coordination: The Medium Access Coordination sub-layer that has been proposed in [57,58] implements the enhanced distributed channel access (EDCA) mechanism used in 802.11e [20]. This sub-layer also provides congestion control, power saving, synchronization, and beacon collision avoidance. Multiple channel

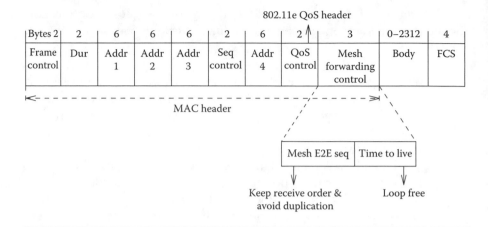

Figure 2.11   The 802.11s mesh data frame.

operations which are based on the common channel framework (CCF) [21] are also supported in multiradio, single radio, or hybrid environments.

■ Mesh Configuration and Management: Mesh networks rely on node self-configuration to accelerate and facilitate the deployment. Therefore, mesh nodes need to implement automatic management modules and association protocols that enable the MPs associating with other MPs neighbors and even external nodes. Management functions should be able to detect the failed nodes to replace them although the mesh network is to a certain extent failure-tolerant. The format of a management frame is shown by Figure 2.12; it includes the DA (destination address) or receiving MP MAC address, the SA (source address) or transmitting MP MAC address, and the BSSID (basic service set ID) field stating for the wildcard value.

It is worth noticing that the interfaces need to implement the 802.11h to enable compliance with dynamic frequency selection

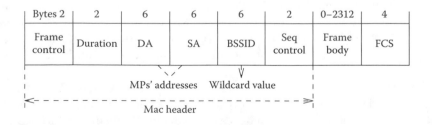

Figure 2.12   The 802.11s mesh management frame.

(DFS) requirements and enhance the efficiency of the multi-hop transmissions, the power saving, and the total capacity.

2.4.5   Routing and QoS Support

Using the mesh network architecture allows a wide coverage, thanks to multi-hop ad hoc communication, but requires a particular QoS management, especially because the mesh nodes act as routers and clients at the same time and do not rely necessarily on a centralized management point. A QoS routing protocol has been proposed for a wireless mesh LAN infrastructure called WMR [21] that supports multimedia applications by guaranteeing minimum bandwidth and maximum end-to-end delay for all intra-BSS and inter-BSS communications.

2.4.5.1   WMR Protocol Overview

The WMR protocol is based on the Ad hoc QoS Routing (AQOR) protocol that has been developed for the MANET context by the authors in [60]. It is based on the following phases: topology discovery, route discovery, admission control with QoS constraints, and route recovery.

■ Topology Discovery: The topology discovery phase consists of exchanging local information with the mesh nodes neighbors to get an updated view of the current topology and estimate the distance to the backhaul. Each mesh node maintains a distance Tag D(I) that indicates the number of hops to the nearest AP; it is set to 0 for APs and to 16 for each newcomer. Moreover, each mesh client and AP within the network should periodically send a Hello message with TTL field set to 1 and a tag field indicating the distance to the nearest AP. This control message is then used to update a list of neighbors N[I] and determine the distance from the nearby AP.
■ Route Discovery: The route discovery is processed on-demand by sending a Route Request for route exploration and then waiting for the correspondent Route Reply enabling the route registration. The traffic addressed to nodes that do not belong to the mesh network is sent to the nearest AP as if it was the final destination.
  ■ Route Exploration: Each node wishing to communicate has to send a Route Request while indicating its QoS requirements in terms of minimal bandwidth and end-to-end delays. The route exploration algorithm differs according to the nature of the destination node. In fact, if the destination is internal to the mesh network, the Route Request is assigned a TTL value and then flooded. However, if the traffic is addressed to an external node (e.g., a node that does not belong to the mesh network such as an Internet destination), the chosen multi-hop wireless path

to the AP should be as short as possible to guarantee good route stability and channel efficiency. Therefore, a distance-constrained discovery algorithm based on the distance tag information stored at every crossed node is proposed. In fact, the source includes its distance tag in the request, and then only the nodes having a smaller value should receive the control packet, update it by setting their own distance tag, and forward it. An initial sequence number equal to zero is set for each Route Request and updated so that only the first accepted packet of a flow is relayed during one round of the control packet propagation, thus minimizing the overhead and reducing the traffic aggregation induced by the multi-hop flow. When a node receives the Route Request, it checks whether its available bandwidth is equal or superior to the required one. If it is the case, the flow is accepted, a new entry is added to the routing table with the status explored, and the packet is forwarded.

■ Route Registration: The destination node should send a Route Reply on the reverse path to the source for every received Route Request. When receiving the reply, intermediate nodes re-estimate their available bandwidth and update the routing table entry by setting the status registered, but the effective bandwidth reservation is only done after receiving the first data packet. All intermediate nodes of all established paths will still be in the registered status for a period of 2 * Tmax, where Tmax is the maximum end-to-end delay of the requesting flow. If no data packet of the correspondent flow arrives within the threshold period, the route will be released.

■ Admission Control: The admission control decision is performed at every node during the exploration phase to discover paths. Thereafter, the route offering the shortest end-to-end delay will be chosen among the paths providing the minimal requested bandwidth.

■ Bandwidth Control: To estimate whether a flow can be transmitted over a path while providing the bandwidth-specified requirements, a correct estimation of the available link capacity and the truly consumed bandwidth is required. As wireless links are shared among all neighboring nodes, the available bandwidth at a node I is determined by the raw data rate of that node and the neighboring transmissions. This available bandwidth value is continuously changing due to the node's mobility. Besides, the bandwidth consumed by a flow (j) is different from the minimal bandwidth required by that flow due to the interference caused by neighbors. To estimate bandwidth values, a half duplex channel and identical data rates and transmission range for all nodes have been assumed [21]. The available bandwidth at a node I is estimated by computing the existing total channel traffic load, which includes the traffic generated by I and its neighbors, I's neighboring traffic, and finally the boundary traffic crossing the boundaries of I's range and exchanged by I's neighbors and nodes that are outside I's range. Finally, to estimate the bandwidth that should be reserved for a flow (j), both the new self traffic and boundary traffic introduced by the requesting flow were considered [21]. After computing the available bandwidth and the required minimum bandwidth, the admission control compares these results to determine whether to accept the flow.

■ End-to-End Delay Control: A proposal was put forth in [21] to estimate the delay from the source to the destination denoted by Tup and the delay back to the source Tdown and verify whether Tup + Tdown < 2Tmax, where Tmax is the maximum tolerated delay. Because many paths may be found, the route on which the route reply arrives first is chosen. If no reply arrives within 2 T max, the source may later retry the route discovery or turn down the flow.

■ Route Recovery: Discovered routes may be broken due to node mobility or channel deterioration, thus leading to QoS violations. To address this issue, the destination node estimates the end-to-end delay experienced by the arriving data packets and triggers the QoS recovery mechanism when needed. With a traditional ad hoc routing algorithm, an intermediate node that does not receive the hello packet from its neighbor after a time-out notifies the source by sending an error packet. Consequently, the path problems cannot be detected at real-time and resolved quickly. WMR detects a QoS violation using the bandwidth reservation information at the destination node. In fact, the destination triggers the recovery mechanism when it does not receive the data packets before the reservation time-out. Besides, an intermediate node may send an error notification back to the source if the next hop cannot be reached to release the reserved resources.

■ Simulation Results: The WMR [21] protocol simulation has been done using OPNET Modeler 7.0, which was modified to support multi-hop communications. The MAC layer module was the default IEEE 802.11 DCF and the WMR was inserted on top of it. The authors have also supposed that all nodes had a transmission range of 200 meters and a raw bandwidth of 2 Mbps. The maximum packet size used in temporary bandwidth reservation was set to 1024 bytes while the sender buffer was set to 64 packets. A source node might retry the route discovery three successive times. Hello messages were sent every second and the neighbor time-out was set to three seconds. Forty nodes were randomly deployed in a 800 m * 800 m range and ten flows were randomly spread among these nodes; the network also included two APs located at diagonal corners of the field. The simulation period was set to 300 seconds. Stream media applications used Constant Bit Rate (CBR) flows with ten packets per second and fixed data packet size of 1024 bytes. All flows tolerated a maximum delay Tmax equal to 0.1 second and required a minimum bandwidth of 80 kbps. The performance metrics that have been considered were (1) the traffic admission ratio, (2) the end-to-end delivery ratio, (3) the average end-to-end delay, (4) the ratio of late packets, and (5) the normalized routing overhead.

The traffic admission ratio is the ratio between the number of data packets sent to the network from the sources and the number of data packets generated at the sources up to time T. The end-to-end delivery ratio is the ratio between the number of data packets that arrive at the destination and the number of data packets sent from the source up to time T. The average end-to-end delay is the average end-to-end delay of data packets received at the destination up to time T, including all possible delays caused by buffering during route discovery, queuing delay at the transmission queue, retransmission delays at the MAC, and propagation delay. The ratio of late packets is the ratio between the number of data packets that exceed the delay bound and the number of data packets that arrive at the destination up to time T. Finally, the normalized routing overhead is the number of control packets transmitted per data packet arrived at the destination up to time T.

The simulation results showed that WMR has succeeded in providing the required QoS while adapting to the network changes and minimizing the control overhead [21]. Nevertheless, we believe that WMR provides QoS within the mesh network; that is, when mesh nodes communicate with external ones (e.g., Internet nodes), the QoS is only provided on the sub-path between the source and the AP. We think that the AP needs to perform a re-estimation of the required QoS in terms of minimum delay by taking into consideration the time already spent when crossing the intermediate mesh nodes Tcross until the AP. It is clear that if the minimum delay is close to Tcross, it will be difficult to provide the required end-to-end QoS. Finally, the WMR did not provide an optimal mechanism for effectively achieving routes recovery in case of paths breakage.

2.4.6 Overview of Available Commercial Systems

■ Strix Systems: The Access/One® Network powered by Strix Systems provides a wireless LAN system that supports multiple radio frequency technology within a scalable network [14]. Access/One Network wireless APs deployed within a mesh architecture can automatically discover their neighbors and route traffic choosing optimal paths according to environment conditions changes. For this purpose, each node identifies the optimal route to the closest and least-congested network server (an Access/One Network module used for control signaling and data registry) and a path to the wired links via mesh nodes. When new nodes integrate the network or congestion occurs on the wireless links, the established routes are automatically re-evaluated to guarantee the maximum performances. Moreover, the network modules scan all available channels in real-time to define a list of potential reachable client modules. Particular radios may be dedicated for particular functionalities (either send or receive) and the least-congested channels are selected to build the mesh. Furthermore, Access/One Network nodes guarantee the authentication by supporting encapsulated RADIUS exchanges, including the MD5, TLS, TTLS, and PEAP mechanisms. Besides, privacy is provided using the supported WEP, including TKIP/MIC enhancements, and AES cipher suites, with either static or dynamic keys. Finally, Access/One Network nodes support the IEEE 802.1q VLAN tagging of wireless frames and assign priorities to them so that they can be processed by a VLAN-aware switch.

■ Tropos® Networks: Tropos Networks propose the MetroMesh™ Networks architecture that provides WiFi clients with a secure access to network services in a coverage area ranging from local to metropolitan [15]. For instance, the Tropos 3210 indoor MetroMesh router implements the proprietary Predictive Wireless Routing Protocol (PWRP) to create a self-organizing and self-healing wireless mesh by searching for the optimal data path to the wired network. The Tropos 3210 indoor MetroMesh router guarantees wireless connectivity to standard 802.11b/g clients. Moreover, it seamlessly meshes with the Tropos 5210 outdoor MetroMesh router to extend the coverage area of the metro-scale WiFi network. The supported MetroMesh OS provides the VLAN technology and implements the auto-discovery and auto-configuration on power-up with a real-time adjustment of the established paths to guarantee optimal performances. Secure management features include AES encryption of wireless routing, MAC address access control lists definition, and a full VPN compatibility. Thanks to such mechanisms, individual users with different privileges and security needs may operate independently while maximizing network economics and performance.

2.5 Wireless Mesh MAN

2.5.1 Purpose

Complex multimedia applications are becoming very popular, leading customers to request the marriage of mobility support with a high bandwidth and an enhanced availability, reliability, and flexibility. As cellular-based technologies have not been satisfactory in many aspects, broadband wireless access is gaining the interest of researchers and network operators while multi-hop communication is expected to become the leading technology. The aim of the mesh metropolitan networks is to provide broadband access everywhere and anytime by increasing reach and coverage through multiple hops, without compromising performance or reliability. Some of the IEEE 802.16 standards have provided the mesh network support and tried to minimize the impact of multipath interference while providing connectivity between network endpoints without direct line-of-sight.

2.5.2 Targeted Services

Compared to wired or cellular networks, wireless mesh MANs are an economic alternative to enable ubiquitous broadband networking with high throughput and multimedia-applications support even for underdeveloped regions. Targeted services are mainly wireless Internet access, public safety, and implementation of intelligent transportation systems.

■ ISP: Internet service providers are searching for integrated solutions that provide public Internet access for residents, enterprises, and travelers with consistent levels of service and pricing, guaranteed scalability, and minimal investments. On the other hand, countries and cities are encouraging the deployment of information technologies to improve government services which will attract business and citizens and boost the economic development. A growing number of ISPs have found in the wireless mesh networks an ideal solution to provide both indoor and outdoor broadband wireless connectivity in urban and rural environments without the need for costly network infrastructure. With a Wireless Internet Service Provider (WISP), users are able to connect to the Internet when they travel outside their home or business, or go to another city that also has a WISP. As examples, the city of Chaska, Minnesota, has formed chaska.net, a WISP that provides low-cost, high-speed Internet connections to more than 7,500 homes and 18,000 residents [22]. The city of Moorhead, Minnesota, has also succeeded in installing a metro-scale broadband WiFi network from Tropos Networks, which provides lower-cost Internet access anywhere in the city [23].

■ Public Safety: Municipal police, fire, and emergency departments have a pressing need for adopting metro-scale mesh networks and the resulting mobile broadband data access. In fact, public safety agents have used mobile data radio systems for years, but the implemented cellular networks offered near-ubiquitous coverage and low data rates (9.6 kbps), thus prohibiting in-field access to multimedia data and applications. Adopting metro-scale mesh networks for mobile broadband data access will improve the effectiveness and efficiency of public safety officers by getting critical information in their hands on the street in a totally secure manner. Furthermore, deploying metro-scale video surveillance (e.g., in high crime areas and strategic targets) will enhance public safety and bring applications such as virtual lineups, fingerprint analysis, and access to detailed mug shots or floor plans out of the station house and into the field where they are needed. Besides, equipping firemen with locator chips and helmet-mounted wireless video cameras can help incident commanders and field crews share knowledge during emergencies.

■ Intelligent Transportation Systems: Mesh networking technology can be adopted by transportation companies to provide intelligent transport systems, if a high-speed mobile backhaul from a vehicle to the Internet is supported. Buses, ferries, and trains equipped with wireless mesh access can provide real-time travel information, allow remote monitoring of in-vehicle security video, permit the addressing of transportation congestion, and help control the pollution.

2.5.3 Architecture

Broadband wireless MAN standards detail two modes of communication: the Point-to-Multipoint (PMP) mode and the mesh mode. With the PMP mode, the subscriber station (SS) can only communicate with a base station (BS) using separate downlink and uplink sub-frames [28]. Consequently, the BS always has to route data between two communicating SSs [29]. The mesh mode adopts a multi-hop communication by allowing every station (subscriber or base station) to directly communicate with other stations in the network, independently of their nature. Thus, traffic can be routed through other SSs and occur directly between SSs while the mesh BS connects the wireless network to the backhaul links. An adaptive scheduling mechanism is used to allocate mini slots and associated channels within the data sub-frame. The assignment of transmission opportunities in the direct links can be controlled by either a centralized or distributed algorithm; furthermore, a three-way handshake is always used to request, grant, and confirm those transmission opportunities.

■ Centralized Scheduling: In centralized scheduling, the BS has to provide the schedule configuration for the SSs within a threshold number of hops after analyzing the transmission requests. Consequently, the BS has the same functionality as in the PMP mode. However, not all the SSs have to be directly connected to the BS because some of them can determine the actual schedule for their direct neighbors from these flow assignments [61]. The centralized scheduling is coordinated because the scheduling packets are transmitted within scheduling control sub-frames without risks of collision. It is particularly adapted for the transmission of persistent traffic streams.

■ Distributed Scheduling: In distributed scheduling, the mesh BS does not coordinate the process in a centralized manner. In fact, all stations (BS and SS) have to coordinate their transmissions with their two-hop neighbors and broadcast their schedules to all their direct neighbors. Each request is analyzed by the granter using a given slot allocation algorithm; then the granter returns a grant message in case of success. In this case, the requester sends back the received message to acknowledge its reception. The distributed scheduling may be coordinated or uncoordinated. The coordinated distributed scheduling uses the scheduling packets transmitted within the control sub-frame. The uncoordinated distributed scheduling fits to occasional or brief traffic over links which have not been considered by the current centralized or coordinated distributed schedule. It is performed in a contention-based manner where scheduling control messages are sent during the data sub-frame while avoiding conflict with the schedules already established using the coordinated procedures [40].

2.5.4 Standards

The IEEE 802.16 standards, also known as WiMAX (Worldwide Interoperability for Microwave Access), is currently viewed as the future technology that will be adopted for the deployment of broadband wireless metropolitan area networks [28]. The physical layer detailed by the IEEE 802.16 standards uses the frequency ranges 2 to 11 GHz and 10 to 66 GHz and supports single carrier (SC), Orthogonal Frequency Division Multiplexing (OFDM) and Orthogonal Frequency Division Multiple Access (OFDMA). The 2 to 11 GHz has no line-of-sight requirements; however, it induces multi-path

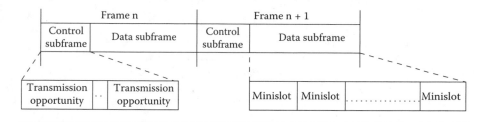

Figure 2.13   802.16 MAC frame in mesh mode.

and requires additional functionalities such as power management, error recovery, and interference mitigation. The MAC layer which manages the share of the common channel resources adopts the Time Division Multiple Access (TDMA) and supports both PMP mode and mesh mode. In the following section, we detail the PHY and MAC extensions needed to support mesh mode.

2.5.4.1   MAC Layer Overview in WiMAX Mesh Mode

The mesh mode defined by the IEEE 802.16 standard supports only Time Division Duplex (TDD), which separates uplink and downlink in time. A MAC frame in mesh mode is made up of two sub-frames fixed in length, the control sub-frame and the data sub-frame, as illustrated by Figure 2.13

The data sub-frame illustrated by Figure 2.14 is used for data transmission in a link connection-oriented basis (there is no end-to-end connection [42]). One link is used for bidirectional data transfers between two SSs without distinction between uplink and downlink sub-frames (per-analysis mesh mode).

Physical bursts vary in length; they are made up of a preamble followed by MAC PDUs. The latter includes a fixed-length MAC header, a fixed-length mesh sub-header, a variable length payload, and an optional CRC field. The control sub-frame is only used for the signaling message transmission transfers. It serves the cohesion, creation, and maintenance between all SSs and to the data scheduling [41]. The parameter MSH_CTRL_LEN determines the number of transmission opportunities that can be carried by one control sub-frame, and ranges between 0 and 15. Besides, each transmission opportunity has the length of 7 OFDM symbols. Consequently, the total length of a control sub-frame is computed by Lcs = 7 ∗ MSH_CTRL_LEN. A control sub-frame can be a network-control sub-frame or a schedule-control sub-frame, as illustrated by Figure 2.15.

The network control sub-frame is useful for new terminals that want to access the network because it is used to advertise network information and

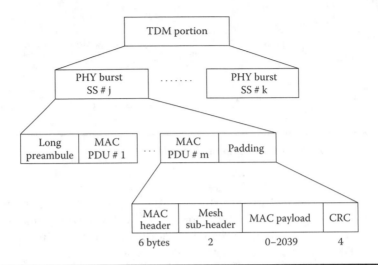

Figure 2.14   802.16 data sub-frame in mesh mode.

synchronization elements [34]. In fact, active nodes periodically broadcast the MSH-NCFG message containing basic configuration information such as the BS identifier and the base channel in current use [35]. A new node that wants to access the mesh starts listening to the MSH-NCFG to pinpoint active networks. Based on the advertised information, it establishes a coarse synchronization and starts the network entry process.

The network entry process begins when a joining node, also called a candidate node, selects one sponsoring node and sends the network entry

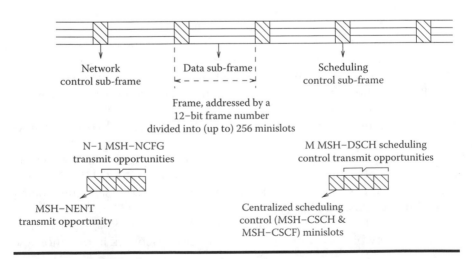

Figure 2.15   The MAC control sub-frame in mesh mode.

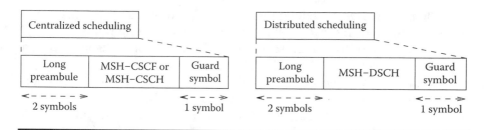

Figure 2.16   The schedule control sub-frame in mesh mode.

message MSH-NENT:Request, including provider configuration data and optional authentication code. The sponsoring node responds by the MSH-NCFG:NetEntryOpen message advertising the candidate's MAC address as being sponsored and including initial schedule. The new node acknowledges by sending a MSH-NENT:Ack; then higher-layer DHCP configuration and authentication are processed. Finally, the new node sends the MSH-NENT:Close and the sponsor responds with the MSH-NCFG:Ack [40]. If the selected sponsor does not advertise the new node's MAC address, then the procedure is repeated MSH-SPONSOR-ATTEMPTS times using a random back-off between attempts. A new sponsor is selected when all attempts fail.

To request bandwidth, SSs send connection-based requests in stand-alone or piggyback messages, including required numbers of bytes. Bandwidth is then allocated on an SS basis. The schedule control sub-frame carries the scheduling information of the data sub-frame transmission opportunities. It is also divided into two parts: the centralized scheduling mechanism (CSCH) and the distributed scheduling mechanism (DSCH), as detailed in Figure 2.16. When centralized scheduling is adopted, the mesh BS periodically collects network information and resources reservation demands while the SS sends its resource allocation request to the BS encapsulated in a CSCH:Request message. The corresponding CSCH:Grant is created by the BS and broadcasted to the SSs within a threshold hop range; then those SSs shall forward the received message to their neighbors that are further away from the BS (i.e., more hops to the BS). The CSCH includes the following parameters [31]:

- Flow Scale: Determines scale of the granted bandwidth
- NumAssignments: Number of 8-bit assignment fields followed
- UpstreamAssignment: Base of the granted bandwidth as bits per second for the ingress traffic of the node in the BS routing tree
- DownstreamAssignment: Base of the granted bandwidth as bits per second for the egress traffic of the node in the BS routing tree

When distributed scheduling is adopted, request and grant of channel resource are delivered by an MSH-DSCH message among nodes.

In coordinated distributed scheduling, all the stations (BS and SS) periodically transmit the MSH-DSCH in a collision-free fashion to inform neighbors with the schedule of transmissions. The mesh distributed election-based scheduling used for scheduling the MSH-NCFG and the coordinated MSH-DSCH control messages guarantees collision-free scheduling within each node's extended neighborhood. The algorithm is run when the local node should transmit (NextXmtTime = now); its inputs are as follows:

- The frame number and the transmit opportunity number within that frame for the type of message being scheduled
- All the node's identifiers within the two or three hops neighborhood
- The XmtHoldoff Time of the local node, which is the node transmit hold-off delay
- As many couples of {node ID, NextXmtTime, XmtHoldoffTime} of nodes within the two or three hops neighborhood as have been recently received, where NextXmtTime is the node's next transmission time of MSH-NCFG

The algorithm processes a pseudo-random mixing function to deduce the NextXmtTime of the current node. In fact if the pseudo-random mix of the local node is superior to all the mixes of eligible competing nodes, the NextXmtTime for the local node is set to CandXmtOpportunityNum and the algorithm returns a success. It is worth noticing that the proposed algorithm is fair and robust because all nodes are treated equally and scheduling seeds are varying pseudo-randomly for each frame leading to non-persistent collisions.

However, in uncoordinated distributed scheduling, the MSH-DSCH message is transmitted to the intended neighbor in the free slots of the data sub-frame without paying attention to possible collusions [10,11,28]. The MSH-DSCH message always includes the following fields [31]:

- Scheduling IE includes the next MSH-DSCH transmission time and hold-off exponent of the node and its neighbor nodes.
- Request IE conveys the resource request of the node.
- Availability IE implies the available channel resource of the node.
- Grants IE conveys grant or confirm information of the channel resource.

Both centralized scheduling and distributed scheduling use the three-way handshake, which principle is given by Figure 2.17. If no MSH-DSCH is received for an uncoordinated distributed scheduling request, the second requestee sends an MSH-DSCH:Grant packet.

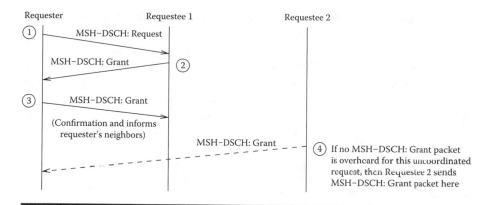

Figure 2.17   The three-way handshake.

Transmission errors are corrected interactively, thanks to the Automatic Repeat Request (ARQ) protocol. The ARQ principle states that when a receiver detects corruptions in a message, it automatically requests a retransmission; then, after getting the correspondent ARQ message, the sender retransmits the message until it is correctly received or until the number of attempts exceeds a configured threshold. The ARQ mechanism is defined at the MAC layer; its implementation is optional and may be per-connection based [47]. However, a connection cannot support ARQ and non-ARQ at once.

2.5.4.2   Hand-Over

An Access Service Network (ASN) includes at least one ASN gateway (GW) and a BS associated with one or more ASN gateway. The BS or ASN GW are called a serving BS or a serving ASN GW, respectively, when they manage the MS before the hand-over and a target BS or a target ASN GW, respectively, if they are associated to the MS after the hand-over. Furthermore, an ASN GW can be an anchoring ASN GW when it used to relay MS data to the serving ASN GW. In this case, the CSN does not carry information about the MS location and the IP address changes become less frequent.

Mobility management needs the implementation of hand-over procedures combined with the SS's context management and data transmissions. For instance, the data path function establishes the correspondent paths and guarantees the data transfers while the SS's context and its exchange in the backbone are handled by the context function. The hand-off functions are responsible for the hand-over signaling and decisions. In fact, the hand-over procedure is first initiated by a request emitted by a serving hand-off function; then the involved targets reply and wait for the correspondent confirmation. Only the entity which receives the confirmation becomes the serving one.

Intra ASN hand-overs which take place between BSs belonging to the same ASN do not result in important delays and data loss; moreover, they do not induce changes in IP addresses because the movement of the SS is transparent outside. However, inter-ASN hand-overs which occur between BSs belonging to different ASNs require a special coordination between the involved ASN GWs where anchoring and re-anchoring are adopted. SSs collect the channel information of the neighboring BSs either by performing ranging or by listening to the current BS's broadcast messages.

2.5.4.3 Physical Layer Overview in WiMAX Mesh Mode

The IEEE 802.16a standard extends the physical layer defined for the 10 to 66 GHz range to support mesh mode operations in the 2 to 11 GHz band of licensed and unlicensed spectrum [36]. In fact, the standard has enabled non-line-of-sight (NLOS) operations while addressing the resulting multipath constraint by adopting the OFDM modulation. Data bits enter the channel coding block to be treated by the Forward Error Correction (FEC) and then interleaved [34]. They are then passed to the constellation map of the modulator. An Inverse Discrete Fourier Transform (IDFT) of length N is then applied to the data sequence, resulting in a frequency domain representation bn composed of N carriers. A digital/analogical conversion is then applied and the resulting signal is low-pass filtered and modulated up to the carrier frequency of choice. The time domain impulse response of a multipath transmission channel approximates that of the Rayleigh distribution [36].

Using the OFDM modulation allows a good average signal-to-noise ratio (SNR), but the SNR of each carrier varies widely. To address this issue, forward error correction codes are used. However, it is important to notice that using OFDM in a noisy environment such as an NLOS air-link simplifies the equalizer design and allows the demodulator estimating the SNR for each carrier and feeds this information to the FEC stage to squeeze the most out of the channel [33].

The IEEE 802.16-2005, also known as IEEE 802.16e or Mobile WiMAX, which was approved in December 2005, is an improvement of the modulation schemes adopted by the original fixed WiMAX standard. In fact, it uses a new modulation method called Scalable OFDMA, which improves NLOS coverage by using advanced antenna diversity schemes and hybrid automatic retransmission request. Moreover, the standard improves indoor penetration and introduces high-performance coding such as Turbo Coding to enhance security and NLOS performance.

2.5.4.4 QoS Support

- QoS Support in WiMAX Mesh Mode: The IEEE 802.16 standard provides QoS for the PMP mode by defining four classes of service:

unsolicited grant, real-time polling, non-real-time polling, and best effort. When examining the MAC header, we find a 16-bits field called CID, which is in charge of distinguishing between unicast and broadcast frames, defining service parameters, and identifying link IDs. Figure 2.18 illustrates the CID of a unicast packet containing the fields Reliability, Priority/Class, and Drop Precedence.

The Reliability field is set to zero when there is no retransmission. It is set to one to indicate retransmit more than four times. The Priority/Class value indicates the priority of the packet and Drop Precedence refers to the probability of the packet when congestion occurs. These three QoS parameters are defined in the protocol despite the lack of a slot allocation algorithm that uses them. To achieve QoS features in the mesh mode, a simple slot allocation algorithm has been proposed in [30]. The principle is to determine a reasonable transmission time by looking up the channel resource table after receiving a request and returning the detail of slot occupation information. For this purpose, the node first computes the number of mini slots (R) requested for transmitting within a frame, according to its Demand Level and Demand Persistence. Then, it deduces the value of the next MSH-DSCH transmission time (T) by consulting the neighbor table, which is stored locally. After that, the node looks up R continuous available mini slots at the same position of the continuous frames (the number is Demand Persistence) starting from time T. In case of success, it returns a grant to the requester; otherwise, failure information is forwarded.

Unfortunately, this simple algorithm is not sufficient for guaranteeing the QoS. To improve it, the authors of [30] have set a checkpoint along the first available time slots and a threshold in the channel resource table. The number of allocated mini slots reflects the utilization of the data sub-frame in a certain degree and the threshold varies between 0 and 256. When the utilization level of the data sub-frame at checkpoint is lower than the threshold, the network state is assumed good and the transmission requests will be treated with the same priority. A utilization level higher than the threshold

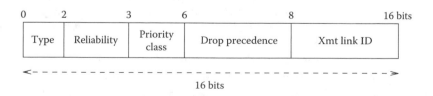

Figure 2.18   The CID field of a unicast packet.

reflects a congested state. In this case, low-priority requests will be answered by failure information.

The drawback of the improved algorithm is that one checkpoint is not enough and may cause mistakes under some circumstances. To address this issue, a second checkpoint is added. When the utilization level at checkpoint 1 is lower than the threshold, the algorithm turns to check the utilization level at checkpoint 2; if exceeded, it searches a frame from checkpoint 2 whose utilization level is below the threshold and allocates mini slots for the frame.

■ QoS Provision on the Backbone: Mesh routers forming the backbone relay traffic between the client nodes and the wireline gateways to communicate with external networks such as the Internet. To increase the coverage area, new wireless routers may be easily added; however, an efficient QoS routing should be provided while addressing scalability issues and taking advantage of the low mobility and power consumption of the nodes. To address these issues, authors in [68] have presented a wireless DiffServ architecture for the wireless mesh backbone. In fact, the DiffServ approach may interconnect heterogeneous wireless/wireline networks; however, its wireless version, which is proposed over the wireless mesh backbone, needs to address the following challenges [68]:

■ Routers need to support both edge and core functionalities as they may collect service requirements from different clients and aggregate them to a unique service level agreement (SLA) requirement or relay traffic to and from the gateways.

■ The centralized bandwidth broker (BB), which collects traffic status at the edge/core router and monitors resource allocation and QoS provision, cannot be defined in the mesh context; therefore, a distributed protocol should be defined to guarantee the BB services in a distributed manner.

■ The wireless DiffServ should handle a large number of gateways. Therefore, the service requirement from a wireless mesh backbone represents the summation of all the aggregating SLAs through all the involved gateways. SLA configuration on each gateway should take into account the wireless mesh backbone topology and the traffic density generated by each router.

■ Wireless links capacity changes constantly. Therefore, the physical and link layers should be taken into account when performing QoS provisioning.

Multi-hop networks generally adopt distributed control and resource allocation protocols. Therefore, the routing protocols are QoS-aware; they search for paths satisfying multiple QoS constraints such as delay and bandwidth. The mesh backbone is a multi-hop network characterized by a low mobility scheme.

The involved routers provide a broadband wireless connectivity and perform the differentiation and classification of the flows generated by their associated networks while optimizing the resources utilization. As a router may monitor multiple ad hoc networks or WLANs within its coverage area, it aggregates flows into classes and routes, the flows of the same class in a single path satisfying that class QoS requirements. Authors in [68] propose a cross-layer routing protocol based on four components: the load classifier, the path selector, the call admission control routine, and the route repair routine. The load classifier determines whether the traffic load of a certain class is low, medium, or high, then triggers the path selector to select the less-congested gateway and select a suitable path to that gateway based on the Greedy Perimeter Stateless Routing protocol [69]. Thereafter, the destination gateway triggers a call admission control procedure which has MAC contention awareness. The route repair routine is started when the route to the destination gateway breaks or when it can no longer meet the QoS requirements. In this case, the path selector should select a new path from the breaking point in order to minimize the overhead.

The wireless mesh backbone can adopt either a CSMA/CA or a reservation-based MAC [68]. The CSMA/CA approach is widely deployed in the WLAN context; however, it suffers from poor throughput and unfairness problems when applied in a multi-hop environment. The reservation-based MAC approach is gaining increasing interest as it guarantees contention-free transmissions, thanks to reservations. Nevertheless the channel reservation is a challenging issue, as it needs to be monitored in a distributed manner [3]. To optimize the MAC resource utilization, resources which are not used by the high-priority traffic class should be assigned to the low-priority traffic class. When reservation-based MAC is used, additional control mechanisms need to be defined to exploit the resources originally reserved for other classes. Controversially, the CSMA/CA MAC approach, which is completely distributed, may become suitable for the wireless DiffServ after addressing the hidden terminal problem, as stated in [68]. To serve the most prior traffic first, the black burst contention scheme is adopted to modify the traditional Enhanced Distributed Control Function (EDCF) proposed by the IEEE 802.11e standard. In fact, each node that wants to transmit should first wait for the channel to be idle for an arbitration interframe period (AIFS) proper to its traffic class. Then, instead of traditionally waiting for the back-off duration, the node should send a black burst, the length of which (in the unit of slot time) equals the back-off timer in order to jam the channel. The node will then wait for the channel to become idle. If it is the case, the node may monitor the channel; otherwise, it will quit the current contention, change the back-off duration, and wait for the channel to be in an idle state for the AIFS again. The node which has high-priority traffic will have a long back-off timer so that the low-priority nodes will sense the black burst of the high-priority node and find the channel busy, thus being obliged to differ the transmissions.

2.5.5   Deployed Solutions

Constructors such as Tropos Networks, Strix, and Nortel have already deployed metropolitan mesh networks in the United States and Taiwan. This section is an overview of the proposed coverage solutions.

2.5.5.1   Tropos® Networks

Tropos Networks tries to offer data communications anywhere, anytime, to anyone that needs it. To achieve this goal, the Tropos MetroMesh architecture combines the ubiquitous coverage of cellular with the ease and speed of WiFi. Thanks to this marriage, effects of interference and multi-path fading across the mesh are overcome while throughput in the range of > 1 Mbps (symmetric) is consistently delivered to standard WiFi client devices.

Many cities in the United States have adopted the MetroMesh architecture to deliver ubiquitous broadband access to their residents. The pioneer case studies of Chaska and Corpus Christi deserve to be investigated.

■ The Chaska Wireless Internet Service Provider: Chaska, Minnesota, has always tried to offer attractive services to its residents. First, the city started its own electricity utility so that its habitants have escaped the pricing demands of a private utility. In 1998, the incumbent telecommunications providers were ignoring the broadband data needs of the schools in the community. To face the problem, the city formed chaska.net, a WISP owned and operated by the city. The WISP implemented wireless point-to-multi point (PMP) technology to replace the traditional T-1 line required by the city's educational institutions.

But the spring of 2004 was the real turning point in Chaska's history. While more and more residents were asking for lower-priced broadband and Internet connectivity that did not tie up phone lines, the city government was struggling to attract new residents and business to Chaska, and to keep them in town rather than going to neighboring Minneapolis. After carefully considering the situation, chaska.net decided to adopt the metro-scale WiFi from Tropos Networks. The city's wireless metropolitan network made use of the city's existing fiber network and was constructed using a combination of Tropos Networks' MetroMesh™ architecture, KarlNet PMP wireless backhaul connections, and an Operations Support System (OSS) from Pronto Networks. The deployment of wireless broadband needed a capital investment of $535,000 and occurred in less than eight weeks, although traditional wireline broadband networks and incumbent wireless (3G) networks can take years and require tremendous investments.

As it uses the 802.11 standard (WiFi) for backhaul and client access, the network requires no proprietary radio frequency (RF) equipment for access devices. Besides, mobile users pay only $15.99 per month with no time-term contracts required and have the ability to freely roam throughout the entire 16 square miles of the city because the 230 deployed Tropos 5110 MetroMesh routers allow transparent roaming. Backhaul was injected at 36 locations around the city using a combination of KarlNet PMP wireless links and connections to the city's fiber network. Scalability was guaranteed because the Tropos 5110 MetroMesh routers automatically reorganize to take advantage of the increased capacity and the additional backhaul.

By using Tropos Networks' metro-scale WiFi technology and existing infrastructure, chaska.net provides broadband access to all 7500 homes in the city as well as city employees, public safety officials, and small businesses at rates up to 60 percent less than competing broadband services, and in many cases at or below the cost of dial-up services. The subscriber management is done using the Tropos Control element manager, which allows chaska.net staff to monitor the WiFi network from a centralized location. When subscribers access the network, the Pronto OSS redirects them to a Web page on the chaska.net Web server. In fact, the Pronto OSS platform and Community Broadband Gateway are in charge of provisioning, authentication, customer billing, administration, customer relationship management (CRM), and roaming agreements. In addition, a global MAC address white list is defined to provide additional security support.

■ The Multi-Use Metro-Scale WiFi, City of Corpus Christi, Texas: Corpus Christi is rated as the largest city on the Texas coast and the nation's sixth largest port. The city always relied on its technology infrastructure to enhance the productivity and efficiency of its municipal services, attract more business, and better serve its residents. However, Corpus Christi was facing permanent problems with meter reading. "Meter readers often have difficulty accessing a property because of fences or dogs," explained Leonard Scott, MIS unit manager and program manager for the WiFi project. "We average several complaints per day, every day, from customers who believe their utility statements are incorrect. If someone wants to buy a house, there is no easy way to check gas and water usage history." To address this issue, Corpus Christi has decided to automate meter reading for municipal gas and water services that supply a 147-square-mile area.

Although a fiber-optic network backbone was covering two-thirds of the city, it did not extend to the third of the area that the Automated Meter Reading (AMR) system would need to cover. To allow coverage of the totality of the zone, Corpus Christi selected Tropos Networks for relaying gas and water meter data from AMR concentrators to the city's utilities business office system. With automated data collection, gas and water customers were able to check daily meter data online and view a property's gas and water consumption history while the municipality was better able to monitor gas usage and water flow.

After living the success story of the AMR application which used a limited portion of the available bandwidth, the city departments soon predicted the potential for hosting new services such as vehicles equipped with laptops for police, fire, and other public safety officers; mobile desktops for field supervisors and managers; and anywhere, anytime access for residents and visitors to city resources such as the library, City Hall, and museums. The only critical question was how to allow broad use of the wireless network while restricting the municipal system to some authenticated users and guarantee the security services for the public safety system. To overcome this problem, the mesh metro-mesh architecture powered by Tropos Networks was combined with the Pronto's OSS, which provides an SSL-encrypted registration and authentication process and supports VPN, which allows secure and encrypted access. Besides, the 300 Tropos 5110 outdoor MetroMesh routers allowed the delivery of multimedia data with automated roaming over the coverage area.

Thanks to the combination of the metro-mesh architecture powered by Tropos Networks with the OSS for subscriber management, Corpus Christi's residents, municipality officers, public safety agents, public works department employees, and building inspectors have been able to get broadband ubiquitous access to vital online information while they are in the field [27].

2.5.5.2 Strix Systems

The Access/One Network Outdoor Wireless System (OWS) of Strix Systems is designed for the deployment of 802.11 networks across large urban areas, rural counties, and entire regions. OWS solutions have been deployed in hundreds of networks worldwide, outdoor and indoor, for the metro, public safety, government, energy, transportation, hospitality, education, enterprise, residential, and carrier access markets. The resulting structured wireless mesh networks provide intelligence, scalability, security, and unrivaled performance. Using Access/One, public safety markets can deploy secure and manageable wireless networks in unlicensed spectrums that support voice, video, and data applications. Furthermore, high-speed Internet access can be provided even in underserved rural areas.

- The Tempe Case Study: The City of Tempe, Arizona, selected the Access/One Network OWS for its citywide WiFi deployment [32]. Tempe will offer secure WiFi access for its residents, businesses, and visitors. Moreover, public safety agents will be provided with WiFi access to their secure private network within all 40 square miles of the city limits. Strix was chosen in partnership with MobilePro for the high throughput and low latency the system offers across large networks. When complete, the citywide network will provide anytime, anywhere access to residents, businesses, and municipal workers, enhancing the way people connect to the Internet, do business, and serve the community.

  The City of Tempe was considered validation of Strix's technology because it was hand-selected from a group of 113 possible proposals. This also speaks very highly of the combined systems and services that the solution is capable of deploying. Some experts affirm that the Access/One Network OWS is an efficient solution that enables customers to dedicate radios for both ingress and egress in the mesh backhaul as well as separate radios for client access.

- The Chittagong Case Study: Strix Wireless Mesh will enable new-generation wireless voice/data/video services for 3.5 million people in Chittagong, the commercial capital of Bangladesh. The deployed mesh network will be based on Strix's Access/One Network OWS and will provide broadband phone and Internet service for residents, businesses, and visitors. Accatel Inc. has partnered with Nextel Telecom to deploy the citywide wireless mesh infrastructure; it is now installing 90 Strix OWS nodes for the initial network deployment, which will support 10,000 voice subscriber lines in an eight-square-mile area. The second phase of the project will add 15 to 20,000 voice subscribers in 12 months. Within three years, the Strix wireless mesh network is expected to include hundreds of OWS nodes and serve 200,000 voice subscriber lines. In the near future, the wireless mesh network will be deployed over the whole area of Chittagong and other cities within the licensing area.

■ Nortel's Case Study: Marshalltown, Iowa, is a rural community with a small population. To encourage economic development and attract businesses and residents, Marshalltown has decided to adopt the wireless technology and launch the first WiFi city network in the state of Iowa. The Marshalltown Economic Development Impact Committee, in conjunction with critical communications system integrator RACOM, has chosen Nortel's wireless mesh solution to initially provide end-user WiFi services to a 20-square-block area in the downtown core. The network infrastructure is based on seven Nortel 7220 WLAN APs supported by a Nortel Wireless Gateway 7250, giving free public WiFi services to local residents and businesses. The new broadband network delivers mobile Internet access at 800 kbps for roaming users within the downtown core. Public safety workers are also supported by the network. Besides, the mesh solution allows the network to differentiate high-priority emergency response traffic from low-priority public Internet access. Marshalltown plans to support the delivery of data communications for emergency response teams, including video surveillance, as well as access to local, state, and national databases for relevant information. In the near future, the wireless mesh network will cover the entire county and support WLAN IP telephony and VPN capabilities [39].

2.6 Wireless Mesh WAN

Mesh WANs intend to provide ubiquitous mobile broadband wireless access in a cellular architecture while supporting mesh networking in indoor and outdoor scenarios. For instance, mobile travelers can enjoy Internet access while passengers information services, remote monitoring of in-vehicle security video, and driver communications may be supported within a complete transportation system. Besides, the guarantee of an NLOS communication enables users to extend the coverage area and to build a wide mesh network that provides Internet-based applications such as streaming and VoIP with enhanced throughput, reliable services, and QoS support.

The Mobile Wireless Broadband Access (MWBA) is a transmission technology that allows important throughput for last-mile wireless connections [43], which is why it has been adopted by both IEEE 802.20 and IEEE 802.16e standards. Broadband services are provided to potential customers with support of multimedia applications. Besides, MBWA systems are resistant to rapid channel variation and address the implications of mobility on the IP layer by maintaining the routability of packets during IP hand-off.

The IEEE 802.20 standard intends to provide wireless access systems with mesh networking support for high-speed mobile subscriber stations within a medium-to-extended metropolitan area. In fact, IEEE 802.20 operates in licensed bands below 3.5 GHz and specifies the MAC and physical layers extensions that offer ubiquitous mobile broadband access for cellular and mesh architectures for mobile users traveling at up to 155 mph with NLOS communications support. In the following sub-sections, what little information is currently available about the 802.20 PHY and MAC layers will be presented and the similarity and differences with respect to 802.16e will be discussed.

2.6.1 IEEE 802.16 Mobility Management

The IEEE 802.16e standard is an amendment of the IEEE 802.16d standard, also known as IEEE 802.16-2004, which supports the mesh mode. IEEE 802.16e adapts the scalable OFDMA (SOFDMA) technique at the physical layer to improve multi-access capabilities while enhancing the MAC layer by addressing mobility issues and particularly hand-over. IEEE 802.16e overlaps with the mandate of IEEE 802.20 and introduces nomadic capabilities allowing mobile users to connect to wireless Internet services providers while moving at a speed of 75 to 93 mph. To manage client mobility, different types of hand-over have been addressed [48]. Following is a brief description of each type.

- MS-Initiated Hand-Over: This hand-over occurs when a node detects degradations in the signal with its serving BS or when it deduces that it can get a higher QoS at another BS. The hand-over decision is taken after collecting gain information from the neighboring nodes which periodically broadcast the mobile neighbor advertisement message specifying frequency of the BS they belong to, its identifier, the types of services it supports, and its available radio resources. The mobile station may also precede a neighbor scanning by synchronizing with some targeted BS's downlink transmissions and estimating the quality of the physical channel. After defining a list of candidate BSs, the MS sends a notification to its serving BS. The serving BS coordinates with the candidates to get a hand-over pre-notification response and define a list of targets. The MS may then choose one target and should inform its serving BS that it is leaving.
- BS-Initiated Hand-Over: A serving BS may decide to exclude some MSs when it detects that the managed nodes are leaving the coverage zone or when it estimates that it can no longer provide the required QoS.
- Soft Hand-Over: Soft hand-over is performed when an MS is able to receive the same MAC/PHY protocol data units from one or more

BSs, thanks to diversity combining at the antenna. Soft hand-over permits the MS to continue receiving real-time data despite the hand-over procedure; however, it requires multiple antennas and it is more complex.

2.6.2 IEEE 802.20

The IEEE 802.20 standard intends to provide a downlink rate of 1 Mbps and an uplink one of 300 kbps for high-speed mobile users while guaranteeing efficient packet-based data services with real-time traffic support [44]. It supports the mesh networking paradigm and the NLOS communications. The architecture of an IEEE 802.20 network guarantees seamless integration of different user domains. In fact, targeted applications are VoIP, financial transactions, online gaming, audio and video streaming, videoconference, WAP, file download, Web browsing, etc. The supported devices (laptops, PDAs, and smart phones), which have different mobility, battery, and storage constraints, will generate different traffic and application models, depending on their characteristics. However, they will benefit from a seamless ubiquitous access.

The IEEE 802.20 standard gives the specifications of the physical and MAC layers that provide enhanced services to the third layer of the OSI model to achieve reliable IP packets routing between external terminals and mobile users or between mobile users. The IEEE 802.20 MWBA system architecture addresses resource allocation, rate management, and authentication issues, and pays specific attention to location management and hand-over.

Table 2.1 summarizes the principal characteristics of the air interface as specified by the IEEE 802.20 standard. In addition to its support for the multimedia applications and QoS requirements, IEEE 802.20 guarantees a seamless hand-over between other network technologies, thanks to the adaptation layer (virtual interface). In fact, the hand-off is implemented at the MAC layer while the virtual interface manages multiple wireless network interfaces on a single host by providing a virtual MAC address to the station. As a result, each mobile node is assigned a unique IP address although it may move between different wireless networks; the station's mobility will be reflected by the changes in the virtual MAC values.

2.6.2.1   802.20 PHY Layer

The PHY layer of the 802.20 standard is typically based on the technologies developed in the 802.16 working groups. The standard for the PHY layer, however, is more heavily angled toward use in a mobile setting and seems to be inclining toward using OFDMA (Orthogonal Frequency Division Multiple Access) in a similar way to 802.16e. This mainly can reduce the

Table 2.1   The IEEE 802.20 Air Interface Specifications

Characteristic	Target Value
Mobility	Vehicular mobility classes up to 250 kmph (as defined in ITU-RM.1034-1)
Peak user data rate (downlink [DL])	> 1 Mbps
Peak user data rate (uplink [UL])	> 300 kbps
Peak aggregate data rate per cell (DL)	> 4 Mbps
Peak aggregate data rate per cell (UL)	> 800 kbps
Airlink MAC frame RTT	< 10 ms
Bandwidth	e.g., 1.25 MHz, 5 MHz
Cell sizes	Appropriate for ubiquitious MANs and capable of reusing existing infrastructure
Maximum operating frequency	< 3.5 GHz
Spectrum (frequency arrangements)	Supports FDD and TDD frequency arrangements
Spectrum allocations	Licensed spectrum allocated to the mobile service
Security	Support AES

development time of products. However, the possibility of using OFDMA (Orthogonal Frequency Division Multiple Access) on the downlink connection and CDMA (Code Division Multiple Access) on the uplink has been mentioned. The reason for using CDMA on the uplink is that using OFMDA somewhat limits the benefits that antenna technologies like spatial multiplexing can provide. CDMA can help to reduce this limitation by assigning the same bandwidth resources to all users in a sector and using spatial processing at base station to recover the signal [26].

Modulation and coding in 802.20 is essentially identical to that of 802.16a/d. Besides, to allow flexible high-speed mobility, the 802.20 standard is expected to support basically all of the advanced transmission options that the 802.16 standards define. These standards include, but are not limited to space–time block code and various forms of spatial multiplexing/MIMO (Multiple-Input Multiple-Output). A wide variety of channel bandwidths from 1.25 to 40 MHz are also expected to be supported with both TDD and FDD multiplexing. Using 1.25 MHz channel speeds (similar to ADSL), while providing 1 Mbps downstream and 300 kbps upstream, are expected to scale with wider channels. This will allow the support of up to 100 users per cell.

2.6.2.2   802.20 MAC Layer

The MAC layer of the 802.20 standard is also loosely based on technologies developed in the 802.16 working groups. Similar to 802.16, the 802.20

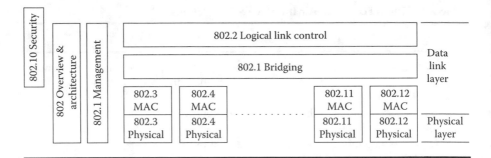

Figure 2.19   LLC functionalities.

MAC is divided into convergence-specific and common-part sub-layers. Furthermore, mobility techniques developed in 802.16e such as hand-off and power management are also implemented in the 802.20 standard. Figure 2.19 details the logical link control (LLC) services that intend to guarantee reliable data transmissions. It also shows that the IEEE 802.20 may support common and specific parts of the physical layer to support various PHY technologies [45]. The connection establishment mechanism to be provided by 802.20 is not yet fully defined, but due to the standard's resemblance to 802.16e, it is expected that the mechanisms will be largely similar. One difference between the two, however, is that CDMA (with respect to OFDM/OFDMA) could be utilized on uplink connections.

Because 802.20 is a fully mobile standard, it will provide support for all types of hand-off mechanisms to enable users to freely roam between cells without interruption. Soft hand-off provision will be entirely integrated. 802.20 also will fully integrate higher-level hand-offs over Mobile IPv4 and Mobile IPv6. Because different forward and reverse-link connection mechanisms may be used, hand-off will need to occur in both directions [26]

The level of QoS support that 802.20 will offer is to some extent undecided at this moment. The common requirements document agrees, however, on the fact that DiffServ and RSVP will be supported for end-to-end compatibility with traditional networks. Finally, note that the 802.20 standard offers performance similar to that provided by usual 2.5G and 3G cellular technologies. 802.20, however, presents the clear advantage of being a fully IP-based, packetized network standard. Consequently, network throughput is enhanced versus a circuit-switched standard, because messages do not have to be encoded from pre-allocated circuits into packets (and back) each time a request is sent or received. Additionally, the 802.20 offers a higher spectral efficiency than any current cellular standard. Thus, the 802.20 is expected do more with less channel bandwidth and would handle a higher number of users per cell.

2.7 Advanced Issues

Factors such as network topology and architecture, traffic nature, and node mobility highly mark the mesh network's capacity and performance, thus affecting protocols development and implementation. All protocols need to be improved or reinvented while considering a cross-layer design. This section gives an overview of the hottest research issues aimed at designing scalable, low-cost, and easily deployable wireless mesh networks.

2.7.1 Physical Layer

WMNs physical layer should be revised to provide important rates and wide coverage while enhancing reliability by solving the fading, multipath, and interference constraints. Traditional modulation techniques such as OFDM and UWB should be replaced by new schemes that allow better data rates in larger areas. For instance, the MIMO technique, which intends to improve the wireless network capacity by adopting antenna diversity and spatial multiplexing, can be exploited. In fact, using multiple antennas for reception provides the receiver with replicas of the transmitted signal, thus reducing fading and interferences. Moreover, adopting spatial multiplexing permits the simultaneous transmission of different data streams by breaking the channel into multiple spatial channels and then using each of them to transmit a differently encoded traffic.

As diversity techniques are inefficient in case of strong interference, smart antennas with beam-forming capability may also be adopted to provide the receiver with high gain in the direction of the desired signal and low gain in all other directions. Cheap directional-antenna implementation and frequency-agile techniques should be further investigated to build a high-capacity wireless backhaul system [62]. The MAC layer design should also be done according to the added values of the physical layer to achieve the expected improvements. Many MAC protocols as stated in [63–66] have been developed to support directional and smart antennas in the ad hoc network context, but an additional effort is required to implement a MAC protocol with multi-antenna-systems support. Moreover, cognitive radios technologies represent a new research field that needs to be investigated.

2.7.2 MAC Layer

Mesh nodes mobility and nature (router or client) combined with power constraints add complexity to the design of a MAC scalable protocol. In fact, existing medium access-control protocols (such as CSMA/CA), which apply to the ad hoc context, suffer from poor performance and frequent collisions when the number of nodes increases; therefore, they should be replaced by TDMA and CDMA schemes while overcoming the induced difficulties. Advanced techniques such as MIMO and cognitive radios, which can be implemented at the physical layer, need a particular MAC design to effectively enhance the throughput and coverage.

The scheduling is also a critical issue because it should address multi-user diversity according to the cross-layer design. In fact, transmission opportunities allocation should be coordinated among all wireless routers to grant transmission to users experiencing peak in their channel quality [67]. Moreover, open research issues related to scheduling should determine how to profit from other diversity techniques implemented at the physical layer, such as spatial diversity and frequency diversity, to enhance the throughput. Besides, interoperability of various wireless technologies requires the definition of particular bridging functions at the MAC level. Furthermore, a multi-channel multi-transceiver MAC can be a promising solution to guarantee reliability and enhance the provided data rates. Finally, a better QoS has to be offered at the MAC level to support multimedia traffic transmissions that are particularly affected by delays, packet loss, and jitter.

2.7.3 Network Layer

Multi-hop communication protocols rapidly lose their performance when the network size increases. Routing schemes designed for WMNs should ensure scalability and enhance network performance without adding complexity and management difficulties. In fact, the destination of mesh traffic may be multiple mobile nodes; furthermore, the same traffic may simultaneously follow multiple paths to reach the same AP. Thus, the routing protocols need to rely on correct link status information provided by the physical and MAC layers to discover high-quality routes. New routing metrics that reflect the loss rate and the available bandwidth of intermediate links need to be developed. Multicast traffic routing can also be a hot research topic. Cross-layer design, which intends to enhance routing performances by considering MAC parameters and feedback, is a promising research issue that needs to be further investigated. Routing protocols should also take into consideration the mesh nodes' nature (which can be routers or clients) to correctly respond to different mobility and power constraints.

2.7.4 Transport Layer

Transport protocols that are used in the ad hoc context are also adopted by the WMNs. These protocols can be classified as reliable TCP variants, entirely new reliable protocols, or protocols designed for real-time delivery. TCP variant protocols aim at overcoming the performance degradations experienced by TCP when it is applied to the ad hoc context. In fact, non-congestion packet losses caused by the transmission over unreliable wireless links are considered by TCP as congestion losses and induce severe throughput decreases. To address this issue, the protocol designed in [50] adopts a feedback mechanism that allows a differentiation between losses caused by congestion and those caused by wireless channels; however, a future study is needed to correctly design a loss differentiation approach and to accordingly modify the TCP protocol for WMNs.

Besides, the connection-oriented TCP protocol which relies on ACK reception is highly affected by mesh network asymmetry in terms of bandwidth, loss rates, and latency [37]. In fact, TCP data and the correspondent ACK may take different routes in the mesh network, thus leading to performance degradations. Some ACK processing schemes have been proposed and a different network architecture has been presented in [15] to solve the asymmetry-related problem, but their effectiveness for WMNs should be further investigated. A cross-layer optimization can also be adopted to enhance the TCP performance because the network asymmetry is closely related to lower-layer protocols. Moreover, the high variation of the RTT caused by node mobility and dynamic path changes has severe consequences on the TCP performance. Adapting TCP to RTT variation in the WMNs is still an open research topic.

To address TCP shortcomings, new protocols have been developed. To this end, the ATP protocol, [12], which is rate-based, differentiates between congestion and non-congestion losses by examining the resulting delays and does not set transmission time-outs while addressing congestion control and reliability separately. However, adopting a brand new transport protocol for the WMNs will result in non-interoperability with existing technologies. More specifically, WMNs should be able to permit network access for conventional and mesh clients and wireless mesh nodes which need to access the Internet and also to be integrated with heterogeneous wireless networks such as IEEE 802.11, 802.16, and 802.15. One solution will be the development of a special adaptive TCP variant for WMNs which addresses traditional TCP performance degradations while being compatible with the traditional TCP protocol. Furthermore, end-to-end real-time transmission guarantees have been addressed by both RTP (Real Time Protocol) and RTCP (Real-Time Transport Protocol) in compliance with an RCP (Rate Control Protocol). However, there has been no RCP proposition specifically designed for the WMNs.

2.7.5 Application Layer

New application layer algorithms need to be developed so that real-time Internet applications can be supported by multi-hop wireless mesh networks. Furthermore, distributed information sharing over WMNs has specific characteristics that need to be addressed by new applications protocols. Finally, new applications that take advantage of the WMN's particularities need to be invented to effectively provide an added value. For example, new tools may be developed for a home networking environment to achieve home automation by allowing the remote monitoring, configuration, and control of all electronic devices.

2.7.6 Network Management

A centralized control of node location and hand-over is not applicable in the mesh context where an LOS with the BS is not required and where the client nodes may constantly roam while mesh routers have restricted mobility. Developing a distributed location management scheme for WMNs is an interesting research topic that needs to be investigated. In the same way, power management procedures vary according to the nature of the mesh nodes. On one hand, mesh routers which do not have power constraints need to manage their transmission power to control the connectivity and reduce interference while increasing the spectrum spatial-reuse efficiency. On the other hand, mesh clients which may be IP phones or sensors require particular power efficiency.

Consequently, power management for the WMNs is an open research topic that needs to be further investigated. Finally, network monitoring protocols need to be developed to effectively manage mesh routers and enhance network performance. In fact, mesh routers have to report statistical data to one or more servers to detect network anomalies and correctly respond to them. Special data processing algorithms need to be developed and network management procedures designed for the ad hoc networks need to be further enhanced to support large-scale mesh networks.

2.7.7 Security

Security schemes designed for WLANs provide authentication, authorization, and accounting services by implementing them at the AP or at special gateways. Besides, VPN techniques are provided over WLANs using standard key encryption algorithms for tunneling, such as IPSEC. Unfortunately, such schemes are not completely suitable for WMNs because the WMNs do not provide a trusted centralized party that ensures a secure key and certificates management. Besides, attackers may easily benefit from the lack of infrastructure to target routing and MAC protocols, leading to congestion and denial of service.

All these security breaches need to be addressed to convince wireless mesh networks customers to subscribe to reliable services. Security mechanisms need to be embedded into the communications protocols of the different layers so that intrusions are detected and tolerated. Designing a cross-layer framework that monitors the security of the communication protocols is a challenging research topic that needs to be investigated.

2.8 Conclusion

The goal of this chapter has been to present the wireless mesh networking fundamentals aimed at designing scalable, low-cost, and easily deployable mesh networks with coverage ranges from PAN to WAN. We may state that, although they inherit from the MANETs characteristics, mesh networks have their own specificities. In fact, scalability issues need to be addressed as the network may integrate a large number of nodes and provide a wide coverage. Besides, distributed protocols need to be implemented to guarantee an efficient network management and control. As multimedia applications support is a must, mesh networks need to rely on QoS-aware routing protocols able to establish the most suitable path while providing the QoS requirements in terms of bandwidth, delay, and jitter. Nodes mobility management and hand-off should also be addressed because clients need to move at different speeds without losing access to the applications they are using (e.g., Internet access, access to a public-safety private network, etc.). Last but not least, mesh networks need to provide advanced security mechanisms to encourage client subscribing to reliable services.

We can state that mesh PANs, LANs, MANs, and WANs share common characteristics and face common communication challenges although their requirements may differ. For instance, when addressing transmission issues, we can conclude that the UWB technique enhances the meshing capabilities, but is only applicable in the short-range communications context. Therefore, different transmission techniques may be used in the MAN and WAN context to support node mobility at medium and high speeds while resisting multipath and fading. To provide QoS, it is possible to adopt the IntServ approach for PANs and LANs because the node number is not very important. However, a DiffServ approach fits the MANs and WANs contexts because it provides a scalable solution and guarantees soft QoS requirements. In addition, mobility constraints highly differ according to the network size. In fact, in the mesh PAN context, it is difficult to maintain QoS-aware paths because both routers and mesh nodes are mobile; however, the average speed is about 5 kmph. Mesh LANs always rely on a fixed infrastructure; nevertheless, they need to address hand-over and roaming issues as the served mobile nodes may move from one ESS to another. Mesh MANs and WANs include a fixed backhaul and a large number of mobile nodes moving at medium or high speed; therefore, guaranteeing QoS and addressing hand-over and roaming becomes a challenging issue, especially when propagation conditions induce multipath and fading.

Many works are currently being conducted on designing robust mesh networks ranging from PAN to WAN, but the finalized standards versions have not yet been released.

References

[1] Raffaele Bruno, Marco Conti, and Enrico Gregori, Mesh networks: Commodity multihop ad hoc networks, *IEEE Communications Magazine*, March 2005, 43(3): 123–131.

[2] Myung J. Lee, Jianliang Zheng, Young-Bae Ko, and Deepesh Man Shrestha, Emerging standards for wireless mesh technology, *IEEE Wireless Communications*, April 2006, Volume 13, Issue 2, pp. 56–63.

[3] Ian F. Akyildiz, Xudong Wang, and Weilin Wang, Wireless mesh networks: A survey, *Computer Networks*, Elsevier, 2005.

[4] David H. Axner, The Up Side and Down Side of Wireless Mesh Networks, available at www.packethop.com/pdf/business_communications_review_january_2006.pdf.

[5] Steven Conner and Roxanne Gryder, Building a Wireless World with Mesh Networking Technology, *Technology@Intel Magazine*, November 2003, available at www.intel.com/technology/magazine/communications/nc11032.pdf.

[6] Bogdan Carbunar, Ioannis Ioannis, Cristina Nita-Rotaru, and Aaron Walters, Building Trustworthy Wireless Mesh Networks, Dependable and Secure Distributed Systems Labs, Purdue University (Microsoft presentation).

[7] Francis daCosta, Managing the Performance of Ad hoc Mesh Networks, MeshDynamics proposal for the IEEE P802.15 Working Group for Wireless Personal Area Networks (WPANs), IEEE P802. 15-04-0211-00-0005, May 2004, available at www.meshdynamics.com/Publications/MDWMAN-OVERVIEW.pdf.

[8] Jianliang Zheng, Yong Liu, Chunhui Zhu, Marcus Wong, and Myung Lee, IEEE 802.15.5 WPAN Mesh Networks, IEEE P802.15-05-0260-00-0005. Available at grouper.ieee.org/groups/802/15/pub/ 05/15-05-0260-00-0005-802-15-5-mesh-networks.pdf.

[9] John Boot, IEEE P802.15.TG5 Call for Applications, IEEE P802.15-05/267r1, May 2004.

[10] Jack Lang, Mesh Network Outline, IEEE P802.15-15-03-0393-00-0030, September 2003, available at grouper.ieee.org/groups/802/15/pub/03/15-03-0393-00-0030-mesh-network-outline.doc.

[11] Jinyun Zhang, Mesh Networking Support for IEEE 802.15 WPAN, Mitsubishi Electric Research Laboratories, 2005, available at http://www.merl.com/projects/wpan-mesh/.

[12] K. Sundaresan, V. Anantharaman, H.-Y. Hsieh, and R. Sivakumar, ATP: A Reliable Transport Protocol for Ad hoc Networks, in *ACM International Symposium on Mobile Ad hoc Networking and Computing* (MOBIHOC), 2003, pp. 6475.

[13] IEEE P802.15.4/D18, Draft Standard: Low Rate Wireless Personal Area Networks, February 2003, available at www.ember.com/pdf/ EM2420datasheet.pdf.

[14] Access One Network, Strix Systems, Product description, available at www. strixsystems.com.

[15] 3210 Indoor MetroMesh Router, Tropos Networks, available at www.tropos. com.

[16] Daniela Maniezzo, Gianluca Villa, and Mario Gerla, A Smart MAC-Routing Protocol for WLAN Mesh Networks, UCLA Technical report #040032, October 2004, available at www.cs.ucla.edu/ST/docs/tech_rep2004.pdf.

[17] Wireless LAN Infrastructure Mesh Networks: Capabilities and Benefits, A Farpoint Group White Paper, Document FPG 2004-185.1, July 2004, available at http://wireless.itworld.com/4260/farpoint_wlanmesh/.

[18] Joseph F. Herzig Jr., An Analysis of the Feasibility of Implementing Ultra Wide-band and Mesh Network Technology in Support of Military Operations, Thesis, Naval Postgraduate School, Monterey, California, March 2005, available at http://www.stormingmedia.us/24/2422/A242234.html.

[19] Are Two Links Better than One? Digitimes interviews Ted Kuo of Accton Technology about the state of mesh standards.

[20] Roxanne Gryder and Steven Conner, Building a Wireless World with Mesh Networking Technology, available at www.intel.com/technology/ magazine/communications/nc11032.pdf.

[21] Qi Xue and Aura Ganz, QoS routing for mesh based wireless LANs, *International Journal of Wireless Information Networks*, Vol. 9, No. 3, July 2002.

[22] Metro-Scale WiFi as City Service, chaska.net, Chaska, Minnesota, A Tropos Networks Case Study, October 2004, available at www.tropos.com/pdf/ chaska_casestudy.pdf.

[23] Anytime, Anywhere Broadband GoMoorhead!, Moorhead, Minnesota, A Tropos Networks Case Study, November 2005, available at www.tropos. com/pdf/casestudy_gomoorehead.pdf.

[24] Metro-Scale WiFi for Public Safety, San Mateo Police Department, A Tropos Networks Case Study, March 2004, available at www.tropos.com/pdf/ SMPD_Casestudy.pdf.

[25] Metro-Scale Video Surveillance: High-Profile Criminal Trial, A Tropos Networks Case Study, August 2004, available at www.tropos.com/pdf/ peterson_casestudy.pdf.

[26] Jim Tomcik, MBFDD and MBTDD Wideband Mode: Technology Overview, http://grouper.ieee.org/groups/802/20/Contribs/C802.20-05-68r1.pdf.

[27] Pioneering Multi-Use Metro-Scale WiFi City of Corpus Christi, Texas, A Tropos Networks Case Study, June 2005, available at www.tropos.com/ pdf/corpus_casestudy.pdf.

[28] Nico Bayer, Dmitry Sivchenko, Bangnan Xu, Veselin Rakocevic, and Joachim Habermann, Transmission timing of signalling messages in IEEE 802.16-based Mesh Networks, University of Applied Sciences, Giessen-Friedberg, Germany; Veselin Rakocevic, City University, UK, available at www.staff.city.ac.uk/~veselin/publications/Bayer_EW2006.pdf.

[29] Simone Redana, Matthias Lott, and Antonio Capone, Performance Evaluation of Point-to-Multi-Point (PMP) and Mesh Air-Interface in IEEE Standard 802.16a, in *Proceedings of IEEE VTC Fall 2004*, Los Angeles, September 2004.

[30] Fuqiang Liu, Zhihui Zeng, Jian Tao, Qing Li, and Zhangxi Lin, Achieving QoS for IEEE 802.16 in Mesh Mode, 8th International Conference on Computer Science and Informatics, Salt Lake City, Utah, July 21–26, 2005.

[31] Mika Kasslin and Dave Beyer, Mesh Mode Text for 802.16ab Standard, Nokia, available at www.ieee802.org/16/tg4/contrib/802164c-01_39.pdf.

[32] City of Tempe Selects Strix for City-wide WiFi Mesh Network, City of Tempe Deploying Mesh Networks, May 2006, available at www.bbwexchange.com/ publications/page1386.asp.

[33] Strix Systems Products and Solutions for Mesh Networks, June 2006, available at www.strixsystems.com.

[34] Simone Redana and Matthias Lott, Performance Analysis of IEEE 802.16a in Mesh Operation Mode, in *Proceedings of IST SUMMIT 2004*, Lyon, France, June 2004.

[35] Hung-Yu Wei, Samrat Ganguly, Rauf Izmailov, and Zygmunt J. Haas, Interference-Aware IEEE 802.16 WiMax Mesh Networks, in *Proceedings 61st IEEE Vehicular Technology Conference* (VTC Spring '05), 2005.

[36] Alan Barry, George Healy, Cian Daly, Joseph Johnson, and Ronan J. Skehill, Overview of Wi-Max IEEE 802.16, in *Proceedings of the 5th Annual ICT Information Technology and Telecommunications*, Cork, October 2005.

[37] D. Aguayo, J. Bicket, S. Biswas, D.S.J. De Couto, and R. Morris, MIT Roofnet Implementation, available at http://pdos.lcs.mit.edu/roofnet/design/.

[38] Michael Carlberg Lax and Annelie Dammander, WiMAX — A Study of Mobility and a MAC-layer Implementation in GloMoSim, Master's thesis, Umea University, Sweden, April 2006.

[39] Nortel Solutions, Products and Experience in the Mesh Networking Field, June 2006, available at www.nortelnetworks.com.

[40] Dave Beyer, Nico van Waes, and Carl Eklund, Tutorial: 802.16 MAC Layer Mesh Extensions Overview, IEEE 802.16 Standard Group Discussions, February 2002.

[41] Harish Shetiya and Vinod Sharma, Algorithms for Routing and Centralized Scheduling in IEEE 802.16 Mesh Networks, IEEE Wireless Communications and Networking Conference 2006, Las Vegas, April 2006.

[42] Proposal for Connection Oriented Mesh, available at www.ieee802.org/16/tgd/contrib/S80216d-03_18.pdf.

[43] Thikrait Al Mosawi, Review of Existing Mobile Broadband Wireless Access (MBWA) Technologies (IEEE 802.16 AND IEEE 802.20), Center for Telecommunications Research, November 2004.

[44] Introduction of the IEEE 802.20 Standard for Mobile Broadband Wireless Access systems, RSAC paper 2/2004, available at www.ofta.gov.hk/en/ad-comm/rsac/paper/rsac2-2004.pdf.

[45] System Requirements for IEEE 802.20 Mobile Broadband Wireless Access Systems, available at www.ieee802.org/20/Contribs/C802.20-04-44.doc.

[46] IEEE C802.20-03/104: IEEE 802.20 Working Group on Mobile Broadband Wireless Access, available at www-sop.inria.fr/planete/qni/C802.20-03-104.pdf.

[47] 802.16 IEEE Standard for Local and Metropolitan Area Networks, Part 16: Air Interface for Fixed Broadband Wireless Access Systems, IEEE Computer Society and IEEE Microwave Theory and Techniques Society.

[48] Stephen Roberts, Development of IEEE 802.16e Functionality in QualNet, Master's thesis, Dublin City University, School of Electronic Engineering, January 2006.

[49] D.B. Johnson, and D.A. Maltz, Dynamic source routing in ad-hoc wireless networks, In T. Imielinski and H. Korth, eds., *Mobile Computing*, Kluwer, 1996, pp. 153–181.

[50] K. Chandran, S. Raghunathan, and S.R. Prakash, A feedback-based scheme for improving TCP performance in ad hoc wireless networks, *IEEE Personal Communications*, 8, 1, 3439, 2001.

[51] C. Perkins, E. Belding-Royer, and S. Das, Ad hoc On-demand Distance Vector (AODV) Routing, IETF RFC 3561, July 2003.

[52] D.S.J. De Couto, D. Aguayo, J. Bicket, and R. Morris, A high-throughput path metric for multi-hop wireless routing, in *ACM Annual International Conference on Mobile Computing and Networking* (MOBICOM), 2003, pp. 134–146.

[53] E.M. Belding-Royer, Multi-level hierarchies for scalable ad hoc routing, *ACM/Kluwer Wireless Networks*, 9, 5, 461–478, 2003.

[54] A.K. Saha and D.B. Johnson, Self-organizing Hierarchical Routing for Scalable Ad hoc Networking, Technical report, TR04-433, Department of Computer Science, Rice University.

[55] K. Xu, X. Hong, and M. Gerla, Landmark routing in ad hoc networks with mobile backbones, in *Proceedings of IEEE GLOBECOM 2000*, San Francisco, November 2000, http://citeseer.ist.psu.edu/gerla00landmark.html.

[56] C.-C. Shen, C. Srisathapornphat, and C. Jaikaeo, An adaptive management architecture for ad hoc networks, *IEEE Communications Magazine*, 41, 2, 108–115, 2003.

[57] Aoki et al., 802.11 TGs Simple Efficient Extensible, Mesh (SEE-Mesh) Proposal, IEEE 802 11-05/0562r01, 2005.

[58] Sheu et al., 802.11 TGs MAC Enhancement Proposal, IEEE 802 11-05/0575r4, 2005.

[59] M. Papadopouli and H. Schulzrinne, Network Connection Sharing in an Ad hoc Wireless Network among Collaborative Hosts, NOSSDAV, June 24th–26th, 1999, Bell Labs, New Jersey.

[60] Q. Xue and A. Ganz, Ad hoc QoS On-demand Routing (AQOR) Algorithm for QoS Support in Mobile Ad hoc Networks, Technical report TR-CSE-02-09, *J. Parallel Distrib. Comput.*, 63, 2, 154–165, 2003.

[61] I.F. Akyildiz, J. Xie, and S. Mohanty, A survey of mobility management in next-generation all-IP-based wireless systems, *IEEE Wireless Communications*, 11, 4, 1628, 2004.

[62] J. Kajiya, Commodity Software Steerable Antennas for Mesh Networks, Microsoft Mesh Networking Summit, June 2004.

[63] T.-S. Yum and K.-W. Hung, Design algorithms for multihop packet radio networks with multiple directional antennas stations, *IEEE Transactions on Communications*, vol. 40, pp. 1716–1724, November 1992.

[64] A. Nasipuri, S. Ye, and R.E. Hiromoto, A MAC protocol for mobile ad hoc networks using directional antennas, in *IEEE Wireless Communications and Networking Conference* (WCNC), 2000, pp. 1214–1219.

[65] Y.B. Ko, V. Shankarkumar, and N.H. Vaidya, Medium access control protocols using directional antennas in ad hoc networks, in *IEEE Annual Conference on Computer Communications* (INFOCOM), 2000, pp. 1321.

[66] R.R. Choudhury, X. Yang, R. Ramanathan, and N.H. Vaidya, Using directional antennas for medium access control in ad hoc networks, in *ACM Annual International Conference on Mobile Computing and Networking* (MOBICOM), 2002, pp. 5970.

[67] X. Quin and R. Berry, Exploiting multiuser diversity for medium access control in wireless networks, in *Proceedings IEEE INFOCOM 2003*, San Francisco, 30 Mar.–3 Apr. 2003, vol. 2, pp. 108–494.

[68] H. Jiang, W. Zhuang, X. (Sherman) Shen, A. Abdrabou, and P. Wang, Differentiated services for wireless mesh backbone, Centre for Wireless Communications (CWC), Department of Electrical and Computer Engineering, University of Waterloo, Canada, *IEEE Communications Magazine*, Volume 44, Issue 7, pp. 113–119, July 2006.

[69] B. Karp and H. Kung, GPSR: Greedy perimeter stateless routing for wireless networks, in *Proceedings ACM MOBICOM00*, pp. 243–254.

[70] S. C. Ergen, ZigBee/IEEE 802.15.4 Summary, September 10, 2004, available at http://www.cs.wisc.edu/~suman/courses/838/readinglist.html.

[71] G.R Hiertz, Y. Zang, S. Max, and H.-J. Reumerman, 802.15.5 MAC Design Proposal, Philips Research Laboratories, ComNets, RWTH Aachen University, http://mpa.comnets.rwth-aachen.de.

[72] Supriya Maheshwari, An E_cient QoS Scheduling Architecture for IEEE 802.16 Wireless MANs. Master's thesis, Indian Institute of Technology, Bombay, India, 2005.

[73] IETF Integrated Services Working Group, available at www.ietf.org/html.charters/IntServ-charter.htm.

[74] IETF Differentiated Services Working Group, available at www.ietf.org/html.charters/DiffServ-charter.htm.

SECURITY
PROTOCOLS
AND TECHNIQUES

Chapter 3

Attacks and Security Mechanisms

Anjum Naveed, Salil S. Kanhere, and Sanjay K. Jha

Contents

The true potential of any network cannot be exploited without considering and adequately addressing the security issues. Wireless mesh networks (WMNs), being multi-hop wireless networks, are prone to most of the security attacks on multi-hop wireless networks. In this chapter, we will discuss the security vulnerabilities in multi-hop wireless networks that are relevant to WMNs. We will consider the attacks in WMNs and the possible solution mechanisms to prevent and counteract these attacks.

3.1 Introduction

In recent years, WiFi (802.11) networks have become pervasive with numerous hotspots being deployed in urban city centers. However, to be connected, the mobile clients need to be within the radio range of the access point. To ensure that the target area is sufficiently covered, ISPs would need to install additional hotspots in strategically placed locations to extend existing coverage. This may not always be possible due to constraints on the terrain, social issues, etc. Further, deploying additional hotspots adds to the installation cost and more importantly to the running costs (subscription cost for Internet connectivity for each access point). A promising, low-cost alternative for providing last-mile wireless connectivity is the concept of WMNs, which are multi-hop wireless networks consisting of mesh routers and mesh clients. Generally, mesh routers have limited mobility and act as access points for the mobile clients to provide the connectivity over multiple hops as well as route the traffic for neighboring mesh routers. Some of the routers are equipped with wired interface and serve the purpose of gateway to provide the connectivity with the Internet. The clients' nodes may also act as intermediate hops for neighboring nodes to extend the connectivity. A typical WMN architecture is shown in Figure 3.1. By enabling multi-hop communication between the mesh nodes, it is possible for

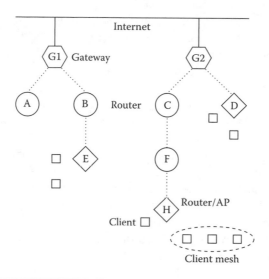

Figure 3.1   Wireless mesh network architecture.

several mobile clients to share a single broadband connection to the Internet. Several WMN deployments have been planned for major cities across the globe (Taipei, Moscow, Philadelphia, etc.) in the near future. However, very little attention has been devoted by the research community to address the security issues in WMNs.

The broadcast nature of transmission and the dependency on the intermediate nodes for routing the user traffic leads to security vulnerabilities making WMNs prone to various attacks. The attacks can be external as well as internal in nature. External attacks are launched by intruders who are not part of the WMN and gain illegitimate access to the network. For example, an intruding node may eavesdrop on the packets and replay those packets at a later stage of time to gain access to the network resources. Attacks from external nodes can be prevented by resorting to cryptographic techniques such as encryption and authentication. On the other hand, the internal attacks are launched by the nodes that are part of the WMN. One example of such an attack is an intermediate node dropping the packets, which it was supposed to forward, leading to a denial-of-service (DoS) attack. Similarly, the intermediate node may keep the copy of all the data that it forwards (internal eavesdropping) for offline processing and meaningful information retrieval without the knowledge of any other node in the network. Such attacks are typically launched either by selfish nodes or by malicious nodes, which may have been possibly compromised by attackers. There is a subtle difference in their motives. The selfish node is seeking to greedily acquire greater than its fair share of the network resources at the expense of other users. On the contrary a malicious attacker's sole aim is to undermine the performance of the entire network. Note that in an internal attack, the misbehaving node is part of the WMN and hence has access to all the keying and authentication information. Consequently, cooperative mechanisms, which enable other nodes within the network to detect and possibly isolate these misbehaving nodes, need to be employed.

It is evident that the true potential of WMN cannot be exploited without considering and adequately addressing the internal as well as the external security issues. In this chapter, we identify the security issues in WMNs, followed by descriptions of attacks on WMNs. The primary focus will be the attacks that affect the MAC layer and the network layer of WMNs. The characteristics of the security solution for WMNs are identified and different solution mechanisms are discussed. The standardization efforts for the security in WMNs are discussed. The chapter is concluded with some open issues yet to be considered in relation to security of WMNs.

3.2 Security Issues in Wireless Mesh Networks

Several vulnerabilities exist in the protocols for WMNs that can be exploited by the attackers to degrade the performance of the network. The WMN nodes depend on the intermediate nodes for connectivity with other nodes in the network and the Internet. Consequently, the MAC layer protocols as well as the routing protocols for WMNs assume that the participating nodes are well behaved with no malicious intentions. Therefore, all the nodes are assumed to follow the MAC protocol and perform the routing and packet forwarding operations as specified by the respective protocols. Based on this assumed trust, the nodes make independent decisions for their transmission, depending on the wireless channel availability. Similarly, the routing protocols require the WMN nodes to exchange their routing information within the neighborhood to make efficient routing decisions. Because the nodes are assumed to be well behaved, each node makes an independent decision based on the routing protocol specifications. The node then informs its neighbors about the decision. The neighbor nodes neither verify the decision nor the information transmitted by the node. In practice, however, some WMN nodes may behave in a selfish manner and other nodes may be compromised by malicious users. The assumed trust and the lack of accountability make the MAC layer protocols and the routing protocols vulnerable to various active attacks, such as black hole attacks, wormhole attacks, and rushing attacks [11–13].

The malicious or selfish nodes can drop data packets selectively or may choose to drop all the packets without forwarding any traffic. Further, because the participating nodes may not be owned by one administrator, specifically in case of community deployment of WMNs, data confidentiality and data integrity can be compromised if the intermediate node keeps the copy of all the data for offline cryptanalysis and information retrieval. The malicious nodes may also inject bad packets in the network, which may lead to a DoS attack. Similarly, passively sniffed packets can be replayed at a later time to gain access to the network resources. All these vulnerabilities render WMNs prone to security attacks. We consider the attacks on WMNs that exploit these vulnerabilities in the next section.

3.3 Attacks in Wireless Mesh Networks

In this section, the details of various attacks on WMNs are given. We consider the attacks affecting the physical layer, MAC layer, and the network layer because these layers form the core of the network. We do not consider the attacks on the transport layer and the application layers because these layers are primarily implemented in the end-user devices, hence the attacks on these layers are independent of the underlying network. Therefore, the attacks and the counter-measures on these layers (application and transport) for WMNs, other wireless networks, or even wired networks would be the same rather than being specific to WMNs.

3.3.1 Physical Layer Attacks

All wireless networks, including WMNs, are vulnerable to radio jamming attacks at the physical layer. The radio jamming attack [14] is a potentially damaging attack which can be launched with relative ease by simply allowing a wireless device to transmit a strong signal, which can cause sufficient interference to prevent packets in the victim network from being received. In its simplest form, the attacker may continuously transmit the jamming signal (constant jammer). Alternately, the attacker may resort to slightly sophisticated strategies whereby the attacker only transmits the radio signal when it senses some activity on the channel and remains quiet otherwise (reactive jammer). However, these types of jamming attacks, where the transmission is an arbitrary signal, can be regarded as noise in the channel and MAC protocols like BMAC [15] can successfully counteract these attacks to a certain degree by adjusting the signal-to-noise ratio (SNR) threshold at the receiving node. More complex forms of radio jamming attacks have been studied in [14], where the attacking devices do not obey the MAC layer protocol. We discuss these attacks in Section 3.3.2 as link layer jamming attacks.

3.3.2 MAC Layer Attacks

3.3.2.1 Passive Eavesdropping

The broadcast nature of transmission of the wireless networks makes these networks prone to passive eavesdropping by the external attackers within the transmission range of the communicating nodes. Multi-hop wireless networks like WMNs are also prone to internal eavesdropping by the intermediate hops, whereby a malicious intermediate node may keep the copy of all the data that it forwards, without knowledge of any other node in the network. Although passive eavesdropping does not affect the network functionality directly, it leads to the compromise in data confidentiality and data integrity. Data encryption is generally employed using strong encryption keys to protect the confidentiality and integrity of data.

3.3.2.2 Link Layer Jamming Attack

Link layer jamming attacks are more complex compared to blind physical layer radio jamming attacks. Rather than transmitting random bits constantly, the attacker may transmit regular MAC frame headers (no payload) on the transmission channel which conform to the MAC protocol being used in the victim network [16]. Consequently, the legitimate nodes always find the channel busy and back off for a random period of time before sensing the channel again. This leads to the denial of service for the legitimate nodes and also enables the jamming node to conserve its energy resources. In addition to the MAC layer, jamming can also be used to exploit the network and transport layer protocols [17]. Intelligent jamming is not a purely transmit activity. Sophisticated sensors can be deployed, which detect and identify victim network activity, with a particular focus on the semantics of higher-layer protocols (e.g., AODV [Ad-hoc On-demand Distance Vector] and TCP). Based on the observations of the sensor, the attacker can exploit the predictable timing behavior exhibited by higher-layer protocols and use offline analysis of packet sequences to maximize the potential gain for the jammer. These attacks can be effective even if encryption techniques such as Wired Equivalent Privacy (WEP) and WiFi Protected Access (WPA) have been employed. This is because the sensor that assists the jammer can still monitor the packet size, timing, and sequence to guide the jammer. Because these attacks are based on carefully exploiting protocol patterns and consistencies across size, timing, and sequence, preventing them will require modifications to the protocol semantics so that these consistencies are removed wherever possible.

3.3.2.3 MAC Spoofing Attack

MAC addresses have long been used as the singularly unique layer-2 network identifiers in both wired and wireless LANs. MAC addresses which are globally unique have often been used as an authentication factor or as a unique identifier for granting varying levels of network privileges to a user. This is particularly common in 802.11 WiFi networks. However, to-day's MAC protocols (802.11) and network interface cards do not provide for any safeguards that would prevent a potential attacker from modifying the source MAC address in its transmitted frames. On the contrary, there is often full support in the form of drivers from manufacturers, which makes this particularly easy. Modifying the MAC address in transmitted frames is referred to as MAC spoofing, and can be used by attackers in a variety of ways. MAC spoofing enables the attacker to evade Intrusion Detection Systems (IDSs) that are in place. Further, today's network administrators of-ten use MAC addresses in access control lists. For example, only registered MAC addresses are allowed to connect to the access points. An attacker can easily eavesdrop on the network to determine the MAC addresses of legitimate devices. This enables the attacker to masquerade as a legitimate user and gain access to the network. An attacker can even inject a large number of bogus frames into the network to deplete the resources (in par-ticular, bandwidth and energy), which may lead to denial of service for the legitimate nodes.

3.3.2.4 Replay Attack

The replay attack, often known as the man-in-the-middle attack [18], can be launched by external as well as internal nodes. An external malicious node (not part of WMN) can eavesdrop on the broadcast communication between two nodes (A and B) in the network, as shown in Figure 3.2. It

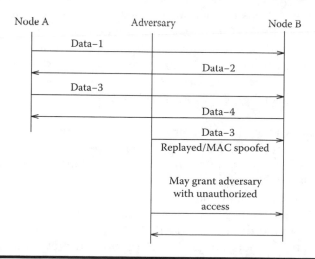

Figure 3.2   Robustness against MAC spoofing and replay attacks.

can then transmit these legitimate messages at a later stage of time to gain access to the network resources. Generally, the authentication information is replayed where the attacker deceives a node (node B in Figure 3.2) to believe that the attacker is a legitimate node (node A in Figure 3.2). On a similar note, an internal malicious node, which is an intermediate hop between two communicating nodes, can keep a copy of all relayed data. It can then retransmit this data at a later point in time to gain the unauthorized access to the network resources. The replay attack, exploiting the IEEE 802.1X [33] authentication mechanism, is discussed in Section 3.6.

3.3.2.5 Pre-Computation and Partial Matching Attacks

In this section we discuss a different form of security attacks. Unlike the above-mentioned attacks where MAC protocol vulnerabilities are exploited, these attacks exploit the vulnerabilities in the security mechanisms that are employed to secure the MAC layer of the network. Pre-computation and partial matching attacks exploit the cryptographic primitives that are used at MAC layer to secure the communication. In a pre-computation attack or Time Memory Trade-Off attack (TMTO), the attacker computes a large amount of information (key, plaintext, and respective cipher text) and stores that information before launching the attack. When the actual transmission starts, the attacker uses the pre-computed information to speed up the cryptanalysis process. TMTO attacks are highly effective against a large number of cryptographic solutions. On the other hand, in a partial matching attack, the attacker has access to some (cipher text, plaintext) pairs, which in turn decreases the encryption key strength and improves the chances of success of the brute force mechanisms. Partial matching attacks exploit the weak implementations of encryption algorithms. For example, in the IEEE 802.11i standard for MAC layer security in wireless networks [30], the MAC address fields in the MAC header are used in the message integrity code (MIC). The MAC header is transmitted as plaintext while the MIC field is transmitted in the encrypted form. Partial knowledge of the plaintext (MAC address) and the cipher text (MIC) makes IEEE 802.11i vulnerable to partial matching attacks.

DoS attacks may also be launched by exploiting the security mechanisms. For example, the IEEE 802.11i standard for MAC layer security in wireless networks is prone to the session hijacking attack and the man-in-the-middle attack, exploiting vulnerabilities in IEEE 802.1X, and DoS attack, exploiting vulnerabilities in the four-way handshake procedure in IEEE 802.11i. Although these attacks are also considered as MAC layer attacks, we pend the discussion on IEEE 802.11i, its vulnerabilities, attacks exploiting these vulnerabilities, and the proposed prevention mechanisms till Section 3.6.

3.3.3   Network Layer Attacks

The attacks on the network layer can be divided into control plane attacks and data plane attacks and can be active or passive in nature. Control plane attacks generally target the routing functionality of the network layer. The objective of the attacker is to make routes unavailable or force the network to choose sub-optimum routes. On the other hand, the data plane attacks affect the packet forwarding functionality of the network. The objective of the attacker is to cause the denial of service for the legitimate user by making user data undeliverable or injecting malicious data into the network. We first consider the network layer control plane attacks, followed by a discussion on network layer data plane attacks.

3.3.3.1   Control Plane Attacks

Rushing attacks [11] targeting the on-demand routing protocols (e.g., AODV) were among the first exposed attacks on the network layer of multi-hop wireless networks. Rushing attacks exploit the route discovery mechanism of on-demand routing protocols. In these protocols, the node requiring the route to the destination floods the Route Request message, which is identified by a sequence number. To limit the flooding, each node only forwards the first message that it receives and drops remaining messages with the same sequence number. The protocols specify a specific amount of delay between receiving the Route Request message by a particular node and forwarding it, to avoid collusion of these messages. The malicious node launching the rushing attack forwards the Route Request message to the target node before any other intermediate node from source to destination. This can easily be achieved by ignoring the specified delay. Consequently, the route from source to destination includes the malicious node as an intermediate hop, which can then drop the packets of the flow resulting in data plane DoS attack.

A wormhole attack has a similar objective albeit it uses a different technique [12]. During a wormhole attack, two or more malicious nodes collude together by establishing a tunnel using an efficient communication medium (i.e., wired connection or high-speed wireless connection, etc.), as shown in Figure 3.3. During the route discovery phase of on-demand routing protocols, the Route Request messages are forwarded between the malicious nodes using the established tunnel. Therefore, the first Route Request message that reaches the destination node is the one forwarded by the malicious nodes. Consequently, the malicious nodes are added in the path from source to destination. Once the malicious nodes are included in the routing path, the malicious nodes either drop all the packets, resulting in complete denial of service, or drop the packets selectively to avoid detection.

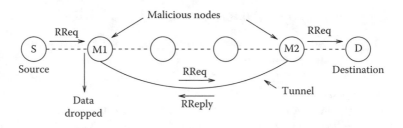

Figure 3.3 Wormhole attack launched by nodes M1 and M2. Nodes use high-speed tunnel to forward routing protocol control messages while data is dropped.

A black hole attack (or sink hole attack) [19] is another attack that leads to denial of service in wireless mesh networks. It also exploits the route discovery mechanism of on-demand routing protocols. In a black hole attack, the malicious node always replies positively to a Route Request although it may not have a valid route to the destination. Because the malicious node does not check its routing entries, it will always be the first to reply to the Route Request message. Therefore, almost all the traffic within the neighborhood of the malicious node will be directed toward the malicious node, which may drop all the packets, resulting in denial of service. Figure 3.4 shows the effect of a black hole attack in the neighborhood of the malicious node where all the traffic is directed toward the malicious node. A more complex form of the attack is the cooperative black hole attack where multiple malicious nodes collude together, resulting in complete disruption

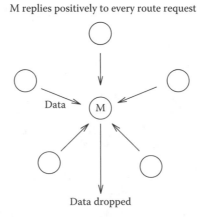

Figure 3.4 Black hole attack. Node M replies positively to every Route Request. Consequently all data is forwarded to the node, which then drops the data.

of routing and packet forwarding functionality of the network. The cooperative black hole attack and the prevention mechanisms have been studied in [13].

A grey hole attack is a variant of the black hole attack. In a black hole attack, the malicious node drops all the traffic that it is supposed to forward. This may lead to possible detection of the malicious node. In a grey hole attack, the adversary avoids the detection by dropping the packets selectively. A grey hole attack does not lead to complete denial of service, but it may go undetected for a longer duration of time. This is because the malicious packet dropping may be considered congestion in the network, which also leads to selective packet loss.

A Sybil attack is the form of attack where a malicious node creates multiple identities in the network, each appearing as a legitimate node [20]. A Sybil attack was first exposed in distributed computing applications where the redundancy in the system was exploited by creating multiple identities and controlling the considerable system resources. In the networking scenario, a number of services like packet forwarding, routing, and collaborative security mechanisms can be disrupted by the adversary using a Sybil attack. Following form of the attack affects the network layer of WMNs, which are supposed to take advantage of the path diversity in the network to increase the available bandwidth and reliability. If the malicious node creates multiple identities in the network, the legitimate nodes, assuming these identities to be distinct network nodes, will add these identities in the list of distinct paths available to a particular destination. When the packets are forwarded to these fake nodes, the malicious node that created the identities processes these packets. Consequently, all the distinct routing paths will pass through the malicious node. The malicious node may then launch any of the above-mentioned attacks. Even if no other attack is launched, the advantage of path diversity is diminished, resulting in degraded performance.

In addition to the above-mentioned attacks, the wireless mesh networks are also prone to network partitioning attacks and routing loop attacks. In a network partitioning attack, the malicious nodes collude together to disrupt the routing tables in such a way that the network is divided into non-connected partitions, resulting in denial of service for a certain network portion. Routing loop attacks affect the packet-forwarding capability of the network where the packets keep circulating in loop until they reach the maximum hop count, at which stage the packets are simply discarded.

3.3.3.2 Data Plane Attacks

Data plane attacks are primarily launched by the selfish and malicious (compromised) nodes in the network and lead to performance degradation or denial of service for the legitimate user data traffic. The simplest of the data plane attacks is passive eavesdropping. Eavesdropping has already been discussed in Section 3.3.2 as a MAC layer attack and we do not discuss it further. Selfish behavior of the participating WMN nodes is a major security issue because the WMN nodes are dependent on each other for data forwarding. The intermediate-hop selfish nodes may not perform the packet-forwarding functionality as per the protocol. The selfish node may drop all the data packets, resulting in complete denial of service, or it may drop the data packets selectively or randomly. It is hard to distinguish between such a selfish behavior and the link failure or network congestion. On the other hand, malicious intermediate-hop nodes may inject junk packets into the network. Considerable network resources (bandwidth and packet processing time) may be consumed to forward the junk packets, which may lead to denial of service for the legitimate user traffic. The malicious nodes may also inject the maliciously crafted control packets, which may lead to the disruption of routing functionality. The control plane attacks are dependent on such maliciously crafted control packets. The malicious and selfish behavior has been studied in [22,23].

3.3.4 Multi-Radio Multi-Channel Wireless Mesh Network Attacks

In this section, we consider the attacks that affect the network layer as well as the MAC layer of WMNs. These attacks exploit the channel assignment and routing algorithms in multi-radio multi-channel wireless mesh networks (MR-MC WMN). Bandwidth capacity is a major limitation for wireless mesh networks. In MR-MC WMN, each WMN node is equipped with multiple radios to increase the available bandwidth. Orthogonal channels are used for each interface of the node, which ensures simultaneous communication using all the wireless interfaces without interference. Dynamic channel assignment is required to assign the channels to the network links. The objective of the channel assignment algorithms is to ensure the minimum interference within a WMN. Various joint routing and channel assignment algorithms have been proposed for MR-MC WMN [1–5]. Readers are encouraged to review the dynamic routing and channel assignment algorithms proposed in [2] for better understanding of the attacks discussed in this section. Note that channel assignment is done at the MAC layer while the routing is a network layer functionality. All the joint routing and channel assignment algorithms assume that the mesh nodes are well-behaved. Hence the nodes make independent decisions about their channel assignment based on the neigbhor channel assignment information and inform neighboring nodes about the decision, which is not verified. The assumed

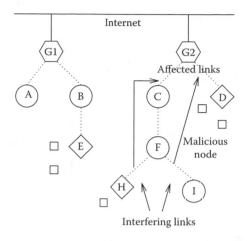

Figure 3.5 Network endo-parasite attack (NEPA). Assuming the node F is within interference domain of node G.

trust among the WMN nodes and the independent decision of the nodes make these algorithms vulnerable to security attacks.

A network endo-parasite attack (NEPA) [21] is launched by the compromised malicious node when it changes the channel assignment of its interfaces in such a way that the interference on heavily loaded high priority channels increases (each interface is switched to a different high-priority channel). This is contrary to the normal operation of the channel assignment algorithm where the node assigns the least loaded channels to its interfaces. Figure 3.5 shows the attack. The malicious node F has switched the channel on link FH to the same channel as the link GC and link FI to the channel used by link GD. The malicious switching by node F will increase the interference on links GC and GD. The malicious node does not inform its neighbors about the change in channel assignment; therefore, the neighboring nodes are unable to adjust their channel assignment to mitigate the effect of increased interference. The increase in interference results in serious performance degradation.

A channel ecto-parasite attack (CEPA) [21] is a special type of NEPA. During CEPA, the malicious node switches all its interfaces to the most heavily loaded highest priority channel. Like NEPA, the malicious node does not inform its interference domain neighbors about the change in channel assignment. The effect of the attack is the hidden usage of the most heavily loaded channel, which increases the interference considerably, resulting in a decrease in performance. The attack is shown in Figure 3.6 where the malicious node has switched both its child links FH and FI to the channel

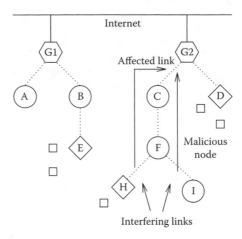

Figure 3.6   Channel ecto-parasite attack (CEPA) (assuming the node F is within interference domain of node G).

that is being used by the high-priority link GC. As the links FH and FI are within the interference range of the link GC, the link GC will experience high interference. However, the malicious node has not informed its neighbors about the change in channel assignment; therefore, the node G will continue to use the same channel on link GC, assuming the external noise or other factors to be the reason for degraded performance.

A low cost ripple effect attack (LORA) [21] is launched when the compromised malicious node transmits misleading channel assignment information about its interfaces to the neighboring nodes without actually changing the channel assignment. The information is calculated in such a way that the neighboring nodes are forced to adjust their channel assignments to minimize the interference, which may generate a series of changes even in the channel assignment of the nodes that are not direct neighbors of the malicious node. The effect of the attack is shown in Figure 3.7 using the arrow. Although most of the dynamic channel assignment algorithms prevent the ripple effect to propagate within the network from the parent nodes (closer to the wired gateway) to the child nodes, the effect can still propagate in the reverse direction. The objective of the attack is to force the network in the quasi-stable state by imposing premature channel adjustment on other nodes repeatedly. Considerable network resources are consumed for channel adjustment and the user data forwarding capability is severely affected. The attack is relatively more severe than NEPA and CEPA because the effect is propagated to a large portion of the network even beyond the neighbors of the compromised node, disrupting the traffic forwarding capability of various nodes for considerable time duration.

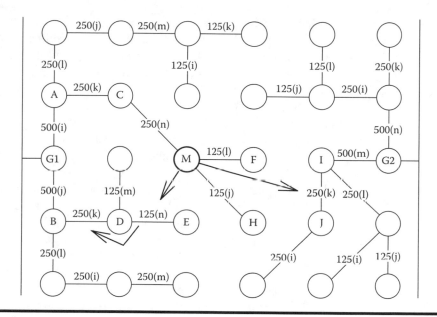

Figure 3.7 **Example WMN with routers physically arranged in grid topology. G1 and G2 are gateways connected to wired network. Edges show routing topology and labels along edges are bandwidth in kbps (channel). For simplicity, $(k + 1)$-hop neighbors include immediate physical neighbors only. Arrows show propagation of ripple effect attack from compromised node M.**

3.4 Characteristics of Security Solutions for Wireless Mesh Networks

In the previous section, we discussed the security attacks that exploit the vulnerabilities in the MAC layer and the network layer protocols for WMN. We now list the essential characteristics that a security mechanism for WMN should have to successfully prevent, detect, and counter these attacks. We only list the characteristics that differentiate WMN security mechanisms from existing security mechanisms for wired and wireless networks.

■ In wired networks, the security services of data confidentiality and data integrity are generally provided on a per-link basis (between two devices). This is based on the assumption that the end devices are secure. However, as discussed in previous sections, the WMN nodes may resort to the selfish and malicious behavior. To counteract the selfish and malicious behavior of the intermediate-hop nodes, the WMN must provide the end-to-end services of data confidentiality and data integrity, in addition to the security services on a per-link basis.

■ The trust establishment mechanism should be robust against internal selfish and malicious behavior. Note that the internal selfish and malicious nodes are part of WMNs, therefore the conventional authentication mechanisms based on cryptographic primitives may not be effective against the internal misbehavior.

■ Section 3.3.3 and Section 3.3.4 indicate that the accountability should be a necessary characteristic for WMNs to ensure that the WMN nodes behave according to the protocol specification even if the nodes make independent decisions about routing and channel assignment.

■ Wireless mesh networks are self-administered networks and lack the centralized administration authority which can respond to the network issues. Therefore, the attack and anomaly detection mechanisms for wireless mesh networks should be self-sufficient and must not be dependent on the administrator to verify the possible attack and anomaly alerts.

■ An important characteristic of wireless mesh networks is the self-healing nature. Therefore, the detection mechanisms must be coupled with adequate automated response to the security attacks and identified anomalies.

Having identified the essential characteristics of the security mechanisms for wireless mesh networks, we now consider different security mechanisms that are employed to counter the attacks identified in Section 3.3.

3.5 Security Mechanisms for Wireless Mesh Networks

ITU-T Recommendation X.800 [29]—Security Architecture for OSI—defines the required security services for communication networks. The security services have been broadly categorized into five groups: authentication, access control or authorization, confidentiality, integrity, and non-repudiation. Security management services have also been defined aimed at ensuring availability, accountability, and event management. The security services can be categorized into two broad categories: intrusion prevention and intrusion detection. In case of intrusion prevention, measures are taken to stop the attacker from intruding into the network and launching the attack on the network. The protection can be from external as well as internal intruders. Security services of authentication, access control, data confidentiality, data integrity, and non-repudiation lead to intrusion prevention. However, intrusion prevention is insufficient to protect the network from all attacks because no prevention technique can ensure complete protection.

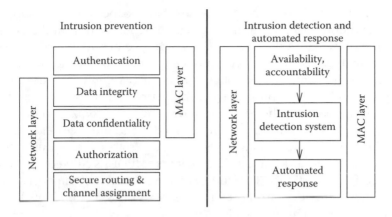

Figure 3.8 Security model for wireless mesh networks.

Therefore, the intrusion prevention mechanisms are complemented by intrusion detection and response mechanisms. The role of intrusion detection is to identify the illegitimate activities which may be the consequence of the attacks or may lead to the attacks. Early detection and timely response can limit the effect of the attack on the network. The intrusion detection and response mechanisms aim at ensuring the accountability and availability of the network services. Figure 3.8 shows how different security services fit together in the security model for wireless mesh networks. We now consider the intrusion prevention mechanisms as well as intrusion detection mechanisms both at the MAC layer and the network layer of wireless mesh networks.

3.5.1 MAC Layer Security Mechanisms

3.5.1.1 Intrusion Prevention Mechanisms

Various security frameworks [30–32] have been proposed for multi-hop wireless networks that are applicable to wireless mesh networks with slight modification. These security frameworks provide the security services of authentication, data confidentiality, and data integrity at the MAC layer of the network on a per-link basis. Most of the security frameworks employ the cryptographic primitives. For example, Soliman and Omari [31] have proposed the security framework based on stream cipher for encryption to provide the services of data confidentiality, data integrity, and authentication. The objective of using stream cipher is to allow the online processing of the data. Consequently, minimum delay is introduced because of the security provisioning. Two secret security keys, Secret Authentication Key (SAK) and Secret Session Key (SSK), are used for authentication of the supplicant and authenticator. SAK is exchanged between the supplicant and the authenticator after initial mutual authentication from the authentication server, whereas the SSK is used for a given communication session between the two nodes. The SAK and SSK pair is used by the communicating nodes to generate the permutation vector (PV), which is used for the encryption and decryption of data. In the strongest mode of security, the data is also involved in the PV generation. The synchronization of the generated permutation vector between the sender and the receiver of the data results in origin authentication of every MAC Protocol Data Unit (MPDU). To minimize the security overhead, plaintext MPDU is XORed with the PV generated for that MPDU. The authors have proved that the encryption of data using PV provides strong security services of data confidentiality, data integrity, and origin authentication.

IEEE 802.11i was ratified in June 2004 as the standard for the security of the MAC layer of the wireless networks. The standard is based on the cryptographic primitives and provides the services of data confidentiality, data integrity, and authentication. The standard is discussed in detail in Section 3.6.

One of the major security requirements in case of multi-hop wireless networks like WMN is the trust establishment between communicating nodes. As mentioned in Section 3.4, conventional cryptography-based mechanisms are generally non-applicable to multi-hop networks like WMN. Consequently, a number of distributed neighbor-collaboration authentication protocols have been proposed by researchers for this purpose [38,39,42]. A comprehensive analysis of the authentication protocols for wireless networks can be found in [41]. Deng et al. [42] have proposed the threshold and identity-based authentication and key management for multi-hop wireless networks. A threshold cryptography-based solution is proposed for the distribution of the master key <public key, private key> and the authentication of the nodes based on the private key. In the proposed scheme, all nodes possess the public key while every node has got a share of the private key. (k,n) Threshold secret sharing is employed to generate the private key for the node which states that "k" out of "n" shares of private key are required to construct the complete private key and less than k shares of the secret key cannot construct the complete private key. Based on this mechanism, whenever a node needs to refresh its private key, it needs k neighbors to send their secret share to the node to reconstruct the private key and no node can construct the private key based on its own information. The process of private key generation is shown in Figure 3.9, where the requesting node broadcasts the request message along with its own share for verification. The neighboring nodes reply to the request message by sending their own share of the secret key to the requesting node. The requesting node is able to generate the private key on receiving k shares of the key. Using this mechanism, the intruding node cannot generate the

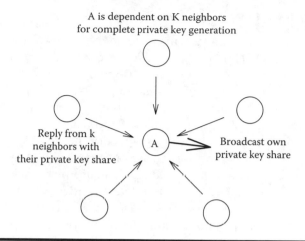

Figure 3.9 Neighbor collaboration for private key generation in wireless mesh networks.

private key unless its own share of private key is verified by k neighboring nodes. Similarly, the private key of the misbehaving node is not refreshed by the neighbors. Therefore, the threshold secret sharing serves as the strong authentication and key management solution.

The security mechanisms discussed above prevent the network from MAC layer attacks as follows. The security service of data confidentiality leads to the protection against passive eavesdropping attack. Although the nodes within the transmission range of the communicating nodes can still overhear the communication, the data is protected using encryption mechanisms provided by the data confidentiality service. Therefore, the received information is useless, unless it is decrypted using brute force methods, which are impractical, keeping in view the value of information retrieved versus the cost of attack. Data and header integrity service provides the protection against MAC spoofing attacks. The message with spoofed MAC address (IP address for IP spoofing) will fail the integrity check at the receiving node and will be discarded. Per-packet authentication and integrity provided by the solutions [30,31] protect the data against replay attacks. These solutions use a fresh key for each message which is synchronously computed by the sender and the receiver. Therefore, a replayed packet, encrypted using an outdated key, will fail the integrity check and will be discarded. Use of a fresh key for each message also protects the data from pre-computation and partial matching attacks because the pre-computed information needs to be applied on every message to decrypt that message. This renders the attack extremely costly compared to the information retrieved.

3.5.1.2 Intrusion Detection Mechanisms

Very few intrusion detection systems have been proposed at the MAC layer of wireless networks. Lim et al. [43] have proposed an intrusion detection system to secure wireless access points coupled with automated active response. The authors have proposed the deployment of specific detection devices closer to wireless access points and the detection is done at the MAC layer. RTS/CTS (Ready To Send/Clear To Send) messages from the black-listed MAC addresses are proposed as detection metrics. As a response to the intrusion, the authors propose the use of the intruder's tactics back onto the intruder by crafting and transmitting the malformed packets back. The proposed idea of deploying dedicated detection devices may not be cost effective. Similarly, the response mechanism may be computation resource extensive. Further, the legitimate nodes may get punished if the detected information is not accurate.

One of the most recent works in this context is from Liu et al. [24]. The authors have proposed the game theoretic approach for selecting the optimum intrusion detection strategy at a given instance from a set of deployed weak intrusion detection mechanisms. The basic idea is that different intrusion detection techniques are very good at detecting certain types of attacks, but do not perform optimally in other cases. The combination of these strategies and the use of optimum strategy in a given scenario can increase the detection accuracy of the resulting system. However, while the idea of selecting the optimum technique at a given instance has strength, basically at a given instance of time, only one weak intrusion detection technique will be used. Consequently, the performance of intrusion detection may not significantly improve as compared to the increase in overhead because of the IDS selection mechanism.

The intrusion detection mechanisms at the MAC layer are used to detect the attacks launched by misbehaving nodes that do not obey the MAC layer protocol. These attacks include the link layer jamming attacks and DoS attacks.

3.5.2 Network Layer Security Mechanisms

3.5.2.1 Intrusion Prevention Mechanisms

Intrusion prevention techniques have been proposed to secure the routing protocols for multi-hop wireless networks. These protocols include Secure Routing Protocol (SRP) [6], Secure AODV (SAODV) [7], Authenticated Routing for Ad hoc Network (ARAN) [8] and Ariadne, a secure on-demand routing protocol [9], to list a few. The most recent work in this domain is described in [10]. All these protocols use cryptographic primitives to establish some form of trust between the network nodes through the process of mutual authentication. For example, SRP [6] is aimed at securing the route discovery process and safeguards the routing functionality from attacks exploiting the routing protocol itself. The Route Request and Route Reply messages are protected by message authentication code (MAC) for authentication of the originating node. The IP address of the intermediate nodes is also added in the Route Request message for cross validation to prevent the network from black hole and wormhole attacks. The authors prove that the protection of Route Request and Route Reply messages ensures protection against multiple attacks except for the case where multiple nodes collude together and launch the attack. SAODV [7] uses digital signatures to authenticate all the fields of Route Request and Route Reply messages except from the hop count field. Digital signatures are used on end-to-end basis between source and destination. The hop count field is secured using hash-chains on per-link basis.

The intrusion prevention mechanisms are primarily used to establish the trust between the participating nodes and providing the control message integrity and confidentiality. These services can provide some protection against wormhole and black hole attacks. However, the problem of malicious and misbehaving nodes cannot be addressed completely using the intrusion prevention mechanisms at the network layer and the support from intrusion detection mechanisms becomes mandatory.

3.5.2.2 Intrusion Detection Mechanisms

Numerous intrusion detection techniques have been proposed at the network layer for wired as well as wireless networks. In this section we briefly discuss some of the recent research efforts in this domain; however, the survey by no means is exhaustive. Most of the intrusion detection systems rely on the knowledge-based systems and data mining techniques [25–28]. For example, Huang et al. [26] have proposed IDS for multi-hop mobile wireless networks based on the cross-feature analysis. The nodes monitor different parameters in the network and, based on values of $(i - 1)$ parameters, predict the value of ith parameter and compare it with monitored value of that parameter to detect routing anomalies in the networks. The authors have also proposed the distributed cluster-based approach as an extension to this work [27], where they propose the division of networks into clusters and only few elected nodes within each cluster perform the monitoring with the intrusion detection probability almost the same as with all the nodes actively monitoring. This scheme is resource efficient, which is the primary design goal for wireless networks.

Yang et al. [28] have proposed the self-organized network layer security solution for mobile ad hoc networks. This is one of the very few solutions which ensure self-healing and self-organized networks. The solution is based on distributed neighbor collaboration and information cross-validation, resulting in self-organized, self-healing networks. The scheme is based on the threshold secret sharing discussed in Section 3.5.1 which is used to refresh the token of the nodes. The authors have proposed a novel token-based crediting scheme. The token of the node expires after a specific time duration. The token expiry time of the node depends upon the credit of the node. The credit of the well-behaving nodes gets accumulated over the period of time. Therefore, the token expiry time of these nodes is longer and is linearly incremented every time the node refreshes its token. The token of malicious or selfish nodes is revoked by neighbor collaboration refraining them to participate in the network. The detection metrics used to differentiate between well-behaving and malicious nodes are based on the routing protocols and consist of hop count distance, packet forwarding ratio, etc.

The intrusion detection mechanisms at the network layer primarily address the issues of malicious, selfish, and misbehaving nodes that are at the heart of almost all the attacks at the network layer. The solutions described in [26–28] identify the anomalies in the control messages to detect the control plane attacks like rushing, wormhole, black hole, grey hole, network partitioning, and routing loop attacks. On the other hand, neighbor monitoring techniques [26,27] are employed to detect the data plane attacks.

3.6 Toward Standardization

IEEE 802.11i [30] is the defined standard for the MAC layer security of the wireless networks. The draft standard for wireless mesh networks, IEEE 802.11s, has proposed the use of IEEE 802.11i for the MAC layer security in wireless mesh networks. Therefore, we dedicate this section to discuss the IEEE 802.11i standard. We first explain the security methods used and the security services provided in the IEEE 802.11i standard, and later we will expose the vulnerabilities in IEEE 802.11i that render the standard prone to security attacks. These attacks include the pre-computation and partial matching, session hijacking, and the man-in-the-middle attacks exploiting vulnerabilities in IEEE 802.1X, and DoS attacks exploiting vulnerabilities in the four-way handshake. We also discuss the proposed prevention mechanisms for these attacks briefly.

IEEE 802.11i provides the security services of data confidentiality, data integrity, authentication, and protection against replay attacks. The standard consists of three components: key distribution, mutual authentication, and data confidentiality integrity and origin authentication. In the following paragraphs, we briefly discuss these components.

IEEE 802.1X [33] is used for key distribution and authentication, entailing the use of Extensible Authentication Protocol (EAP) [34] and an authentication, authorization, and accounting server (AAA server) like RADIUS or DIAMETER [35,36]. IEEE 802.1X is a port-based access control protocol which operates in client–server architecture. When the router/access point (authenticator) detects a new client (supplicant), the port on the authenticator is enabled and set to the "unauthorized" state for that client. In this state, only 802.1X traffic (EAP messages) is allowed and all other traffic is blocked from that client. The authenticator sends out the EAP-Request message to the supplicant, and the supplicant replies with the EAP-Response message. The authenticator forwards this message to the AAA server. If the server authenticates the client and accepts the request, it generates a Pairwise Master Key (PMK), which is distributed to the authenticator and the supplicant using EAP messages. After authentication from the server, the authenticator sets the port for the client to the "authorized" state and normal traffic is allowed. Note that the same protocol can be used to authenticate and distribute keys between two peer routers or two peer clients in case of wireless mesh networks.

Encryption key distribution and authentication using 802.1X is followed by mutual authentication of supplicant (client or peer router) and authenticator (router/AP or peer router), which is based on the four-way handshake. The four way handshake is initiated when the two nodes intend to exchange data. The encryption key distribution makes the shared secret key PMK available to the supplicant as well as the authenticator. However, this key is designed to last the entire session and should be exposed as little as possible. Therefore the four-way handshake is used to establish two more keys called the Pairwise Transient Key (PTK) and Group Temporal Key (GTK). PTK is generated by the supplicant by concatenating the PMK, Authenticator nonce (ANonce), Supplicant nonce (SNonce), Authenticator MAC address, and Supplicant MAC address. The product is then put through a cryptographic hash function. GTK is generated by the authenticator and transmitted to the supplicant during the four-way handshake. PTK is used to generate a Temporal Key (TK), which is used to encrypt unicast messages while the GTK is used to encrypt broadcast and multicast messages. The four-way handshake (shown in Figure 3.10) involves generation and distribution of these keys between supplicant and authenticator, resulting in mutual authentication. The first message of the four-way handshake is transmitted by the authenticator to the supplicant, which consists of ANonce. The supplicant uses this ANonce and readily available fields with itself to generate the PTK. The second message of the handshake is transmitted by the supplicant to the authenticator consisting of SNonce and Message Integrity Code (MIC), which is encrypted using PTK. The authenticator is then able to generate the PTK and GTK. The attached MIC is decrypted using the generated PTK. If the MIC is successfully decrypted, then the authenticator and the supplicant have successfully authenticated each other (Mutual Authentication). This is because the authenticator's generated PTK will only match the PTK transmitted by the supplicant if the two share the same PMK. A third message is transmitted by the authenticator consisting of GTK and MIC.

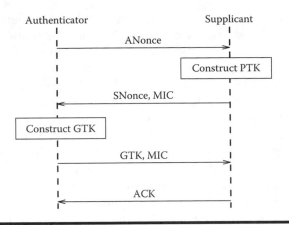

Figure 3.10   Four-way handshake.

The last message of the four-way handshake is the acknowledgment transmitted by the supplicant. The two nodes can exchange the data after a successful four-way handshake.

IEEE 802.11i provides two methods for the security services of data confidentiality, data integrity, origin authentication, and protection against replay attacks. The first method, Temporal Key Integrity Protocol (TKIP), is the enhanced version of WEP and has been provided for backward compatibility with the hardware that was designed to use WEP. RC4 encryption has been used as the encryption algorithm; however, the implementation of the algorithm is weak, rendering the protocol vulnerable to numerous security attacks. We do not discuss this method in detail. Interested readers are referred to Section 8.3.2 of the standard [30] for further details of the method.

The second method is the Counter mode (CTR) with CBC-MAC Protocol (CCMP). CCMP is based on the Counter mode with CBC-MAC (CCM) [37] of the AES encryption algorithm. CCM combines Counter (CTR) for confidentiality and the Cipher Block Chaining (CBC) Message Authentication Code (MAC) for origin authentication and integrity. As shown in Figure 3.11, CCM encryption takes four inputs: the encryption key, Additional Authentication Data (AAD), a unique Nonce for every frame, and the plaintext. CCM requires a fresh TK (generated from PTK) for every session which is used as the encryption key. AAD is constructed from the MAC header, and consists of the following fields: Frame Control field FC (certain bits masked), Address A1, Address A2, Address A3, Sequence Control field SC (certain bits masked), Address A4 (if present in the MAC header), and quality-of-service Control field QC (if present). CCMP uses the A2 and the priority fields of the MAC header along with a 48-bit packet number (PN) to generate the unique nonce value for each frame protected by a given TK. PN is

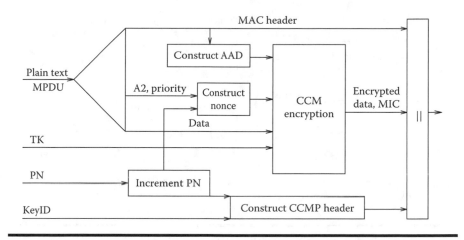

Figure 3.11   CCMP encryption process and encrypted frame generation [30].

incremented for each MPDU, resulting in a fresh value of nonce for each MPDU. The output of the encryption is the cipher text and the MIC. The frame to be transmitted is constructed by concatenating the MPDU header, CCMP header, cipher text, and MIC. CCM encryption is explained in RFC 3610.

3.6.1   Vulnerabilities in IEEE 802.11i and Security Attacks

The IEEE 802.11i standard successfully provides a number of security services; however, a number of security vulnerabilities have been identified in recent years. We discuss these vulnerabilities, the attacks exploiting these vulnerabilities, and the available prevention mechanisms in this sub-section.

3.6.1.1   IEEE 802.1X Vulnerabilities

IEEE 802.1X [33] is used by IEEE 802.11i standard for key distribution and authentication. Three entities, Authenticator, Supplicant, and the Authentication server, participate in the process. The basic assumption underlying the protocol is that the authenticator is always trusted. Therefore, the supplicant does not verify the messages received from the authenticator and unconditionally responds to these messages. However, in practice the adversary can also act as authenticator, which renders the protocol vulnerable to session hijacking and man-in-the-middle attacks as exposed in [45]. Figure 3.12 shows how an adversary can exploit the above-mentioned vulnerability to launch a session hijacking attack. The adversary waits until the authenticator and the supplicant complete the authentication process and the authenticator sends the EAP success message to the supplicant. Following this, the adversary sends a disassociate message to the

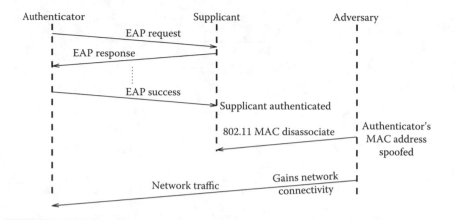

Figure 3.12   Session hijacking attack on 802.1X authentication mechanism.

supplicant with the spoofed IP of the authenticator. The supplicant assumes its session has been terminated by the authenticator as the message is not verified for integrity. The adversary gains access to the network by spoofing the MAC address of the supplicant and proceeds with a mutual authentication procedure using the four-way handshake.

Figure 3.13 shows a man-in-the-middle attack launched by the adversary exploiting the vulnerability in IEEE 802.1X. After the initial exchange of EAP request and response messages between the supplicant and the authenticator, the adversary sends an EAP success message to the supplicant using its own MAC address. Because the IEEE 802.1X protocol suggests unconditional transition upon receiving the EAP success message by the supplicant, the supplicant assumes it is authenticated by the authenticator

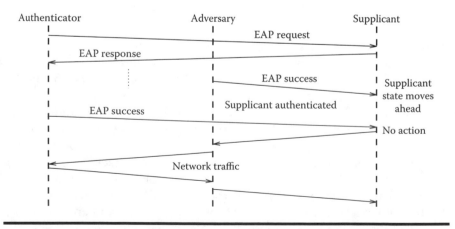

Figure 3.13   Man-in-the-middle attack on 802.1X authentication mechanism.

and changes the state. When the authenticator sends the EAP success message, the supplicant has already passed the stage where it was waiting for the success message, and hence no action is taken for this message. The supplicant assumes the adversary as the legitimate authenticator while the adversary can easily spoof the MAC address of the supplicant to communicate with the actual authenticator. Therefore, the adversary will become the intermediatory between the supplicant and the authenticator. The proposed solutions to prevent these attacks [45] recommend the authentication and integrity check for the EAP messages between the authenticator and the supplicant. The solution also proposes that the peer-to-peer based authentication model be adopted where the authenticator and the supplicant should be treated as peers and the supplicant should verify the messages from the authenticator during the process of trust establishment. The peer-to-peer model is suitable for WMNs where both the authenticator and the supplicant are WMN routers.

3.6.1.2 Four-Way Handshake Vulnerabilities

Four-way handshake is the mechanism used for the mutual authentication of the supplicant and the authenticator in IEEE 802.11i. Vulnerabilities in the four-way handshake have been identified and the DoS attack exploiting these vulnerabilities proposed in [44]. The handshake starts after PMK is distributed to the supplicant and the authenticator. The supplicant waits for a specific interval of time for message 1 of the handshake from the authenticator. If the message is not received, the supplicant disassociates itself from the authenticator. Note that this is the only timer used by the supplicant. If message 1 is received by the supplicant, it is then bound to respond to every message from the authenticator and wait for the response until it is received. On the other hand, the authenticator will time out for every message if it does not receive the expected response within a specific time interval. Further, the supplicant is de-authenticated if the response is not received after several retries. Also note that both the authenticator and the supplicant drop the message silently if the MIC of the message cannot be verified.

This mechanism of four-way handshake is vulnerable to the DoS attack by the adversary. Consider Figure 3.14 where the authenticator sends message 1 to the supplicant. Note that message 1 is not encrypted. Supplicant generates a new SNonce and then generates PTK using the ANonce, SNonce, and other relevant fields and responds with the message 2, which is encrypted using PTK. At this point, the adversary sends the malicious message 1 with the spoofed MAC address of the authenticator. The supplicant is bound to respond to the message. It assumes that the message 2 that it sent to the authenticator is lost so the authenticator has retransmitted the message 1. Therefore, it calculates PTK' (different from PTK and overwriting PTK) based on the ANonce sent by the adversary and sends

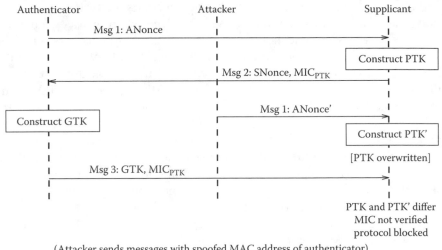

Figure 3.14 DoS attack on four-way handshake. Attacker sends messages with spoofed MAC address of authenticator.

message 2 again which is encrypted using PTK'. Meanwhile, the authenticator responds to the first message 2 of the supplicant by sending the message 3 which is encrypted using PTK. The integrity check performed by the supplicant on message 3 fails because the supplicant is now using PTK' while the authenticator encrypted the message using PTK. Consequently the four-way handshake process is blocked until the authenticator de-authenticates the supplicant after several retries, denying the supplicant of the service.

Three solutions have been proposed in [44] to prevent the above attack. We only discuss the most effective solution here. Note that every time the supplicant receives message 1, it generates a new SNonce which is concatenated with ANonce (transmitted by authenticator in message 1) and other relevant information to generate new PTK. The proposed solution suggests that the supplicant should store the SNonce created in response to the first message 1 that it receives from authenticator. The same SNonce should be used for all subsequent message 1s until the supplicant receives message 3 from the authenticator. Upon receiving the message 3, supplicant should use the newly transmitted ANonce in message 3 and the stored SNonce to generate PTK again, which should be used to decrypt message 3. Use of the same SNonce and ANonce will generate the same PTK if other information remains unchanged. Therefore, the supplicant will be able to respond to the legitimate message 3 even if it receives multiple message 1s from the adversary. Note that the adversary cannot send a malicious message 3 because message 3 is encrypted using PTK, which is dependent on PMK (only known to the supplicant and the authenticator).

3.6.1.3 CCMP Encryption Vulnerabilities

Although CCMP (employed by IEEE 802.11i) uses the CCM encryption, the strength of which is time tested, the protocol is vulnerable to the partial matching and pre-computation attacks. The vulnerabilities of the protocol implementation and the resulting attacks have been exposed in [40]. The research shows that the address field A2 and the priority field of the MAC header and the PN field in the CCMP header are transmitted as plaintext in the headers as well as in the encrypted form as part of the MIC. This leads to the partial matching attack and the researchers have shown that the key strength of the 128-bit encryption key used in CCMP decreases. The decrease in the key strength exposes the protocol to pre-computation attack, resulting in the compromise of data confidentiality and data integrity. Further, the CCM encryption is a two-phase process. During the first phase, the MIC is calculated, and in the second phase, the encryption of the frame takes place. Similarly, the decryption is done in two phases where first the message integrity is verified from MIC and then the decryption takes place. The two-phase processing of the frame at each wireless link may lead to considerable delay in case of multi-hop wireless networks like wireless mesh networks where the data traverses a number of intermediate wireless hops before reaching the wired Internet. The delay introduced by the security services leads to the impracticability of the CCMP protocol for large wireless mesh networks consisting of several intermediate hops.

3.7 Open Issues

A number of security solutions have been discussed aimed at solving different security issues, preventing, detecting, and countering the security attacks; however, a number of open issues still require considerable attention.

- Quite a few intrusion detection systems exist for multi-hop wireless networks; however, very few solutions actually comply with the characteristics of the security solution for WMN (listed in Section 3.4). For example, very few solutions lead to the self-healing and self-organized WMN, primarily because of the lack of appropriate response mechanism to the detected anomalies and possible attacks in the network.
- A number of authentication mechanisms have been proposed for multi-hop wireless networks. However, either the solutions incur unacceptable overheads to cater for mobility or the solutions are

non-robust in an effort to accommodate the trade-off between available resources and the achievable security level. Neither high mobility nor the resource limitation is a major design constraint for WMN. Therefore, the authentication mechanisms for WMN can be more robust with limited overhead and need to be redefined keeping in view the characteristics of WMN.

■ Although efforts have been made to address the security issues originating from colluding malicious nodes that can launch the attacks like wormhole and black hole, no solution has successfully addressed the issue of colluding malicious nodes. The malicious and misbehaving nodes pose serious threats to WMN, specifically if the network has to be self-healing and self-organized.

■ No security mechanism has so far been proposed to address the security vulnerabilities in the joint channel assignment and routing algorithms for multi-radio multi-channel WMN. These algorithms are crucial for the performance of multi-radio multi-channel WMN and a security loophole in these algorithms can lead to severely degraded performance and, in some cases, the complete DoS.

■ IEEE 802.11i, the standard for security in wireless networks, needs to address the issues identified in Section 3.6 before it can be integrated into IEEE 802.11s (draft standard for WMN) as the security component.

3.8 Conclusion

In this chapter, we considered the security issues in wireless mesh networks that render these networks vulnerable to security attacks. Different security attacks on the MAC layer and network layer of wireless mesh networks have been considered in detail. Security mechanisms used to detect, prevent, and counteract these attacks have been discussed briefly. The intrusion prevention and detection mechanisms used in various multi-hop wireless networks and applicable to wireless mesh networks have been considered. The IEEE 802.11i standard for security in wireless networks has been discussed in detail along with a note on the vulnerabilities rendering the protocol impractical for use in wireless mesh networks.

References

[1] Ashish Raniwala, Kartik Gopalan, and Tzi-cker Chiueh. Centralized channel assignment and routing algorithms for multi-channel wireless mesh networks. In *ACM SIGMOBILE Mobile Computing and Communications Review* (MC2R), April 2004.

[2] Ashish Raniwala and Tzi-cker Chiueh. Architecture and algorithms for an IEEE 802.11-based multi-channel wireless mesh network. In *Proceedings of IEEE InfoCom*, March 2005.

[3] Murali Kodialam and Thyaga Nandagopal. Characterizing the capacity region in multi-radio multi-channel wireless mesh networks. In *Proceedings of Mobile Computing and Networking*, August 2005.

[4] Mansoor Alicherry, Randeep Bhatia, and Li (Erran) Li. Joint Channel Assignment and Routing for Throughput Optimization in Multi-radio Wireless Mesh Networks. In *Proceedings of Mobile Computing and Networking*, August 2005.

[5] Krishna N. Ramachandran, Elizabeth M. Belding-Royer, Kevin C. Almeroth, and Milind M. Buddhikot. Interference-aware channel assignment in multi-radio wireless mesh networks. In *Proceedings of IEEE Infocom 2006*, April 2006.

[6] P. Papadimitratos and Z. Haas. Secure routing for mobile ad hoc networks. In *SCS Communication Networks and Distributed Systems Modeling and Simulation Conference* (CNDS 2002), January 2002.

[7] Manel Guerrero Zapata and N. Asokan. Securing ad hoc routing protocols. In *Proceedings of the ACM Workshop on Wireless Security* (WiSe 2002), September 2002.

[8] Kimaya Sanzgiri, Bridget Dahill, Brian Neil Levine, Clay Shields, and Elizabeth Belding-Royer. A secure routing protocol for ad hoc networks. In *Proceedings of the 10th IEEE International Conference on Network Protocols* (ICNP '02), November 2002.

[9] Yih-Chun Hu, Adrian Perrig, and David B. Johnson. Ariadne: A secure on-demand routing protocol for ad hoc networks. In *Proceedings of the 8th Annual International Conference on Mobile Computing and Networking* (MobiCom 2002), pp. 12–23, September 2002.

[10] Huaizhi Li and Mukesh Singhal. A secure routing protocol for wireless ad hoc networks. In *Proceedings of the 39th Hawaii International Conference on System Sciences*, January 2006.

[11] Yih-Chun Hu, Adrian Perrig, and David B. Johnson. Rushing attacks and defense in wireless ad hoc network routing protocols. In *Proceedings of the 2003 ACM Workshop on Wireless Security* (WiSe 2003), in conjunction with MobiCom, pp. 30–40, September 2003.

[12] Yih-Chun Hu, Adrian Perrig, and David B. Johnson. Packet leashes: A defense against wormhole attacks in wireless ad hoc networks. In *Proceedings of IEEE INFOCOM 2003*, April 2003.

[13] Sanjay Ramaswamy, Huirong Fu, Manohar Sreekantaradhya, John Dixon, and Kendall E. Nygard. Prevention of cooperative black hole attacks in wireless ad hoc networks. *International Conference on Wireless Networks*, pp. 570–575, June 2003.

[14] W. Xu, W. Trappe, Y. Zhang, and T. Wood. The feasibility of launching and detecting jamming attacks in wireless networks. In *Proceedings of ACM MOBIHOC*, 2005.

[15] J. Pollastre, J. Hill, and D. Culler. Versatile low power media access for wireless sensor networks. In *Proceedings of ACM Sensys*, 2004.

[16] Y. Law, L. Hoesel, J. Doumen, P. Hartel, and P. Havinga. Energy-efficient link-layer jamming attacks against wireless sensor network MAC protocols. In *Proceedings of the 3rd ACM Workshop on Security of Ad Hoc and Sensor Networks* (SASN 2005).

[17] T. Brown, J. James, and A. Sethi. Jamming and sensing of encrypted wireless ad hoc networks. In *Proceedings of ACM MOBIHOC*, May 2006.

[18] Arunesh Mishra and William A. Arbaugh. An Initial Security Analysis of the IEEE 802.1X Standard, Technical report, University of Maryland, February 2002.

[19] M. Al-Shurman, S. Yoo, and S. Park. Black hole attack in mobile ad hoc networks. In *Proceedings of the 42nd Annual Southeast Regional Conference*, Huntsville, Alabama, April 2004.

[20] J. Newsome, E. Shi, D. Song, and A. Perrig. The Sybil attack in sensor networks: Analysis and defenses, *3rd International Symposium on Information Processing in Sensor Networks*, IPSN 2004, pp. 259–268, April 2004.

[21] Anjum Naveed and Salil S. Kanhere. Security vulnerabilities in channel assignment of multi-radio multi-channel wireless mesh networks. In *Proceedings of IEEE GLOBECOM*, November 2006.

[22] S. Zhong, L.E. Li, Y.G. Liu, and Y.R. Yang. On designing incentive-compatible routing and forwarding protocols in wireless ad-hoc networks: An integrated approach using game theoretical and cryptographic techniques. In *Proceedings of IEEE MOBICOM*, pp. 117–131, August 2005.

[23] N.B. Salem, L. Buttyan, J.-P. Hubaux, and M. Jakobsson, A charging and rewarding scheme for packet forwarding in multi-hop cellular networks. In *Proceedings of IEEE MobiHoc*, p. 1324, June 2003.

[24] Y. Liu, H. Man, and C. Comaniciu. A game theoretic approach to efficient mixed strategies for intrusion detection. In *IEEE International Conference on Communications* (ICC), 2006.

[25] Ana Paula R. da Silva, Marcelo H.T. Martins, Bruno P.S. Rocha, Antonio A.F. Loureiro, Linnyer B. Ruiz, and Hao Chi Wong. Decentralized intrusion detection in wireless sensor networks. In *Proceedings of the 1st ACM International Workshop on Quality of Service and Security in Wireless and Mobile Networks* (Q2SWinet 2005), pp. 16–23, October 2005.

[26] Yi-an Huang, Wei Fan, Wenke Lee, and Philip S. Yu. Cross-feature analysis for detecting ad-hoc routing anomalies. *Proceedings 23rd International Conference on Distributed Computing Systems*, pp. 478–487, May 2003.

[27] Yi-an Huang and Wenke Lee. A cooperative intrusion detection system for ad hoc networks. *Proceedings of the 1st ACM Workshop on Security of Ad Hoc and Sensor Networks*, pp. 135–147, October 2003.

[28] Hao Yang, J. Shu, Xiaoqiao Meng, and Songwu Lu. SCAN: Self-organized network-layer security in mobile ad hoc networks. *IEEE Journal on Selected Areas in Communications*, Volume 24, Issue 2, pp. 261–273, February 2006.

[29] Security Architecture for Open Systems Interconnection for CCITT Applications, *ITU-T Recommendation X.800*, March 1991.

[30] IEEE Std. 802.11i-2004, Wireless Medium Access Control (MAC) and Physical Layer (PHY) Specifications: Medium Access Control (MAC) Security Enhancements, July 2004, http://standards.ieee.org/getieee802/dwnload/802.11i-2004.pdf.

[31] Hamdy S. Soliman and Mohammed Omari. Application of synchronous dynamic encryption system in mobile wireless domains. In *Proceedings of the 1st ACM International Workshop on Quality of Service and Security in Wireless and Mobile Networks* (Q2SWinet '05), pp. 24–30, October 2005.

[32] Kui Ren, Wenjing Lou, and Yanchao Zhang. LEDS: Providing location-aware end-to-end data security in wireless sensor networks. In *Proceedings of IEEE International Conference on Computer Communication* (INFOCOM '06), April 2006.

[33] IEEE Std. 802.1X-2004, *IEEE Standard for Local and Metropolitan Area Networks— Port-Based Network Access Control*, June 2001. http://standards.ieee.org/getieee802/download/802.1X-2004.pdf.

[34] B. Aboba, L. Blunk, J. Vollbrecht, J. Carlson, and H. Levkowetz, Eds., *Extensible Authentication Protocol (EAP)*, RFC 3748, June 2004.

[35] C. Rigney, S. Willens, A. Rubens, and W. Simpson, *Remote Authentication Dial In User Service (RADIUS)*, RFC 2865, June 2000.

[36] P. Calhoun, J. Loughney, E. Guttman, G. Zorn, and J. Arkko, *Diameter Base Protocol*, RFC 3588, September 2003.

[37] D. Whiting, R. Housley, and N. Ferguson, *Counter with CBC-MAC (CCM)*, RFC 3610, September 2003.

[38] S.L. Keoh and E. Lupu. Towards flexible credential verification in mobile ad-hoc networks. In *Proceedings of the 2nd ACM International Workshop on Principles of Mobile Computing*, POMC '02. Toulouse, France, October 2002.

[39] J. Kong, P. Zerfos, H. Luo, S. Lu, and L. Zhang. Providing robust and ubiquitous security support for MANET. In *Proceedings of IEEE ICNP*, 2001, pp. 251–260.

[40] M. Junaid, Muid Mufti, and M. Umar Ilyas. Vulnerabilities of IEEE 802.11i wireless LAN CCMP protocol, *Transactions on Engineering, Computing and Technology*, Volume 11, February 2006.

[41] N. Aboudagga, M.T. Refaei, M. Eltoweissy, L.A. DaSilva, and J. Quisquater. Authentication protocols for ad hoc networks: Taxonomy and research issues. In *Proceedings of the 1st ACM International Workshop on Quality of Service and Security in Wireless and Mobile Networks* (Q2SWinet '05). Montreal, Quebec, Canada, October 2005.

[42] D. Hongmei, A. Mukherjee, and D.P. Agrawal. Threshold and identity-based key management and authentication for wireless ad hoc networks, In *Proceedings of International Conference on Information Technology: Coding and Computing* (ITCC 2004), pp. 107–111, Vol. 1, April 2004.

[43] Y.-X. Lim, T.S. Yer, J. Levine, and H.L. Owen. Wireless intrusion detection and response. Information assurance workshop 2003. *IEEE Systems, Man and Cybernetics Society*, pp. 68–75, June 2003.

[44] Changhua He and John C. Mitchell, Analysis of the 802.11i 4-way hand-shake, WiSE₁04, Philadelphia, October 2004.

[45] Arunesh Mishra and A. William Arbaugh, An Initial Security Analysis of the IEEE 802.1X Standard, Technical report CS-TR-4328, Department of Computer Science, University of Maryland, February 2002, https://drum. umd.edu/dspace/handle/1903/1179?mode=full.

Chapter 4

Intrusion Detection in Wireless Mesh Networks

Thomas M. Chen, Geng-Sheng Kuo, Zheng-Ping Li, and Guo-Mei Zhu

Contents

Wireless mesh networks are potentially vulnerable to a broad variety of attacks. Hence security is an important consideration for the practical operation of wireless mesh networks. Within security, intrusion detection is the second line of defense in wireless networks as well as wired networks. Unfortunately, wireless mesh networks present additional challenges due to their decentralized nature, dynamic network topology, and easy access to the radio medium. Due to these unique challenges, intrusion detection techniques cannot be borrowed straightforwardly from wired networks. New distributed intrusion detection schemes must be designed for wireless mesh networks. This chapter describes the basics of intrusion detection and gives a survey of intrusion detection schemes proposed for wireless mesh networks. The schemes share some common concepts, but differ in the details which are compared. This chapter describes the difficulties with each scheme and ongoing challenges. Due to the difficult challenges presented by the wireless environment, intrusion detection in wireless mesh networks is still an open research problem.

4.1 Introduction

The main goal of networks is to relay data between their users. Usability in terms of quality of service, availability, and reliability is a typical design objective. The value of a network is perceived by the services it provides to its users. Unfortunately, security is often a secondary consideration and somewhat contradictory to usability because it usually imposes access restrictions and usage policies. Consequently, many networks are inadequately safeguarded against a variety of attacks. Attackers may use the network to direct attacks at hosts (e.g., to access or control a host), or attackers may aim to damage the network itself.

Attacks are commonplace and readily seen on the Internet today [1]. The average PC user must be aware of good security practices, such as keeping up with operating system patches, running anti-virus software, turning on a personal firewall, and avoiding suspicious e-mail attachments. Many of these attacks will eventually cross over to wireless networks as well. For example, many attacks exploit vulnerabilities (weaknesses) in operating systems and applications; these are effective in wired or wireless networks. Also, new types of attacks are evolving constantly.

Typical examples of attacks against hosts include:

- Probing for vulnerabilities
- Exploiting vulnerabilities to gain unauthorized access
- Eavesdropping on communications
- Theft or alteration of data
- Installation of malicious software (e.g., viruses, worms, Trojan horses, spyware)
- Denial of service
- Social engineering
- Session hijacking

Some common attacks against the network include:

- Denial of service against a router or server
- Interception or modification of packets
- Interference with routing protocols
- Unauthorized tampering of Web, DNS (Domain Name System), or other servers.

Wireless networks are more vulnerable than wired networks because the wireless medium is shared and accessible through the air. In a wired network, an attacker needs to physically access the network to sniff or inject traffic. In a wireless network, an attacker can listen to or transmit packets on a radio link at a distance (and possibly not in visible sight). Thus, the radio medium makes wireless networks both more attractive as targets and harder to defend.

In addition, the mobility afforded by wireless networks is great for users but has certain implications for security. First, mobile devices tend to travel to different, perhaps unfriendly locations. A mobile device is harder to physically secure than a stationary device in a controlled environment. Without adequate physical protection, mobile devices could be physically compromised. Second, mobile users are more difficult to authenticate. A stationary user will always access the network at a known location, so authentication can be based at least in part on location (e.g., a landline phone is identified by its location). A mobile user may access the network at unpredictable locations at different times.

A mobile ad hoc network (MANET) without any fixed infrastructure presents even more challenges for security. With a fixed infrastructure, mobile users could be authenticated with an authentication server that is always accessible regardless of the user's location. However, in a MANET with a dynamic network topology, nodes may be disconnected from other nodes for periods of time. A centralized authentication server would not work because it may not be always reachable from a mobile user's location.

Without the capability for authentication, impersonation attacks are a major concern in wireless mesh networks. By impersonation, a malicious attacker could participate in the dynamic routing protocol and affect the choice of routes. Wireless mesh networks depend on the cooperation of all nodes to relay packets across the network, so the integrity of the routing protocol is paramount. The effect of an attack on routing could be degradation of network performance, denial of service, or funneling traffic through malicious nodes. Not surprisingly, a great deal of attention has been given to secure routing protocols [2–8].

A unique type of attack called a wormhole has been identified [9]. In physics, a wormhole is theoretically a direct shortcut between two distant points in the space–time continuum. The idea of a wormhole attack is that packets at one location in the network could be tunneled and quickly replayed at another location. A wormhole could be exploited in various ways. For example, it has been hypothesized that routing update packets could go through a wormhole and cause a routing protocol to avoid certain routes [9].

Despite the popular stereotype of a misfit teenage "hacker," there is no "typical" attacker or single motive for malicious attacks. An attacker could be almost anyone — a youth looking for fame, a criminal looking for profit, an acquaintance seeking revenge, a competitor attempting industrial espionage, or a hostile foreign military agency. One of the difficulties in network security (both wired and wireless) is the wide range of types of attackers and attack methods.

On the defense side, network security consists of a variety of protective measures usually deployed in a defense-in-depth strategy. Defense-in-depth refers to multiple lines of defense, such as encryption, firewalls, intrusion detection systems, access controls, anti-virus and anti-spyware programs, combined together to increase the barriers and costs for attackers. The common belief is that a single perfect defense is not feasible. Instead, an effective deterrent can be constructed from multiple lines of defense, even though each individual element of defense is imperfect. Intrusion detection is one of the most fundamental elements in a defense-in-depth strategy.

4.2 Intrusion Detection

Intrusion detection can be viewed as a passive defense, similar to a burglar alarm in a building. Unlike firewalls or access controls, intrusion detection systems (IDSs) are not intended to deter or prevent attacks. Instead, their purpose is to alert system administrators about possible attacks, ideally in time to stop the attack or mitigate the damage [10]. Because wireless networks are easier to attack than wired networks, intrusion detection is more critical in wireless networks as a second line of defense.

4.2.1 Goals of Intrusion Detection

Intrusion detection is generally a difficult problem [11]. An IDS attempts to differentiate abnormal activities from normal ones, and identify truly malicious activities (attacks) from the abnormal but non-malicious activities. Unfortunately, normal activities have a wide range, and attacks may appear similar to normal activities. For example, a ping is a common utility to discover if a host is operating and online, but a ping can also be used for attack reconnaissance to learn information about potential targets. Even if unusual activities can be distinguished from normal activities, an unusual activity may not be truly malicious in intent.

The accuracy of intrusion detection is generally measured in terms of false positives (false alarms) and false negatives (attacks not detected). IDSs attempt to minimize both false positives and false negatives. However, this goal is complicated by the likelihood that a skillful attacker will try to evade detection. Thus, detection must be done in adversarial conditions where the attacker may be intelligent and resourceful.

IDSs also attempt to raise alarms while an attack is in progress, so that the attack can be stopped to minimize damage or the attacker can be identified "in the act." This goal is difficult considering that attacks may consist of a sequence of inconspicuous steps; many events (e.g., packets) must be analyzed in real-time, and an attack may be new and different from past experiences.

4.2.2 Host-Based and Network-Based Monitoring

An IDS essentially consists of three functions, as shown in Figure 4.1 [12]. First, an IDS must collect data by monitoring some type of events. IDSs can be classified into two types depending on the monitored events: host-based or network-based IDSs. Host-based IDSs are installed on hosts and monitor their internal events, usually at the operating system level. These internal events are the type recorded in the host's audit trails and system logs.

In contrast, network-based IDSs monitor packets in the network [13–16]. This is usually done by setting the network interface on a host to promiscuous mode (so all network traffic is captured, regardless of packet addresses). Alternatively, there are also specialized protocol analyzers designed to capture and decode packets at full link speed.

A popular network-based IDS is the open-source Snort [17]. In its simplest mode, Snort can function as a packet sniffer to view packets traversing

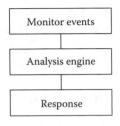

Figure 4.1   Functions of IDS.

a transmission link. In packet logging mode, Snort is able to sniff and dump complete packets into a log for later analysis. Alternatively, Snort configured with a ruleset can function as a real-time IDS. A Snort ruleset is a file of attack signatures that are matched to captured packets. A match to a signature means that an attack is recognized. It is essentially a pattern matching technique. Other popular network-based IDSs are Tcpdump and Ethereal®.

The second functional part of an IDS is an analysis engine that processes the collected data. It is programmed with certain intelligence to detect unusual or malicious signs in the data (elaborated below).

The third functional part of an IDS is a response, which is typically an alert to system administrators. A system administrator is responsible for follow-up investigation of an event after receiving an alert.

4.2.3   Misuse Detection and Anomaly Detection

As mentioned above, the second functional part of an IDS is an analysis engine. Analysis can be done manually by a security expert, but automated analysis is much faster and efficient. The problem with automated analysis is programming the analysis engine with a level of intelligence equivalent to the knowledge and experience of a security expert.

Currently, there are two basic approaches to analysis: misuse detection and anomaly detection. Misuse detection is also called signature-based detection because the idea is to represent every attack by a signature (pattern or rule of behavior). Rules can be divided into single part (atomic) signatures or multi-part (composite) signatures. It is essentially a problem of matching the observed traffic to signatures. If a matching signature is found, that attack is detected.

A common implementation of misuse detection is an expert system. An expert system consists of a knowledge base containing descriptions of attack behavior based on past experiences and rules that allow matching of packets against the knowledge base. These rules are often structured as "if-then-else" statements.

An advantage of misuse detection is its accuracy. If a signature matches, that signature identifies the specific attack. Knowledge of the specific type of attack means that an appropriate response can be determined immediately. For its accuracy, misuse detection is widely preferred in commercial systems.

There are two major drawbacks to misuse detection. First, new signatures must be developed whenever a new attack is discovered. Currently, new attacks are evolving constantly. This means that signatures for IDSs must be updated frequently. Second, an attack is recognized only if a matching signature exists. A signature will not exist for new attacks that are significantly different from known attacks. Thus, misuse detection could have a high rate of false negatives (missed attacks).

Anomaly detection, sometimes called behavior-based detection, is the opposite of misuse detection, as shown in Figure 4.2. Although they are opposite approaches, they can be used together to realize the advantages of both approaches. Misuse detection tries to characterize attacks, and everything else is assumed to be normal. In contrast, anomaly detection tries to characterize normal behavior, and everything else is assumed to be anomalous (although not necessarily malicious). The underlying premise is that malicious activities will deviate significantly from normal behavior.

The characterization of normal behavior is called a normal profile. A normal profile is usually constructed by statistical analysis of training data. Training data is typically obtained from observations of past normal behavior. Thus, a normal profile is a statistical picture of past normal behavior. Significant deviations from the normal behavior are deemed to be anomalous.

An underlying assumption is that normal behavior will remain the same or at least not change quickly. Because real behavior does change over time, practical anomaly detection systems should adapt the normal profile to track normal behavior changes. This means practical systems should have a capability for automated learning.

A major advantage of anomaly detection is the potential to detect new attacks without prior experience. That is, a signature for a new attack is not required; a new attack will be recognized if it significantly deviates from normal behavior.

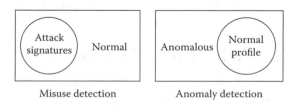

Misuse detection   Anomaly detection

Figure 4.2   Misuse detection and anomaly detection.

There are at least three drawbacks to anomaly detection. First, it has proven to be extremely difficult in practice to accurately characterize normal behavior because normal activities can have wide deviations. The choices of statistical metrics for an accurate profile is still an open research problem. Second, anomalous behavior is not necessarily malicious. In fact, a small fraction of anomalous activities may turn out to be an attack. Thus, anomaly detection often shows a high rate of false negatives. These false alarms must be investigated by system administrators, which is time consuming. Third, a detected anomaly does not identify a specific attack, unlike a signature. The lack of specific information means that system administrators must perform a follow-up investigation to determine whether an actual attack is occurring.

4.2.4 IDS Response

As mentioned above, the third functional component of an IDS is the response. Detection of an intrusion must lead to some type of output. Generally, responses can be passive or active. An example of a passive response is to log the intrusion information and raise an alert to system administrators. The IDS does not attempt to impede or stop the intrusion. An IDS response is usually passive because it is widely believed that human judgment (by a trained administrator) is required to formulate the most appropriate course of action. Also, a system administrator often needs to perform further investigation to identify the root cause of an IDS alert.

Active responses attempt to limit the damage of an attack or stop an attack in progress. Damage can be mitigated by protecting the valuable assets or the specific target of the attack. Another active response could be to track the source of the attack, which might be difficult if the attack is being carried out through intermediaries. For example, a distributed denial-of-service (DDoS) attack is essentially a flooding attack. The flooding traffic usually comes from innocent computers that were surreptitiously compromised by the real attacker. A DDoS attack might be traced to the flooding computers, but it is difficult to trace the attack further back.

There is a risk in tying active responses to intrusion detection, an approach called intrusion prevention. In the event of false positives, normal traffic is mistakenly identified as malicious. This would trigger an active response which could cause damage to an innocent user.

4.3 Unique Challenges of Wireless Mesh Networks

Intrusion detection is a common practice in wired networks. Deployment of IDS is well understood and relatively straightforward because the network environment is static. Traffic is relayed by stationary routers. Normally, there are natural points of traffic concentration which are logical candidates for monitoring. For example, private organizations usually connect to the public Internet through a gateway and firewall. All incoming and outgoing traffic go through this point. An IDS just outside of the firewall will be able to see attacks coming from the untrusted Internet. This is informative for understanding the external threats that the firewall is intended to block. Another IDS inside the firewall would monitor the traffic in the private network. If the firewall is effective, no attacks from the outside should be detected. Obviously any detected intrusion means either an insider attack or an external attack penetrated the firewall.

In comparison with wired networks, wireless mesh networks present difficulties for intrusion detection. As a review, wireless mesh networks have sprung from MANETs. MANETs have no fixed infrastructure. All nodes are mobile and the network topology is dynamic. Nodes are simultaneously user devices and routers. The requirements for MANETs have been driven largely by military or specialized civilian applications [18].

Wireless mesh networks relax the requirement of no fixed infrastructure, and can have a mix of fixed and mobile nodes interconnected by wireless links. As in MANETs, mesh nodes can be simultaneously user devices and routers. Nodes might also be fixed wireless routers, e.g., IEEE 802.11 access points or 802.16 subscriber stations [19]. These nodes could constitute a backbone infrastructure [20,21]. A principal characteristic is multi-hop routing, where packets traverse the network by opportunistic relaying from node to node. Multi-hop routes through a wireless mesh network are computed by MANET dynamic routing protocols.

4.3.1 Wireless Medium

The wireless medium is one of the major factors affecting intrusion detection. In wired networks, traffic is forced to travel along links, and there are natural points of traffic concentration which are convenient locations for intrusion detection. This is not as valid in a wireless mesh network, particularly if it is entirely ad hoc. However, there might be a backbone of fixed wireless routers. In that case, the traffic through access points should be monitored. In practice, this is difficult because access points typically do not have "SPAN ports" that mirror the traffic.

Monitoring traffic by promiscuously eavesdropping on the radio medium is not ideal. Nodes in a wireless mesh network may have relatively short radio ranges (just long enough to reach the next node), so sensors are able to see only limited amounts of traffic. Multiple sensors need to be deployed around the entire network for a comprehensive view of traffic.

Another difficulty presented by the wireless medium is the mobility afforded to nodes. As mentioned earlier, mobile devices might travel to hostile environments. A mobile device without adequate protection could be physically compromised. Therefore, nodes in a wireless mesh network are more vulnerable to compromise and cannot be entirely trusted even if their identity is authenticated.

4.3.2 Dynamic Network Topology

Again, the dynamic topology of wireless mesh networks means that there are no natural fixed points of traffic concentration which would be good choices for monitoring.

A possile approach is to run an IDS on certain hosts to monitor their local neighborhoods. However, a node cannot be expected to monitor the same area for a long time due to its mobility. A node may be unable to obtain a large sample of data for accurate intrusion detection.

4.4 Intrusion Detection for Wireless Mesh Networks

Not surprisingly, most of the research in intrusion detection pertains to MANETs because wireless mesh networks are a relatively recent development. However, virtually all of the intrusion detection schemes for MANETs are relevant to wireless mesh networks.

This section reviews intrusion detection schemes in chronological order to show the evolution of ideas over time; also, see the survey [22].

4.4.1 WATCHERS

Nodes in a wireless mesh network relay data in a cooperative way similar to the way that Internet routers relay IP packets. Therefore, intrusion detection in the Internet environment has direct relevance to intrusion detection in wireless mesh and ad hoc networks. One of the earliest intrusion detection schemes proposed for the Internet environment was WATCHERS (Watching for Anomalies in Transit Conservation: a Heuristic for Ensuring Router Security) [23]. Although WATCHERS was not specifically intended for ad hoc networks, all nodes in ad hoc networks function as routers, so the WATCHERS approach is easily applicable. Later intrusion detection schemes for ad hoc networks have followed similar ideas from WATCHERS.

WATCHERS assumes a wired mesh network of routers where individual routers may be compromised by an attacker or malfunctioning due to a fault or misconfiguration. In either case, it is assumed that an intrusion or malfunction will be manifested in the router's misbehavior (selectively dropping or misrouting packets) that can be observed by other routers.

One of the important ideas of WATCHERS is a totally distributed intrusion detection scheme running concurrently and independently in every router. Each router checks incoming packets to detect any routing anomalies. Also, each router keeps track of the amount of data going through neighboring routers. The objective is to detect misbehaving routers in a distributed way.

A link-state routing protocol is assumed. This assumption is necessary so that each router is aware of other routers and the overall network topology. Each router counts any packets that are misrouted by neighboring routers, based on knowledge of their neighbors' routing tables from the link-state routing protocol. Each router also keeps count of the amount of data received and transmitted on all interfaces.

Routers periodically share their respective data by a flooding protocol, and then start a diagnostic phase. Flooding is necessary to overcome any malicious nodes that might try to interfere in the information sharing by blocking packets. In the diagnostic phase, the counts collected from all routers are compared to determine if any routers (1) have misrouted too many packets, (2) have not participated correctly in the WATCHERS scheme, (3) broadcasted counts that have discrepancies with the counts from their neighbors, and (4) have appeared to drop more packets than a given threshold. If a router is found to exhibit any of these misbehaviors, it is deemed to be a bad router (but it is impossible to determine if the cause is an intrusion or malfunction, based solely on the router's external behavior). In response to any routers deemed to be misbehaving, routing tables at good routers are changed to avoid forwarding packets through those misbehaving routers.

The counts are compared to thresholds. In an ideal world, the thresholds would be zero, but in practice, the thresholds should be chosen to be more than zero. For example, even good routers may drop a significant number of packets if the router is congested. Therefore, the threshold for number of dropped packets could be high. The choice of proper thresholds can be difficult. If the thresholds are too high, misbehaving routers could be undetectable. On the other hand, if thresholds are too low, the rate of false alarms could be significant.

There are costs involved in the WATCHERS scheme. Each router must use memory to keep counts and a routing table for each neighboring router. Also, all routers are involved in a flooding protocol to share information before each diagnostic phase. Moreover, the scheme requires certain conditions to work: (1) each good or bad router must be directly connected to at least one good router, (2) each good router must be able to send a packet to each other good router through a path of good routers, and (3) the majority of routers must be good.

4.4.2 Cooperative Anomaly Detection

One of the earliest intrusion detection schemes for ad hoc networks was proposed by Zhang and Lee [24,25]. One of the basic ideas is distributed monitoring and cooperation among all nodes, similar to the basic idea in

WATCHERS. Each node independently observes its neighborhood (within its radio range) looking for signs of intrusion. Each node runs an IDS agent which keeps track of internal activities on that node and packet communications within its local neighborhood.

A second idea in the scheme is to rely mainly on anomaly detection because of perceived difficulties with misuse detection. Misuse detection is limited to the set of known attacks with existing signatures. Also, signatures must be constantly updated, which would be a difficult process in a wireless ad hoc network. Because anamaly detection does not require the distribution of signatures, it is easier to implement in independent nodes. Each node develops a normal profile during a training period, and looks for significant deviations from the normal profile to detect anomalies.

A third idea in the scheme is cooperation among nodes to cover a broader area. If a node has strong evidence of an anomaly, it can raise an alert itself. However, if a node has weak or inconclusive evidence of an anomaly, it can request a global investigation. The requesting node shares its data about the suspected intrusion with its neighboring nodes. The neighboring nodes share their relevant data, and each participating node follows a consensus algorithm to determine whether to raise an alarm. Any node that comes to the conclusion that an intrusion exists can raise an alarm.

The response to an alarm might be recomputation of routing tables to avoid compromised nodes, or communication links between nodes are forced to re-initialize (re-authenticate each other). The latter would not be effective if an attack has compromised a node and captured its authentication credentials.

4.4.3   Watchdogs and Pathraters

The idea of nodes monitoring the packet forwarding behavior of neighboring nodes was also proposed by Marti et al. [26]. Dynamic source routing is assumed. When a packet is ready to be sent, a path to the destination is discovered on demand, and the addresses of the nodes along the path are encapsulated in the packet header. Two new ideas are introduced: watchdogs and pathraters.

A watchdog is a process running on a node to monitor the behavior of neighboring nodes. After a node forwards a packet, the watchdog monitors the next node to see that the packet is forwarded again. With source routing assumed, the watchdog has knowledge of the proper route for a tracked packet. If a neighboring node is observed to drop more packets than a given threshold, that node is deemed to be misbehaving.

The pathrater works to avoid routing packets through misbehaving nodes. Each node maintains a rating for every other node in the range from 0 to 1. It calculates a path metric by averaging the node ratings in the path. Node ratings are initialized to a neutral value of 0.5. Actively used paths are incremented periodically, but nodes suspected of misbehaving will have their rating lowered severely.

Because the watchdog is a rather simple monitoring process, several limitations were noted. First, the scheme is limited to source routing because the watchdog needs knowledge of the proper route for each packet. Second, it is vulnerable to interference by a malicious node falsely reporting other nodes as misbehaving. Third, multiple misbehaving nodes could collectively interfere with the watchdog process. Lastly, a misbehaving node could escape detection by dropping packets just below the threshold level.

4.4.4 TIARA

TIARA (Techniques for Intrusion-resistant Ad hoc Routing Algorithms) was actually a set of mechanisms to ensure an ad hoc network could continue to operate under hostile adversarial conditions, rather than an intrusion detection scheme [27]. However, a flow monitoring mechanism in TIARA is designed to detect path failures from misbehaving nodes.

The basic idea is for source nodes to periodically send special "flow status" messages to destination nodes. Flow status messages contain information about the number of packets that have been sent from the source to destination since the previous flow status message. To prevent interference with flow status messages, each message is numbered sequentially (to detect loss) and encrypted with a digital signature (for authentication).

Upon receiving a flow status message, the destination node compares the carried number to the actual number of packets received since the last flow status message. A path failure is notified to the source node if (1) a flow status message has been lost or not received by a specified time interval, (2) the actual number of received packets is less than a threshold fraction of the number indicated by the source, or (3) the actual number of received packets is much more than the number indicated by the source.

There are two obvious disadvantages of this scheme for intrusion detection. First, a path failure does not identify which specific nodes could be compromised. Second, the flow status messages incur a cost in additional traffic that is proportional to the number of source-destination pairs in the network.

4.4.5 Malcounts

Another distributed intrusion detection system proposed by Bhargava and Agrawal [28] is essentially an enhancement of Zhang and Lee's approach. As before, it is assumed that each node is independently and concurrently monitoring its local neighborhood (nodes within its radio range). AODV

(Ad hoc On-demand Distance Vector) routing is assumed. When a packet is ready to be sent, the source node will flood a request through the network; a request successfully reaching the destination will be acknowledged back to the source.

The central idea in the intrusion detection scheme is that each node maintains a "malcount" for neighboring nodes, which is the number of observed occurrences of misbehavior. When the malcount for a node exceeds a given threshold, an alert is sent out to other nodes. The other nodes then check their malcounts for the suspected node and may support the initial alert with secondary alerts. If a suspected node triggers two or more alerts, it is deemed to be malicious and a "purge" message is broadcasted. In response, the suspected node is avoided by the other nodes.

A problem with the proposed scheme is that it is not clear if malcounts are only cumulative, so they can increase but not decrease. The ability to decrease malcounts would be useful for nodes with unusual but not malicious behavior that might be falsely identified as malicious. Their unusual behavior might cause their malcount to increase, but then a period of good behavior would result in their malcount returning to a normal value. This could avoid false alerts.

Naturally, this scheme works only if at least two trustworthy nodes are observing a suspected node, and can be defeated by malicious nodes sending out false alerts. Also, the scheme depends on a threshold for malcounts. A compromised node could avoid detection by keeping its misbehavior under the threshold.

4.4.6 CONFIDANT

The CONFIDANT (Cooperation of Nodes: Fairness in Dynamic Ad hoc Networks) scheme was proposed by Buchegger and Le Boudec [29]. Like previous schemes, it is highly distributed with each node monitoring its local neighborhood and cooperatively sharing information with other nodes. Source routing is assumed so that nodes have knowledge of the correct route for tracked packets. In each node, the CONFIDANT system includes four components: the monitor, reputation system, trust manager, and path manager.

Similar to Zhang and Lee's approach, the monitor in each node observes the activities of neighboring nodes (within radio range) to look for misbehavior. With source routing assumed, the monitor has knowledge of the next hop for each packet. When the node forwards a packet to a neighbor, it watches the neighbor to see whether the packet is forwarded correctly to the next hop. A copy of the entire packet is also stored temporarily to detect any suspicious modifications to the forwarded packet. If a misbehavior is observed, the reputation system is called.

The reputation system is similar in concept to Bhargava and Agrawal's malcount and Marti et al.'s node ratings. The reputation system consists of a table listing all observed nodes and their reputation ratings. If a node is observed to be misbehaving (deviating from expected routing behavior), the node's rating is changed by a weighting function, depending on the confidence in the accuracy of the new observation. To reduce the chance of false alarms, a node's rating can be improved after a specified period of good behavior. If a node's rating falls below a threshold, the path manager is called.

The path manager has a number of responsibilities. It keeps track of a security rating for paths, depending on the reputations of nodes in the path. Paths containing a malicious node are deleted. If a received packet is going on a path containing a malicious node, the packet is ignored and the source is alerted. If a received packet comes from a malicious node, the packet is ignored.

The last component, the trust manager, is responsible for receiving and sending "alarm" messages. Alarm messages contain information about observed misbehaviors to warn about suspected nodes. Alarm messages are sent to other nodes on a "friends" list, although the maintenance of the friends list has not been described. When a node receives an alarm message, the trust manager looks up the source of the message. If the source is trusted, the alarm message is added to a table of alarms. If there is enough evidence that a reported node is indeed malicious, the information is passed to the reputation manager.

A number of details in the CONFIDANT scheme remain to be developed. For example, misbehaviors besides incorrect packet forwarding are not yet specified. Other missing details are the values for thresholds, time-out for improving reputations, and who qualifies for the friends list. Also, the scheme is currently limited to source routing.

4.4.7 MobIDS

MobIDS (Mobile Intrusion Detection System) proposed by Kargl et al. [30] is generally similar to the previous distributed IDS schemes. Multiple sensors in the network keep track of observed instances of cooperative and non-cooperative behavior of nodes. Cooperative instances are given positive values whereas non-cooperative instances are given negative values. All instances observed for a suspected node are combined to calculate a sensor rating for that node, where older instances are given less weight. Then all sensor ratings for a suspected node are averaged, with a weight reflecting the credibility of each sensor, into a "local rating" for that node.

The local ratings are distributed periodically by broadcasting them to neighboring nodes within a given range. Each node averages the local ratings that it receives (including its own rating) into global ratings for other nodes. But global ratings are accepted only when at least a prespecified minimum number of nodes have contributed to the rating. Nodes are deemed to be misbehaving if their ratings drop below a given threshold. Routes are changed to avoid misbehaving nodes, and packets related to those nodes are dropped.

4.4.8   Mobile Agents

Puttini et al. [31] propose a distributed IDS scheme that is similar architecturally to previous proposals except that mobile agents are used for interactions between nodes (instead of data). Mobile agents are software programs that can autonomously suspend execution at one node, transfer their code and state to another node, and resume execution at the second node. Mobile agents are usually implemented in Java™ because the Java Virtual Machine is widely supported on a broad variety of operating systems.

Each node independently runs a process called local IDS (LIDS). The LIDS includes a sensor that is essentially an SNMP (Simple Network Management Protocol) agent to retrieve data from the node's MIB (management information base). The LIDS includes a file of signatures and performs misuse detection to detect attacks.

Information is shared among nodes by dispatching mobile agents, although implementation details about this procedure are lacking. Also, the performance and costs of the mobile agents have not been evaluated. Mobile agents have been studied for many years and proposed for fields such as network management and electronic commerce. However, the theoretical advantages of mobile agents have been elusive.

Mobile agents have never seen much commercial success. Part of the reason is the need for universal adoption of a mobile agent platform (e.g., Java Virtual Machine) which supports the execution and migration of mobile agents. Another reason is that mobile agents do not seem to perform any applications that static agents cannot. Finally, mobile agents introduce additional security concerns because they involve the installation of new (possibly untrusted) code on a host. Special security mechanisms must be installed on hosts to ensure that mobile agents do not cause damage. Because they require higher security, mobile agents are probably poor choices as a solution to security problems such as intrusion detection.

4.4.9   AODVSTAT

AODVSTAT is an extension of STAT (state transition analysis technique) to intrusion detection in wireless networks that use AODV routing [32]. STAT is a stateful signature-based detection technique proposed earlier for wired networks [33]. The premise is that computer attacks can be characterized as sequences of actions taken by an intruder. States represent a snapshot of a host's volative, semi-permanent, and permanent memory.

A complete representation of a successful attack starts from a safe initial state, proceeds through a number of intermediate states, and ends in a compromised state. States are characterized by assertions, which are functions with arguments returning Boolean values. These assertions describe aspects of the security state of the system. Transitions between states are associated with signature actions, which are actions by the intruder necessary for a successful attack. Omission of a signature action would prevent successful completion of the attack.

AODVSTAT applies the ideas of STAT to AODV-routed wireless networks. As mentioned earlier, AODV discovers routes on demand when a packet is ready to be sent. The source node floods a request through the network, and a reply is returned by the destination or an intermediate node that has a route to the destination. A malicious node could interfere with the control packets of the routing protocol, or interfere with the forwarding of data packets.

AODVSTAT sensors are placed on a subset of nodes for promiscuous sensing of radio channels. A sensor has two modes of operation. In stand-alone mode, a sensor looks for signs of attack only within its local neighborhood. In distributed mode, sensors periodically exchange "update" packets containing information about the neighboring nodes of each sensor. The purpose for sharing information is to detect attacks in a distributed way.

As in STAT, AODVSTAT works by stateful signature-based analysis of the observed traffic. Each sensor has a file of attack signatures and looks for a signature match with the traffic. A match triggers a response, usually an alert.

AODVSTAT would have largely the same strengths and weaknesses as STAT. As a misuse detection technique, AODVSTAT could accurately detect types of attacks that consist of sequential actions. A practical issue of how to update the attack signature files at all sensors in an ad hoc network has not been addressed. Also, AODVSTAT has the same limitations as all misuse detection techniques, i.e., the inability to detect attacks without an existing signature. However, in a real implementation, it should be straightforward to combine AODVSTAT with anomaly detection for the best of both techniques.

4.4.10 Trust Model

Pirzada and McDonald [34] described an approach to building trust relationships between nodes in an ad hoc network, but the method is essentially intrusion detection. It is assumed that nodes in the network passively monitor the packets received and forwarded by other nodes, called events.

Events are observed and given a weight, depending on the type of application requiring a trust relationship with other nodes. The weights reflect the significance of the observed event to the application. The trust values for all events from a node are combined using weights to compute an aggregate trust level for another node.

Trust values could be viewed as link weights for the computation of routes. Links with smaller weights would be links to more trusted nodes. A shortest-path routing algorithm would compute the most trustworthy paths.

The similarities between this scheme and previous IDS schemes are clear. Both approaches involve nodes observing the behavior of other nodes and making independent judgments about them. The only difference is that intrusion detection attempts to decide whether a node has been compromised (misbehaving) or not, whereas Pirzada and McDonald's trust model decides on the trustworthiness of a node.

4.4.11   RESANE

RESANE (REputation-based Security in Ad hoc NEtworks) [35] takes a view similar to Pirzada and McDonald's trust model. RESANE is not an IDS scheme per se, but uses intrusion detection techniques for a trust model. It assumes that nodes are running an IDS scheme to identify nodes that are misbehaving. The problem addressed is how to make use of the IDS information.

The goal of RESANE is to calculate reputations for nodes and leverage reputations to motivate cooperation between nodes and good behavior throughout the network. The idea is that a bad reputation will motivate a node toward good behavior. If the node continues misbehavior, its reputation will continue to suffer and the node will become isolated from the rest of the network.

A node calculates a reputation rating for a suspected neighbor from the neighbor's misbehaviors observed by the node. The node can also gather reputation ratings for that suspected neighbor from other neighboring nodes that have observed it. If a node detects a misbehavior by a suspected neighbor, the node can proactively broadcast its information to other neighbors to help them protect themselves. Thus, the overall network is protected by cooperative information sharing.

4.4.12   Critical Nodes

Karygiannis et al. advocated the concept of critical nodes [36]. These critical nodes are worth monitoring at the expense of more resources because they have considerable effect on network performance. In other words, if a critical node is malicious or misbehaving or fails, it would significantly degrade network performance. Non-critical nodes are not as important to monitor when resources are limited (the usual case in ad hoc networks).

The notion of critical nodes may aid the problem of intrusion detection, but the work does not address specifically how intrusions may be detected.

4.4.13 SCAN

SCAN attempts to address two problems simultaneously: routing misbehavior (control plane) and packet forwarding misbehavior (data plane) [37]. Routing misbehavior is exhibited by a node that does not participate properly in the routing protocol, e.g., false route advertisements. Packet forwarding misbehavior refers to any intentional interference with the proper relaying of packets, e.g., packet dropping and packet misrouting.

SCAN is based on two central ideas that are similar to previous IDS schemes. First, each node monitors its neighbors independently. Different from a watchdog, which looks only for packet forwarding misbehavior, nodes in SCAN observe their neighbors for both routing misbehavior and packet forwarding misbehavior. The second idea is information cross validation. Each node monitors its neighbors by cross-checking the overhead transmissions with other nodes. Nodes in a neighborhood collaborate with each other through a distributed consensus protocol. A suspected node can be eventually convicted of being malicious only after multiple neighbors have reached that consensus. This assumes that the network density is sufficiently high that a node can promiscuously overhear the packets sent and received by its neighbors, and nodes have multiple neighbors within range.

For routing misbehavior, SCAN requires two modifications to the usual AODV routing protocol. The usual routing update messages do not contain enough information for nodes to make judgments about routing misbehavior. First, an additional field for "previous hop" is needed in route request messages. Second, an additional field for "next hop" is needed in route reply messages. This additional information in routing messages allows nodes to maintain part of the routing tables of its neighbors. The redundant routing information enables a node to examine the trustworthiness of future routing updates from its neighbors.

The distributed consensus protocol is based on an "m out of N" algorithm, where N neighbors have been independently observing a suspected node. The suspected node is convicted as malicious if at least m out of the N nodes votes for that decision (based on observed misbehaviors). Various strategies for choosing the value of m as a function of N are proposed: a fixed fraction of N, a constant value k, or a value depending on a probability of correct detection and probability of false alarm.

If a node is convicted of being malicious, it is blocked from access to the network. In SCAN, each node must present a valid token to interact with other nodes. Tokens for convicted nodes are revoked, and revoked tokens are tracked by each node by means of a token revocation list. Asymmetric cryptography is used to prevent forged tokens. Each token is signed by the same secret key so it can be verified by a systemwide public key known to all nodes. Tokens are issued and renewed by a distributed algorithm. A token can be signed by a group of collaborating nodes, but not by a single node. A token possessed by a node can be renewed by its neighbors if it expires.

SCAN has limitations and involves some overhead in terms of communications and memory. The current SCAN scheme is limited to AODV, but may be extended to other routing protocols if they are appropriately modified (just as AODV messages must be modified with additional fields). Another limitation of SCAN is a requirement for a dense ad hoc network because multiple neighbors must collaborate to form a consensual judgment about a suspected node. Lastly, there is a requirement that collusion among attackers is limited.

4.4.14 Dempster–Shafer

Chen and Venkataramanan [38] addressed the specific problem of combining the observations of multiple neighbors to form a consensual judgment about a suspected node. Dempster–Shafer evidence theory [39] is proposed to be better than simple majority voting or a Bayesian approach. Essentially, Dempster-Shafer theory allows observers to specify a level of uncertainty in their observation. In the context of intrusion detection, if each node has a reputation or trustworthiness rating, that will be reflected by weighting their vote with a corresponding level of uncertainty. In other words, the votes from untrusted nodes will be discounted, in comparison with votes from trusted nodes, in forming a consensual judgment.

4.4.15 Optimization of Limited Resources

In wireless networks, nodes may have limited resources to spend on intrusion monitoring and detection. On the other hand, intrusion detection is more effective when more traffic is monitored. The selection of nodes to operate IDS should consider the trade-off between detection efficiency and usage of limited resources. This trade-off was formulated as an integer linear problem, where detection efficiency is maximized subject to a set of resource constraints [40].

The authors also considered a related problem where sensors could be unreliable due to faults, power savings, or compromise [41]. Again, the problem was formulated as an integer linear problem to minimize resource consumption subject to keeping a desired detection probability and the possibility that sensors could be inactive.

4.5 Open Research Issues

For reasons mentioned earlier, intrusion detection is more difficult in wireless mesh networks than wired networks. Intrusion detection continues to be a difficult and open problem even in wired networks. In wired networks, it is relatively easy to collect traffic data, but the main challenge is detection accuracy. Neither of the two current analysis approaches, misuse detection or anomaly detection, is perfect. Fundamentally, misuse detection needs an attack signature to recognize an attack. New attacks without an existing signature will be missed, resulting in a high rate of false negatives. Also, it takes significant time to develop and distribute a new signature for a new attack. A new attack has a window of opportunity after its first detection where IDSs have not received a new signature yet. A new attack will not be recognizable in the window of opportunity. Anomaly detection has a different challenge: how to construct a normal behavior profile that will yield a low rate of false positives. Detection accuracy will continue to be the main research issue in wireless mesh networks.

4.5.1 Lack of Experience with Wireless Mesh Networks

Another open issue is the lack of experience with incidents in wireless mesh networks. In contrast, security incidents have been occurring in the Internet over the past 30 years. Although no comprehensive database of attacks exists, 30 years of experience have yielded a wealth of information about Internet-based attacks. This wealth of information has helped the Internet security industry grow to considerable size, and a broad range of security products are available.

On the other hand, wireless mesh networks are a recent development, and there is little real experience with security incidents. Attacks are mostly conjectured and theoretical at this point in time. Hence, it is really unknown how to measure the progress or success of research. More real experience is needed, but will not be obtainable until wireless mesh networks are deployed more widely in the field.

4.5.2 Evaluation Difficulties

Different IDSs will detect and miss different attacks. A long-standing problem has been how to fairly evaluate and compare different IDS. In the past, experiments for wired networks have used test sets of various attacks and measured the detection rate. However, the results will obviously depend on the types of attacks in the test set because different IDS methods will have different strengths and weaknesses. Experimental comparisons of IDSs may always be controversial. Also, considering the lack of experience with real wireless mesh networks, it is difficult to know what types of attacks will be important or realistic.

4.5.3 Intrusion Tolerance

An indirectly related issue is the concept of intrusion tolerance. Intrusion detection attempts to discover the occurrence of attacks and mostly leaves the response to system administrators. Intrusion tolerance recognizes that attacks are inevitable and some attacks will be successful. The idea is to design networks from the beginning to maintain robust operation even in the face of adversarial actions. For example, redundant paths can guarantee that packets will still be delivered if an attacker brings down nodes. Clearly, intrusion tolerance is related to fault tolerance, except that fault tolerance assumes that faults are random and caused by equipment failures. Intrusion tolerance assumes an intelligent attacker capable of strategic actions. Intrusion tolerance for wireless mesh networks is virtually unexplored.

4.6 Conclusion

This chapter has reviewed the basic concepts of intrusion detection and surveyed a number of proposals for intrusion detection in wireless mesh networks. The proposals are mostly for MANETs because wireless mesh networks are a relatively recent development, but the intrusion detection schemes are directly relevant to wireless mesh networks.

A common theme in the research is the notion that nodes should independently and concurrently monitor their local neighborhoods. This is a necessity due to the decentralized nature of wireless mesh networks. A second common theme is the combination of observations from multiple nodes to form a consensual judgment about a suspected node. With these common themes, the various proposed intrusion detection schemes differ mainly in their details and not in their ideas.

At this point, a number of things are clear about the future of intrusion detection. First, there is much room for improvement. The primary measure of effective intrusion detection is low false positives and false negatives. This "proof" has not been convincingly offered by any scheme so far. Second, the challenges imposed by wireless mesh networks imply that the intrusion detection problem will continue to be open for the foreseeable future. Finally, breakthrough progress may not be expected until wireless mesh networks are deployed more widely in the field. At this time, attacks and therefore intrusion detection are largely speculative and theoretical. More real experience with wireless mesh networks will certainly help to catalyze research progress.

References

[1] S. McClure, J. Scambray, and G. Kurtz, *Hacking exposed*, 3rd ed., McGraw-Hill, 2001.

[2] L. Zhou and Z. Haas, Securing ad hoc networks, *IEEE Network*, vol. 13, November/December 1999, pp. 24–30.

[3] H. Deng, W. Li, and D. Agrawal, Routing security in wireless ad hoc networks, *IEEE Communications Magazine*, vol. 40, October 2002, pp. 70–75.

[4] K. Sanzgiri et al., Authenticated routing for ad hoc networks, *IEEE J. on Sel. Areas in Commun.*, vol. 23, March 2005, pp. 598–610.

[5] N. Salem and J-P. Hubaux, Securing wireless mesh networks, *IEEE Wireless Communications*, vol. 13, April 2006, pp. 50–55.

[6] C. Basile, Z. Kalbarczyk, and R. Iyer, Neutralization of errors and attacks in wireless ad hoc networks, Int. Conf. on Dependable Systems and Networks (DSN), 2005, pp. 518–527.

[7] N. Milanovic, M. Malek, A. Davidson, and V. Milutinovic, Routing and security in mobile ad hoc networks, *Computer*, vol. 37, February 2004, pp. 61–65.

[8] H. Yang et al., Security in mobile ad hoc networks: Challenges and solutions, *IEEE Wireless Communications*, vol. 11, February 2004, pp. 38–47.

[9] Y-C. Hu, A. Perrig, and D. Johnson, Wormhole attacks in wireless networks, *IEEE J. on Sel. Areas in Communications*, vol. 24, February 2006, pp. 370–380.

[10] J. McHugh, Intrusion and intrusion detection, *Int. J. of Information Security*, vol. 1, August 2001, pp. 14–35.

[11] S. Axelsson, Intrusion Detection Systems: A Survey and Taxonomy, Technical report 99–15, Department of Computer Engineering, Chalmers University of Technology, Sweden, March 2000.

[12] R. Bace, *Intrusion Detection*, MacMillan Technical Publishing, 2000.

[13] D. Marchette, *Computer Intrusion Detection and Network Monitoring: A Statistical Viewpoint*, Springer-Verlag, 2001.

[14] R. Bejtlich, *The Tao of Network Security Monitoring: Beyond Intrusion Detection*, Addison-Wesley, 2005.

[15] S. Northcutt and J. Novak, *Network Intrusion Detection*, 3rd ed., Pearson Education, 2003.

[16] S. Northcutt, M. Cooper, M. Fearnow, and K. Frederick, *Intrusion Signatures and Analysis*, New Riders Publishing, 2001.

[17] K. Cox and C. Gerg, *Snort and IDS Tools*, O'Reilly Media, 2004.

[18] R. Bruno, M. Conti, and E. Gregori, Mesh networks: Commodity multihop ad hoc networks, *IEEE Communications Magazine*, vol. 43, March 2005, pp. 123–131.

[19] M. Lee, J. Zheng, Y-G. Ko, and D. Shrestha, Emerging standards for wireless mesh networks, *IEEE Wireless Communications*, vol. 13, April 2006, pp. 56–63.

[20] I. Akyildiz, X. Wang, and W. Wang, Wireless mesh networks: A survey, *Computer Networks*, vol. 47, 2005, pp. 445–487.

[21] I. Akyildiz, X. Wang, A survey on wireless mesh networks, *IEEE Communications Magazine*, vol. 43, September 2005, pp. S23–S30.

[22] A. Mishra, K. Nadkarni, and A. Patcha, Intrusion detection in wireless ad hoc networks, *IEEE Wireless Communications*, vol. 11, February 2004, pp. 48–60.

[23] K. Bradley et al., Detecting disruptive routers: A distributed network monitoring approach, *IEEE Network*, vol. 12, September/October 1998, pp. 50–60.

[24] Y. Zhang and W. Lee, Intrusion Detection in Wireless Ad-hoc Networks, 6th Annual ACM Int. Conf. on Mobile Computing and Networking, Boston, 2000, pp. 275–283.

[25] Y. Zhang, W. Lee, and Y-A. Huang, Intrusion detection techniques for mobile wireless networks, *Wireless Networks*, vol. 9, 2003, pp. 545–556.

[26] S. Marti, T. Giuli, K. Lai, and M. Baker, Mitigating Routing Misbehavior in Mobile Ad hoc Networks, 6th Annual ACM Int. Conf. on Mobile Computing and Networking, Boston, 2000, pp. 255–265.

[27] R. Ramanujan, A. Ahamad, J. Bonney, R. Hagelstrom, and K. Thurber, Techniques for Intrusion-resistant Ad hoc Routing Algorithms (TIARA), IEEE MILCOM 2000, Los Angeles, 2000, pp. 660–664.

[28] S. Bhargava and D. Agrawal, Security Enhancements in AODV Protocol for Wireless Ad hoc Networks, 2001 IEEE Vehicular Technology Conf. (VTC 2001), 2001, pp. 2143–2147.

[29] S. Buchegger and J-Y. Le Boudec, Performance Analysis of the CONFIDANT Protocol (Cooperation of Nodes: Fairness in Dynamic Ad-hoc Networks), 3rd ACM Int. Symp. on Mobile Ad hoc Networks and Computing, Switzerland, 2002, pp. 226–236.

[30] F. Kargl, A. Klenk, M. Weber, and S. Schlott, Sensors for Detection of Misbehaving Nodes in MANETs, Detection of Intrusion and Malware and Vulnerability Assessment (DIMVA 2004), Dortmund, Germany, 2004.

[31] R. Puttini, J-M. Percher, L. Me, and R. de Sousa, A Fully Distributed IDS for MANET, 9th Int. Symp. on Computers and Commun. (ISCC 2004), 2004, pp. 331–338.

[32] G. Vigna et al., An Intrustion Detection Tool for AODV-based Ad hoc Wireless Networks, Annual Computer Security Applications Conf. (ACSAC 2004), Tucson, 2004, pp. 16–27.

[33] K. Ilgun, R. Kemmerer, and P. Porras, State transition analysis: A rule-based intrusion detection approach, *IEEE Trans. on Software Engineering*, vol. 21, 1995, pp. 181–199.

[34] A. Pirzada and C. McDonald, Establishing Trust in Pure Ad hoc Networks, 27th Australian Conf. on Computer Science, Dunedin, New Zealand, 2004, pp. 47–54.

[35] Y. Rebahi, V. Mujica, and D. Sisalem, A Reputation-based Trust Mechanism for Ad hoc Networks, 10th IEEE Symp. on Computers and Communications (ISCC 2005), 2005, pp. 37–42.

[36] A. Karygiannis, E. Antonakakis, and A. Apostolopoulos, Detecting Critical Nodes for MANET Intrusion Detection, 2nd Int. Workshop on Security, Privacy, and Trust in Pervasive and Ubiquitous Computing (SecPerU 2006), 2006, pp. 7–15.

[37] H. Yang, J. Shu, X. Meng, and S. Lu, SCAN: Self-organized network-layer security in mobile ad hoc networks, *IEEE J. on Sel. Areas in Communications*, vol. 24, February 2006, pp. 261–273.

[38] T. Chen and V. Venkataramanan, Dempster–Shafer theory for intrusion detection in ad hoc networks, *IEEE Internet Computing*, vol. 9, November/December 2005, pp. 35–41.

[39] G. Shafer, *A Mathematical Theory of Evidence*, Princeton University Press, 1976.

[40] D. Subhadrabandhu, S. Sarkar, and F. Anjum, A framework for misuse detection in ad hoc networks — Part I, *IEEE J. on Sel. Areas in Communications*, vol. 24, February 2006, pp. 274–289.

[41] D. Subhadrabandhu, S. Sarkar, and F. Anjum, A framework for misuse detection in ad hoc networks — Part II, *IEEE J. on Sel. Areas in Communications*, vol. 24, Feb. 2006, pp. 290–304.

Chapter 5

Secure Routing in
Wireless Mesh Networks

Manel Guerrero Zapata

Contents

Most routing protocols for client wireless mesh networks (WMNs) were designed without having security in mind. In most of their specifications it is assumed that all the nodes in the network are friendly. The security issue was postponed and there used to be the common feeling that it would be possible to make those routing protocols secure by retrofitting pre-existing cryptosystems.

Nevertheless, securing network transmissions without securing the routing protocols is not sufficient. Unless fixed networks (where one might assume that routers are trusted nodes) in a wireless network (where all the nodes are also routing nodes) are secure, malicious nodes might attack routing protocols to impersonate other nodes and inject forged routing information. Moreover, by retrofitting cryptosystems (like IPSec [19]) security is not necessarily achieved.

Therefore, in client WMNs with security needs, there must be two security systems: one to protect the data transmission and one to make the routing protocol secure. There are already well-studied, point-to-point security systems that can be used for protecting network transmissions. But there was not much work to make wireless routing protocols discover routes in a secure manner [18,31,37] until recently.

5.1 Introduction

Some aspects of wireless and ad hoc networks have interesting security problems [2,33,37]. Routing is one such aspect. Several routing protocols for these kind of networks have been developed, particularly in the MANET Working Group of the Internet Engineering Task Force (IETF). Surveys of routing protocols for ad hoc wireless networks are presented in [29,30] and, more recently, in [15] and [34].

5.2 Related Work

By the year 2000 there was very little published work on the security issues in ad hoc and wireless network routing protocols. Neither the survey by Ramanathan and Steenstrup [29] in 1996, nor the survey by Royer and Toh [30] in 1999 mention security. None of the draft proposals in the IETF MANET Working Group had a non-trivial "security considerations" section. Actually, most of them assumed that all the nodes in the network are friendly, and a few declare the problem out-of-scope by assuming some canned solution like IPSec may be applicable.

Security issues with routing in general have been addressed by several researchers (e.g., [13,32]) at the end of the 20th century. And, later, some work has been done to secure ad hoc networks by using misbehavior detection schemes (e.g., [23]). This approach has two main problems: first, it is quite likely that it will not be feasible to detect several kinds of misbehavior (especially because it is very hard to distinguish misbehaving from transmission failures and other kind of failures); and second, it has no real means to guarantee the integrity and authentication of the routing messages.

Hash chains had being used as an efficient way to obtain authentication in several approaches that tried to secure routing protocols. In [5,13,28] they use them to provide delayed key disclosure. In [36], hash chains are used to create one-time signatures that can be verified immediately. The main drawback of all the above approaches is that they require clock synchronization.

In their paper on securing ad hoc networks [37] in 1999, Zhou and Haas primarily discuss key management. They devote a section to secure routing, but essentially conclude that "nodes can protect routing information in the same way they protect data traffic." They also observe that denial-of-service attacks against routing will be treated as damage and routed around.

Dahill et al. [7] proposed ARAN in 2001, a routing protocol for ad hoc networks that uses authentication and requires the use of a trusted certificate server. In ARAN, every node that forwards a route discovery or a route reply message must also sign it (which is very computing-power-consuming and causes the size of the routing messages to increase at each hop). In addition, it is prone to reply attacks using error messages unless the nodes have time synchronization.

In October 2001, the first draft of SAODV [10] was sent to the MANET mailing list. SAODV [11,12] is an extension of the AODV routing protocol that can be used to protect the route discovery mechanism providing security features like integrity and authentication, and it only requires originators of routing messages to sign the routing messages (as opposed to ARAN, in which all the forwarding nodes sign the messages).

In 2002, Papadimitratos and Haas [27] proposed a protocol (SRP) that can be applied to several existing routing protocols (in particular DSR [17]). SRP requires that, for every route discovery, source and destination must have a security association between them. Furthermore, the paper does not even mention route error messages. Therefore, they are not protected, and any malicious node can just forge error messages with other nodes as source.

In SEAD [16], hash chains are also used in combination with DSDV-SQ [3] (this time to authenticate hop counts and sequence numbers). At every given time each node has its own hash chain. The hash chain is divided into segments; elements in a segment are used to secure hop counts in a way similar to SAODV. The size of the hash chain is determined when it is generated. After using all the elements of the hash chain, a new one must be computed.

SEAD can be used with any suitable authentication and key distribution scheme. But finding such a scheme is not straightforward.

Ariadne [16] is based on DSR [17] and TESLA [27] (on which its authentication mechanism is based). It also requires clock synchronization, which is, arguably, an unrealistic requirement for ad hoc networks.

In principle, the same approach that SAODV takes to protect AODV could be used to create a "secure version" of other routing protocols: signing the non-mutable routing information by the node to which the route will be processed, and securing the hop count by hash chains. In case there are some other mutable fields, how to protect each of them should be studied.

Nevertheless, if the routing protocol has some other mutable information than the hop count (and it does not mutate in a predictable way), protecting this information might end up being quite complex. It will probably require that the intermediate nodes that mutate part of the message also have to sign it. This will, typically, imply a reduction of performance (due to all the additional cryptographic computations) and also a possible decrease of the overall security.

If the routing protocol to be secured is DSR for mobile ad hoc networks [17], then the main problem will be that DSR includes in its routing messages the IP addresses of all the intermediate nodes that have forwarded the packet.

Intermediate nodes could sign the routing message after adding its own IP address, and verify all the signatures in every routing message. But this would greatly decrease the performance of the routing discovery, and it is not really worthwhile, taking into account that the routes to the intermediate nodes are going to be used very seldom. Anyway, hash chains should be used to avoid that a malicious node would eliminate intermediate nodes and their signatures from the routing message (a very similar technique is also used in [16]).

Another solution would be that intermediate nodes would sign the routing message, but that a node would only verify the signature of an intermediate node when it needs to send a packet to this route. But it still requires all intermediate nodes to sign the message (which is not good when the message is a route request).

Therefore, maybe a better solution would be that intermediate nodes do not sign the message. Later on, if a node that received that routing message wants to use a route to one of those intermediate nodes, it should request a signature from the intermediate node with a unicast message.

Obviously, a much more detailed analysis should be made to study the different attacks that can be performed against DSR and against this "secure DSR" to see if there are new attacks as a consequence of differences between AODV and DSR.

SRP [24] and Ariadne [16] also attempt to secure DSR. Nevertheless, SRP requires that, for every route discovery, source and destination must have a security association between them, and does not protect error messages. Ariadne requires clock synchronization, which can be considered unrealistic for ad hoc networks.

More recently and more focused on mesh networks, a paper by Asherson and Hutchison [1] has as a starting point the concern that routing algorithms designed for ad hoc networks might not be applied straightforward to WMNs. Nevertheless, it concludes giving as a solution to use different routing protocols for the infrastructure part and for the ad hoc part (which would use a routing protocol for ad hoc networks), therefore adopting the same approach as the one used in the Internet.

In the area of routing metrics for mesh networks, Yang, Wang, and Kravets [35] have studied how the use of different routing metrics affects the performance of the routing protocol in mesh networks. Nevertheless, they leave as future work the problem of how to transmit routing metrics in a secure manner.

Finally, the recent standardization efforts of the IEEE 802.11s (the IEEE standard for mesh networking) are considering MANET routing protocols like AODV [25,26] and OLSR [6] as their mesh routing protocols, as noted in the performance comparison paper by Chen, Lee, Maniezzo, and Gerla [4].

5.3 Designing a Secure Routing Protocol

When designing a secure routing protocol, as with any secure protocol, things need to be kept as simple and neat as possible, so they can be properly analyzed.

Ferguson and Schneier, in their paper "A Cryptographic Evaluation of IPsec" [8], conclude that the complexity of IPsec results in inefficiencies and weaknesses which make it weaker and very hard to analyze how secure it is. The bottom line is that creating a too-complex solution makes it unfeasible to verify if it is a good solution.

To keep the design of a secure routing protocol as neat as possible, it is convenient to make a clear distinction of the following items:

- The scenario (or scenarios) it is going to protect
- The security features that this scenario requires
- The security mechanisms that will fulfill those security features

Once the design of the secure routing protocol is done, it is time to analyze whether it indeed works, and, because the three items listed above are clearly separated in the design, it is much easier to perform such analysis because it can be split into the following parts:

- The analysis of requirements: Whether the security features are enough for the targeted scenario.
- The analysis of mechanisms: Whether the security mechanisms are indeed fulfilling all the security requirements. When doing this, it will be found that there are still some attacks that can be performed against your system. Some of them typically will not be completely avoided because of a trade-off between security and feasibility.
- The analysis of feasibility: Whether the security mechanisms have requirements that are not feasible in the targeted scenario.

5.4   Security Requirements

In most domains, the primary security service is authorization. Routing is no exception. Typically, a router needs to make two types of authorization decisions. First, when a routing update is received from the outside, the router needs to decide whether to modify its local routing information base accordingly. This is import authorization. Second, a router may carry out export authorization whenever it receives a request for routing information. Import authorization is the critical service.

In traditional routing systems, authorization is a matter of policy. For example, gated, a commonly used routing program,[1] allows the administrator of a router to set policies about whether and how much to trust routing updates from other routers, e.g., statements like "trust router X about routes to networks A and B." In mobile wireless networks, such static policies are not sufficient (and unlikely to be relevant).

Authorization requires other security services such as authentication and integrity. Techniques like digital signatures and message authentication codes are used to provide these services.

In the context of routing, confidentiality and non-repudiation are not necessarily critical services [13]. Zhou and Haas [37] argue that non-repudiation is useful in an ad hoc network for isolating misbehaving routers: a router A which received an "erroneous message" from another router B may use this message to convince other routers that B is misbehaving. This would indeed be useful if there is a reliable way of detecting erroneous messages. This does not appear to be an easy task.

The problem of compromised nodes is not addressed here because it would probably require some sort of mechanism to allow the owner to confirm its presence. Availability is considered to be outside of scope. Although of course it would be desirable, it does not seem to be feasible to prevent denial-of-service attacks in a network that uses wireless technology

---

[1] http://www.gated.org

(where an attacker can focus on the physical layer without bothering to study the routing protocol).

Therefore, in this research work the following requirements were considered:

- Import authorization: It is important to note that this is not referring to the traditional meaning of authorization. What it means is that the ultimate authority about routing messages regarding a certain destination node is that node itself. Therefore, route information will only be authorized in a routing table if that route information concerns the node that is sending the information. In this way, if a malicious node lies about it, the only thing it will cause is that others will not be able to route packets to the malicious node.
- Source authentication: Nodes need to be able to verify that the node is the one it claims to be.
- Integrity: In addition, nodes need to be able to verify that the routing information has arrived unaltered.
- The two last security services combined build data authentication, and they are requirements derived from the import authorization requirement.

Finally, it is quite likely that, for a small team of nodes that trust each other and that want to create an ad hoc network where the messages are only routed by members of the team, the simplest way to keep secret their communications is to encrypt all messages (routing and data) with a "team key." Every member of the team would know the key and, therefore, it would be able to encrypt and decrypt every single packet. Nevertheless, this does not scale well and the members of the team have to trust each other. So it can be used only for a very small subset of the possible scenarios. That renders asymmetric cryptography as the most suitable option for most wireless scenarios.

5.5 Securing Wireless Mesh Network Routing Protocols

If we agree with the idea reflected in the paper by Asherson and Hutchison [1], that the best approach is to use different routing protocols for the infrastructure part and for the ad hoc part (which would use a routing protocol for ad hoc networks), then the problem of securing WMN routing protocols becomes a much simpler one. The mesh network is composed by the infrastructure part and by the ad hoc networks that are connected to the infrastructure network through the access points.

The infrastructure part can use a routing protocol suitable for fixed networks, the ad hoc networks can use a secure routing protocol suitable for MANET networks, and the access points play as gateways of both the infrastructure and the ad hoc networks.

Because the access points act as gateways between two networks that use different routing protocols, they will use "administrative distances" to prioritize the use of routes of the infrastructure part. Remember that, in case there is a route to the same destination provided by two different routing protocols, the one with lowest "administrative distance" is used.

Routing protocols for fixed networks are relatively easy to secure. Therefore, the real challenge is to secure the routing protocol of the ad hoc part of the mesh network.

5.6   Securing Ad hoc Network Routing Protocols

In an ad hoc network, from the point of view of a routing protocol, there are two kinds of messages: the routing messages and the data messages. The routing protocol uses routing messages to establish the routes that are needed to transmit data messages, and, in the case of a reactive routing protocol, it sees the data messages and refreshes the lifetimes of the routes that those data messages use.

The two kinds of messages are different in nature and security needs. Data messages are end-to-end and can be protected with any end-to-end security system (like IPSec). On the other hand, routing messages are sent to neighbors, processed, possibly modified, and re-sent. Moreover, as a result of the processing of the routing message, a node might modify its routing table. This creates the need for the intermediate nodes to be able to authenticate the information contained in the routing messages (a need that does not exist in end-to-end communications) to be able to apply their import authorization policy.

Another consequence of the nature of the transmission of routing messages is that, in many cases, there will be some parts of those messages that will change during their propagation. This is very common in distance-vector routing protocols, where the routing messages usually contain a hop count of the route they are requesting or providing. Therefore, in a routing message, two types of information could be distinguished: mutable and non-mutable. It is desired that the mutable information in a routing message is secured in such a way that no trust in intermediate nodes is needed. Otherwise, securing the mutable information will be much more expensive in computation, plus the overall security of the system will greatly decrease.

If the security system being used to secure the data messages in a wireless network is IPSec, it is necessary that the IPSec implementation can use as a selector the TCP and UDP port numbers. This is because it is necessary that the IPSec policy will be able to apply certain security mechanisms to the data packets and just bypass the routing packets (that can be identified because they use a reserved Transport layer port number).

5.7 Ad hoc On-Demand Vector Routing

The Ad hoc On-Demand Vector Routing (AODV) protocol [25, 26] is a reactive routing protocol for ad hoc and mobile networks that maintains routes only between nodes which need to communicate. The routing messages do not contain information about the whole route path, but only about the source and the destination. Therefore, routing messages do not have an increasing size. It uses destination sequence numbers to specify how fresh a route is (in relation to another), which is used to grant loop freedom.

Whenever a node needs to send a packet to a destination for which it has no "fresh enough" route (i.e., a valid route entry for the destination whose associated sequence number is at least as great as the ones contained in any RREQ that the node has received for that destination), it broadcasts a route request (RREQ) message to its neighbors. Each node that receives the broadcast sets up a reverse route toward the originator of the RREQ, unless it has a "fresher" one (Figure 5.1).

When the intended destination (or an intermediate node that has a "fresh enough" route to the destination) receives the RREQ, it replies by

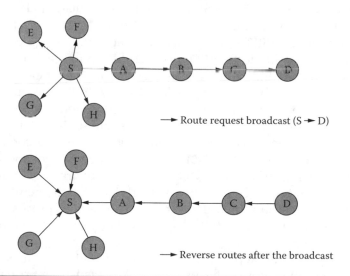

Route request broadcast (S → D)

Reverse routes after the broadcast

Figure 5.1   Route Request. After the RREQ broadcast, D has in its routing table that the next hop to S is D. The rest of the nodes also have in their routing table which is the next hop to S.

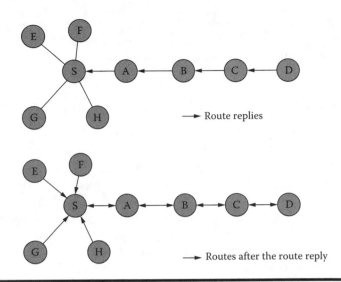

Figure 5.2   Route Reply. After S receives the RREP, all the nodes between S and D know which are the next hops to S and D. The rest of the nodes (E, F, G, and H) also have in their routing table which is the next hop to S. If they do not use that route, it will expire.

sending a Route Reply (RREP). It is important to note that the only mutable information in an RREQ and in an RREP is the hop count (which is being monotonically increased at each hop). The RREP is unicast back to the originator of the RREQ (Figure 5.2). At each intermediate node, a route to the destination is set (again, unless the node has a "fresher" route than the one specified in the RREP). In the case that the RREQ is replied to by an intermediate node (and if the RREQ had set this option), the intermediate node also sends an RREP to the destination. In this way, it can be granted that the route path is being set up bidirectionally. In the case that a node receives a new route (by an RREQ or by an RREP) and the node already has a route "as fresh" as the received one, the shortest one will be updated.

If there is a subnet (a collection of nodes identified by a common network prefix) that does not use AODV as its routing protocol and wants to be able to exchange information with an AODV network, one of the nodes of the subnet can be selected as the "network leader." The network leader is the only node of the subnet that sends, forwards, and processes AODV routing messages. In every RREP that the leader issues, it sets the prefix size of the subnet.

Optionally, a Route Reply Acknowledgment (RREP-ACK) message may be sent by the originator of the RREQ to acknowledge the receipt of the RREP. An RREP-ACK message has no mutable information.

In addition to these routing messages, a Route Error (RERR) message is used to notify the other nodes that certain nodes are not reachable anymore

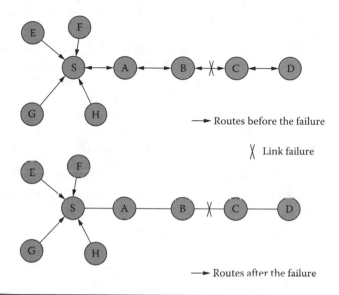

Figure 5.3 **Route Error. When a link failure is detected, all the nodes between S and D get notified about it by a Route Error (RERR) message and erase their routes to S and D.**

due to a link breakage (Figure 5.3). When a node rebroadcasts an RERR, it only adds the unreachable destinations to which the node might forward messages. Therefore, the mutable information in an RERR is the list of unreachable destinations and the counter of unreachable destinations included in the message. It is predictable that, at each hop, the unreachable destination list may not change or become a subset of the original one.

5.8 Security Flaws of AODV

Because AODV has no security mechanisms, malicious nodes can perform many attacks just by not behaving according to the AODV rules. A malicious node M can carry out the following attacks (among many others) against AODV:

1. Impersonate a node S by forging an RREQ with its address as the originator address.
2. When forwarding an RREQ generated by S to discover a route to D, reduce the hop count field to increase the chances of being in the route path between S and D so it can analyze the communication between them. A variant of this is to increment the destination sequence number to make the other nodes believe that this is a "fresher" route.

3. Impersonate a node D by forging an RREP with its address as a destination address.

4. Impersonate a node by forging an RREP that claims that the node is the destination and, to increase the impact of the attack, claims to be a network leader of the subnet SN with a big sequence number and send it to its neighbors. In this way it will become (at least locally) a black hole for the whole subnet SN.

5. Selectively, not forward certain RREQs and RREPs, not reply to certain RREPs, and not forward certain data messages. This kind of attack is especially hard even to detect because transmission errors have the same effect.

6. Forge an RERR message pretending it is the node S and send it to its neighbor D. The RERR message has a very high destination sequence number dsn for one of the unreachable destinations (U). This might cause D to update the destination sequence number corresponding to U with the value dsn and, therefore, future route discoveries performed by D to obtain a route to U will fail (because U's destination sequence number will be much smaller than the one stored in D's routing table).

7. According to the AODV specification [25], the originator of an RREQ can put a much bigger destination sequence number than the real one. In addition, sequence numbers wrap around when they reach the maximum value allowed by the field size. This allows a very easy attack, where an attacker is able to set the sequence number of a node to any desired value by just sending two RREQ messages to the node.

5.9 Secure Ad hoc On-Demand Distance Vector

Assume that there is a key management sub-system that makes it possible for each ad hoc node to obtain public keys from the other nodes of the network. Further, each ad hoc node is capable of securely verifying the association between the identity of a given ad hoc node and the public key of that node. How this is achieved depends on the key management scheme. Do not worry about how key management is achieved at this point.

SAODV uses two mechanisms to secure the AODV messages: digital signatures (why we need the key management sub-system) to authenticate the non-mutable fields of the messages, and hash chains to secure the hop count information (the only mutable information in the messages). For the non-mutable information, authentication is performed in an end-to-end manner, but the same kind of techniques cannot be applied to the mutable information.

The information relative to the hash chains and the signatures is transmitted with the AODV message as an extension message that will be referred to as Signature Extension.

5.9.1 SAODV Hash Chains

SAODV uses hash chains to authenticate the hop count of RREQ and RREP messages in such a way that allows every node that receives the message (either an intermediate node or the final destination) to verify that the hop count has not been decremented by an attacker. This prevents an attack of type 2. A hash chain is formed by applying a one-way hash function repeatedly to a seed.

Every time a node originates an RREQ or RREP message, it performs the following operations:

■ Generates a random number (seed).
■ Sets the Max Hop Count field to the TimeToLive value (from the IP header).

$$Max\_Hop\_Count = TimeToLive$$

■ Sets the Hash field to the seed value.

$$Hash = seed$$

■ Sets the Hash_Function field to the identifier of the hash function that it is going to use. The possible values are shown in Table 5.1.

$$Hash\_Function = h$$

■ Calculates Top_Hash by hashing seed Max_Hop_Count times.

$$Top\_Hash = h^{Max\_Hop\_Count}(seed)$$

where h is a hash function, and $h^i(x)$ is the result of applying the function h to x i times.

In addition, every time a node receives an RREQ or RREP message, it performs the following operations to verify the hop count:

■ Applies the hash function h Maximum_Hop_Count minus Hop_Count times to the value in the Hash field, and verifies that the resultant value is equal to the value contained in the Top_Hash field.

$$Top\_Hash = h^{Max\_Hop\_Count-Hop\_Count}(Hash)$$

where $a = b$ reads: to verify that a and b are equal.

Table 5.1   Possible Values of the Hash Function Field

Value	Hash Function
0	Reserved
1	MD5HMAC96 [21]
2	SHA1HMAC96 [22]
3–127	Reserved
128–255	Implementation dependent

■ Before rebroadcasting an RREQ or forwarding an RREP, a node applies the hash function to the Hash value in the Signature Extension to account for the new hop.

$$Hash = h(Hash)$$

The Hash_Function field indicates which hash function has to be used to compute the hash. Trying to use a different hash function will just create a wrong hash without giving any advantage to a malicious node. Hash_Function, Max_Hop_Count, Top_Hash, and Hash fields are transmitted with the AODV message in the Signature Extension, and as it will be explained later, all of them but the Hash field are signed to protect its integrity.

Figure 5.4 shows the mechanisms to do the hash chain initialization, hop count verification, and hop count incrementation.

5.9.2   SAODV Digital Signatures

Digital signatures are used to protect the integrity of the non-mutable data in RREQ and RREP messages. That means that they sign everything but the Hop_Count of the AODV message and the Hash from the SAODV extension.

The main problem in applying digital signatures is that AODV allows intermediate nodes to reply to RREQ messages if they have a "fresh enough" route to the destination. While this makes the protocol more efficient, it also makes it more complicated to secure. The problem is that an RREP message generated by an intermediate node should be able to sign it on behalf of the final destination; in addition, it is possible that the route stored in the intermediate node would be created as a reverse route after receiving an RREQ message (which means that it does not have the signature for the RREP).

To solve this problem, SAODV offers two alternatives. The first one (and also the obvious one) is that, if an intermediate node cannot reply to an RREQ message because it cannot properly sign its RREP message, it just

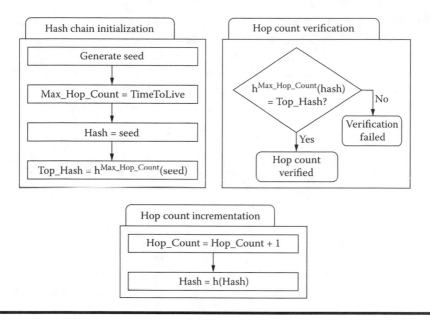

Figure 5.4   Protection of the hop count through hash chains.

behaves as if it did not have the route and forwards the RREQ message. The second is that, every time a node generates an RREQ message, it also includes the RREP flags, the prefix size, and the signature that can be used (by any intermediate node that creates a reverse route to the originator of the RREQ) to reply to an RREQ that asks for the node that originated the first RREQ. Moreover, when an intermediate node generates an RREP message, the lifetime of the route has changed from the original one. Therefore, the intermediate node should include both lifetimes (the old one is needed to verify the signature of the route destination) and sign the new lifetime. In this way, the original information of the route is signed by the final destination and the lifetime is signed by the intermediate node.

To distinguish the different SAODV extension messages, the ones that have two signatures are called RREQ and RREP Double Signature Extensions.

When a node receives an RREQ, it first verifies the signature before creating or updating a reverse route to that host. Only if the signature is verified will it store the route. If the RREQ was received with a Double Signature Extension, then the node will also store the signature for the RREP and the lifetime (which is the "reverse route lifetime" value) in the route entry. An intermediate node will reply to an RREQ with an RREP only if it fulfills the AODV's requirements to do so and the node has the corresponding signature and old lifetime to put into the Signature and Old Lifetime fields of the RREP Double Signature Extension. Otherwise, it will rebroadcast the RREQ.

When an RREQ is received by the destination itself, it will reply with an RREP only if it fulfills the AODV's requirements to do so. This RREP will be sent with an RREP Single Signature Extension.

When a node receives an RREP, it first verifies the signature before creating or updating a route (also called direct route) to that host. Only if the signature is verified, will it store the route with the signature of the RREP and the lifetime.

Both in the case of reverse and direct routes, routes are stored because they meet the import authorization requirement. That is, the route information that is being authorized in the routing table is about the node that is sending the information. In the case of reverse routes, it is about the originator of the RREQ (which is the node toward which the reverse route points). In the case of direct routes, it is about the originator of the RREP (which is the node towards which the direct route points).

In this way, if either the originator of the RREQ or the originator of the RREP messages gives fake information in those messages, the only thing that they might cause is that others will not be able to route packets to them.

Using digital signatures prevents attack scenarios 1 and 3.

5.9.3 Securing Error Messages

Concerning RERR messages, someone could think that the right approach to secure them should be similar to the way the other AODV messages are (signing the non-mutable information and finding out a way to secure the mutable information). Nevertheless, RERR messages have a large amount of mutable information. In addition, it is not relevant which node started the RERR and which nodes are just forwarding it. The only relevant information is that a neighbor node is informing another node that it is not going to be able to route messages to certain destinations anymore.

SAODV's proposal is that every node (generating or forwarding an RERR message) will use digital signatures to sign the whole message and that any neighbor that receives it will verify the signature. In this way it can verify that the sender of the RERR message is really the one that it claims to be. Because destination sequence numbers are not signed by the corresponding node, a node should never update any destination sequence number of its routing table based on an RERR message (this prevents a malicious node from performing attack type 6). Implementing a mechanism that will allow the destination sequence numbers of an RERR message to be signed by their corresponding nodes would add too much overhead compared with the advantage of the use of that information.

Although nodes will not trust destination sequence numbers in an RERR message, they will use them to decide whether or not they should invalidate a route. This does not give any extra advantage to a malicious node.

5.9.4 *Persistence of Sequence Numbers*

The attack type 7 was based on the fact that the originator of the RREQ can set the sequence number of the destination. This should have not been specified in AODV because it is not needed. In the case where everybody behaves according to the protocol, the situation in which the originator of a RREQ will put a destination sequence number bigger than the real one will never happen, not even in the case that the destination of the RREQ has rebooted. After rebooting, the node does not remember its sequence number anymore, but it waits long enough before being active, so that when it wakes up, nobody has stored its old sequence number anymore.

To avoid this attack, in the case that the destination sequence number in the RREQ is bigger than the destination sequence number of the destination node, the destination node will not take into account the value in the RREQ. Instead, it will realize that the originator of the RREQ is misbehaving and will send the RREP with the right sequence number.

In addition, if one of the nodes has a way to store its sequence number every time it modifies it, it might do so. Therefore, when it reboots it, will not need to wait long enough so that everybody deletes routes toward it.

5.10 Open Issues

The digital signature $Digital\_signature_X$ $(routing\_message)$ can be created only by X. Thus, it serves as proof of validity of the information contained in the routing message. This prevents attack scenarios 1, 3, 4, and 6.

The hop authenticator reduces the ability of a malicious intermediate hop to mount the attack type 2 by arbitrarily modifying the hop count without detection. A node that is n hops away from T will know the nth element in the hash chain $(h^n(x))$, but it will not know any element that comes before this because of the one-way property of $h(\,)$. However, the malicious node could still pass on the received authenticator and hop count without modifying it. Thus, the effectiveness of this approach is limited.

In addition, there is another type of attack that cannot be detected by SAODV: tunneling attacks. In that type of attack, two malicious nodes simulate that they have a link between them (that is, they can send and receive messages directly to each other). They achieve this by tunneling AODV messages between them (probably in an encrypted way). In this way they could achieve having certain traffic through them.

No security scheme has been able, so far, to detect this attack. Misbehaving detection schemes could, in principle, detect the so-called tunnel attacks. If the monitor sees a routing message with $Hop\_Count = X + 1$ being sent by a node, but does not see a routing message with $Hop\_Count = X$ being sent to the same node, then the node is either fabricating the routing message or there is a tunnel. In either case, it is cause for raising the alarm. Nevertheless, this kind of scheme has as main problems that there is no way for any node to validate the authenticity of the misbehavior reports and there is the possibility of falsely detecting misbehavior nodes. Therefore, it is not a feasible solution so far.

The way the hop count is authenticated could be changed to a more secure one. For instance, intermediate nodes forwarding the routing messages could include the address of the next hop to which the message is forwarded and sign it [32]. Another possibility would be to use forward-secure signature schemes [20]. A forward-secure signature scheme is like a hash chain, except that to prove that you are *n* hops away from the target, you should sign the routing message with the key corresponding to the *n*th link. Unlike in the hash chain case, the same signing key is not given to the next hop. Only the next signing key is given. This prevents the attack based on the possibility that a malicious node does not increase the hop count when it forwards a routing message. With this scheme, at any time the routing message has only one signature. The problem is, of course, efficiency. There are schemes where the message sizes are reasonably small, but signing and verification are quite expensive. Then there are other schemes where RSA signing could be used, but the public key needed to verify the signatures is size $O(m)$, where m is the diameter of the network. All those approaches would be very expensive (probably not even feasible), and still, they would not prevent tunneling attacks at all. Therefore, the use of hash chains might be, so far, the option that deals best with the trade-off between security and performance.

The use of sequence numbers should prevent most of the possible reply attacks. A node will discard a replied message if it has received an original message because the replied message will not be "fresh enough." To make the prevention of reply attacks stronger, a node could increase its sequence number in more situations than what AODV mandates (or even periodically).

Papadimitratos and Haas suggest in [27] that it is possible to mount an attack by maliciously modifying the IP header of the SAODV messages. This is not true because SAODV does not trust the contents of the IP header, and all the information that needs to operate is inside the AODV message and the SAODV extension.

5.11   AODV Message Formats

Figures 5.5 through 5.8 show the structure of the AODV messages and indicate what the mutable fields of the messages are.

```
 0                   1                   2                   3
 0 1 2 3 4 5 6 7 8 9 0 1 2 3 4 5 6 7 8 9 0 1 2 3 4 5 6 7 8 9 0 1
+-+-+-+-+-+-+-+-+-+-+-+-+-+-+-+-+-+-+-+-+-+-+-+-+-+-+-+-+-+-+-+-+
|     Type          |J|R|G|       Reserved              |  Hop count   |
+-+-+-+-+-+-+-+-+-+-+-+-+-+-+-+-+-+-+-+-+-+-+-+-+-+-+-+-+-+-+-+-+
|                            PREQ ID                              |
+-+-+-+-+-+-+-+-+-+-+-+-+-+-+-+-+-+-+-+-+-+-+-+-+-+-+-+-+-+-+-+-+
|                      Destination IP address                    |
+-+-+-+-+-+-+-+-+-+-+-+-+-+-+-+-+-+-+-+-+-+-+-+-+-+-+-+-+-+-+-+-+
|                   Destination sequence number                  |
+-+-+-+-+-+-+-+-+-+-+-+-+-+-+-+-+-+-+-+-+-+-+-+-+-+-+-+-+-+-+-+-+
|                      Originator IP address                     |
+-+-+-+-+-+-+-+-+-+-+-+-+-+-+-+-+-+-+-+-+-+-+-+-+-+-+-+-+-+-+-+-+
|                   Originator sequence number                   |
+-+-+-+-+-+-+-+-+-+-+-+-+-+-+-+-+-+-+-+-+-+-+-+-+-+-+-+-+-+-+-+-+
```

Figure 5.5   Route request (RREQ) message format. Mutable fields: Hop count.

```
 0                   1                   2                   3
 0 1 2 3 4 5 6 7 8 9 0 1 2 3 4 5 6 7 8 9 0 1 2 3 4 5 6 7 8 9 0 1
+-+-+-+-+-+-+-+-+-+-+-+-+-+-+-+-+-+-+-+-+-+-+-+-+-+-+-+-+-+-+-+-+
|     Type          |R|A|    Reserved     | Prefix Sz |  Hop count   |
+-+-+-+-+-+-+-+-+-+-+-+-+-+-+-+-+-+-+-+-+-+-+-+-+-+-+-+-+-+-+-+-+
|                      Destination IP address                    |
+-+-+-+-+-+-+-+-+-+-+-+-+-+-+-+-+-+-+-+-+-+-+-+-+-+-+-+-+-+-+-+-+
|                   Destination sequence number                  |
+-+-+-+-+-+-+-+-+-+-+-+-+-+-+-+-+-+-+-+-+-+-+-+-+-+-+-+-+-+-+-+-+
|                      Originator IP address                     |
+-+-+-+-+-+-+-+-+-+-+-+-+-+-+-+-+-+-+-+-+-+-+-+-+-+-+-+-+-+-+-+-+
|                            Lifetime                             |
+-+-+-+-+-+-+-+-+-+-+-+-+-+-+-+-+-+-+-+-+-+-+-+-+-+-+-+-+-+-+-+-+
```

Figure 5.6   Route reply (RREP) message format. Mutable fields: Hop count.

```
 0                   1                   2                   3
 0 1 2 3 4 5 6 7 8 9 0 1 2 3 4 5 6 7 8 9 0 1 2 3 4 5 6 7 8 9 0 1
+-+-+-+-+-+-+-+-+-+-+-+-+-+-+-+-+-+-+-+-+-+-+-+-+-+-+-+-+-+-+-+-+
|     Type          |N|       Reserved              |  Dest count  |
+-+-+-+-+-+-+-+-+-+-+-+-+-+-+-+-+-+-+-+-+-+-+-+-+-+-+-+-+-+-+-+-+
|              Unreachable destination IP address (1)            |
+-+-+-+-+-+-+-+-+-+-+-+-+-+-+-+-+-+-+-+-+-+-+-+-+-+-+-+-+-+-+-+-+
|           Unreachable destination sequence number (1)          |
+-+-+-+-+-+-+-+-+-+-+-+-+-+-+-+-+-+-+-+-+-+-+-+-+-+-+-+-+-+-+-+-+
|       Additional unreachable destination IP address (if needed) |
+-+-+-+-+-+-+-+-+-+-+-+-+-+-+-+-+-+-+-+-+-+-+-+-+-+-+-+-+-+-+-+-+
|    Additional unreachable destination sequence numbers (if needed) |
+-+-+-+-+-+-+-+-+-+-+-+-+-+-+-+-+-+-+-+-+-+-+-+-+-+-+-+-+-+-+-+-+
```

Figure 5.7   Route error (RERR) message format. Mutable fields: None.

```
 0                   1
 0 1 2 3 4 5 6 7 8 9 0 1 2 3 4 5
+-+-+-+-+-+-+-+-+-+-+-+-+-+-+-+-+
|     Type          |    Reserved     |
+-+-+-+-+-+-+-+-+-+-+-+-+-+-+-+-+
```

Figure 5.8   Route reply acknowledgment (RREP-ACK) message format. Mutable fields: None.

5.12   Secure AODV Extensions

Figure 5.9 and Figure 5.10 and Table 5.2 show the format of the SAODV signature extensions.

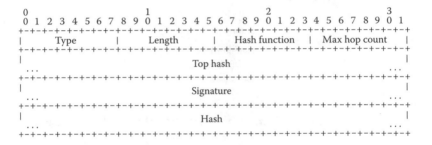

Figure 5.9   RREQ (single) signature extension.

```
 0                   1                   2                   3
 0 1 2 3 4 5 6 7 8 9 0 1 2 3 4 5 6 7 8 9 0 1 2 3 4 5 6 7 8 9 0 1
+-+-+-+-+-+-+-+-+-+-+-+-+-+-+-+-+-+-+-+-+-+-+-+-+-+-+-+-+-+-+-+-+
|     Type      |    Length     | Hash function | Max hop count |
+-+-+-+-+-+-+-+-+-+-+-+-+-+-+-+-+-+-+-+-+-+-+-+-+-+-+-+-+-+-+-+-+
|                                                               |
...                         Top hash                          ...
+-+-+-+-+-+-+-+-+-+-+-+-+-+-+-+-+-+-+-+-+-+-+-+-+-+-+-+-+-+-+-+-+
|                                                               |
...                         Signature                         ...
+-+-+-+-+-+-+-+-+-+-+-+-+-+-+-+-+-+-+-+-+-+-+-+-+-+-+-+-+-+-+-+-+
|                                                               |
...                           Hash                            ...
+-+-+-+-+-+-+-+-+-+-+-+-+-+-+-+-+-+-+-+-+-+-+-+-+-+-+-+-+-+-+-+-+
```

Figure 5.10   RREP (single) signature extension.

Table 5.2   RREQ and RREP Signature Extension Fields

Field	Value
Type	64 in RREQ-SSE and 65 in RREP-SSE
Length	The length of the type-specific data, not including the Type and Length fields of the extension.
Hash Function	The hash function used to compute the Hash and Top Hash fields.
Max Hop Count	The Maximum Hop Count supported by the hop count authentication.
Top Hash	The top hash for the hop count authentication. This field has variable length, but it must be 32-bits aligned.
Signature	The signature of all the fields in the AODV packet that are before this field but the Hop Count field. This field has variable length, but it must be 32-bits aligned.
Hash	The hash corresponding to the actual hop count. This field has variable length, but it must be 32-bits aligned.

Figure 5.11 and Table 5.3 show the format of the RREQ double signature extension.

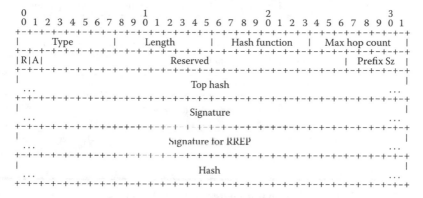

Figure 5.11   RREQ double signature extension.

Table 5.3   RREQ Double Signature Extension Fields

Field	Value
Type	66
Length	The length of the type-specific data, not including the Type and Length fields of the extension.
Hash Function	The hash function used to compute the Hash and Top Hash fields.
Max Hop Count	The Maximum Hop Count supported by the hop count authentication.
R	Repair flag for the RREP.
A	Acknowledgment required flag for the RREP.
Reserved	Sent as 0; ignored on reception.
Prefix Size	The prefix size field for the RREP.
Top Hash	The top hash for the hop count authentication. This field has variable length, but it must be 32-bits aligned.
Signature	The signature of all the fields in the AODV packet that are before this field but the Hop Count field. This field has variable length, but it must be 32-bits aligned.
Signature for the RREP	The signature that should be put into the Signature field of the RREP Double Signature Extension when an intermediate node (that has previously received this RREQ and created a reverse route) wants to generate an RREP for a route to the source of this RREQ. This field has variable length, but it must be 32-bits aligned. Both signatures are generated by the requesting node.
Hash	The hash corresponding to the actual hop count. This field has variable length, but it must be 32-bits aligned.

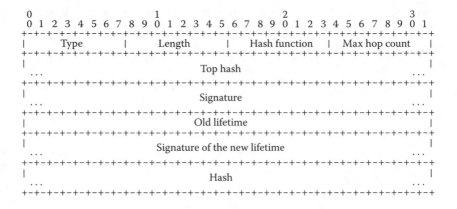

Figure 5.12   RREP double signature extension.

Table 5.4   RREP Double Signature Extension Fields

Field	Value
Type	67
Length	The length of the type-specific data, not including the Type and Length fields of the extension.
Hash Function	The hash function used to compute the Hash and Top Hash fields.
Max Hop Count	The Maximum Hop Count supported by the hop count authentication.
Top Hash	The top hash for the hop count authentication. This field has variable length, but it must be 32-bits aligned.
Signature	The signature of all the fields of the AODV packet that are before this field but the Hop Count field, and with the Old Lifetime value instead of the Lifetime. This signature is the one that was generated by the final destination. This field has variable length, but it must be 32-bits aligned.
Old Lifetime	The lifetime that was in the RREP generated by the final destination.
Signature of the New Lifetime	The signature of the RREP with the actual lifetime (the lifetime of the route in the intermediate node). This signature is generated by the intermediate node. This field has variable length, but it must be 32-bits aligned.
Hash	The hash corresponding to the actual hop count. This field has variable length, but it must be 32-bits aligned.

Figure 5.12 and Table 5.4 show the format of the RREP double signature extension.

Finally, Figure 5.13 and Figure 5.14 and Table 5.5 show the format of the RERR and RREP-ACK signature extensions.

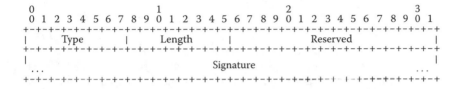

Figure 5.13 RERR signature extension.

```
 0                   1                   2                   3
 0 1 2 3 4 5 6 7 8 9 0 1 2 3 4 5 6 7 8 9 0 1 2 3 4 5 6 7 8 9 0 1
+-+-+-+-+-+-+-+-+-+-+-+-+-+-+-+-+-+-+-+-+-+-+-+-+-+-+-+-+-+-+-+-+
|     Type      |     Length    |            Reserved           |
+-+-+-+-+-+-+-+-+-+-+-+-+-+-+-+-+-+-+-+-+-+-+-+-+-+-+-+-+-+-+-+-+
|                                                               |
| ...                        Signature                      ... |
+-+-+-+-+-+-+-+-+-+-+-+-+-+-+-+-+-+-+-+-+-+-+-+-+-+  +-+-+-+-+-+-+-+
```

```
 0                   1
 0 1 2 3 4 5 6 7 8 9 0 1 2 3 4 5
+-+-+-+-+-+-+-+-+-+-+-+-+-+-+-+-+
|     Type      |    Reserved   |
+-+-+-+-+-+-+-+-+-+-+-+-+-+-+-+-+
```

Figure 5.14 RREP-ACK signature extension.

Table 5.5 RERR and RREP-ACK Signature Extension Fields

Field	Value
Type	68 in RERR-SE and 69 in RREP-ACK-SE
Length	The length of the type-specific data, not including the Type and Length fields of the extension.
Reserved	(Only in RERR-SE). Sent as 0; ignored on reception.
Signature	The signature of all the fields in the AODV packet that are before this field. This field has variable length, but it must be 32-bits aligned.

References

[1] S. Asherson and A. Hutchison. Secure Routing for Wireless Mesh Networks. In *Proceedings of the Southern African Telecommunication Networks and Applications Conference (SATNAC) 2006*, September 2006.

[2] N. Asokan and P. Ginzboorg. Key agreement in ad-hoc networks. *Computer Communication Review*, 23(17): 1627–1637, November 2000.

[3] J. Broch, D. A. Maltz, D. B. Johnson, Y. C. Hu, and J. Jetcheva. A Performance Comparison of Multi-hop Wireless Ad hoc Network Routing Protocols. In *Proceedings of the 4th Annual International Conference on Mobile Computing and Networking*, pp. 85–97, 1998.

[4] J. Chen, Y.-Z. Lee, D. Maniezzo, and M. Gerla. Performance Comparison of AODV and OFLSR in Wireless Mesh Networks. In *Proceedings of the The Fifth Annual Mediterranean Ad hoc Networking Workshop (Med-Hoc-Net 2006)*, pp. 271–278, June 2006.

[5] S. Cheung. An Efficient Message Authentication Scheme for Link State Routing. In *13th Annual Computer Security Applications Conference*, pp. 90–98, 1997.

[6] T. Clausen, P. J. (Editors), C. Adjih, A. Laouiti, P. Minet, P. Muhlethaler, A. Qayyum, and L. Viennot. Optimized link state routing protocol (olsr). RFC 3626, October 2003. Network Working Group.

[7] B. Dahill, B. N. Levine, E. Royer, and C. Shields. A Secure Routing Protocol for Ad hoc Networks. Technical report UM-CS-2001-037, University of Massachusetts, Departament of Computer Science, August 2001.

[8] N. Ferguson and B. Schneier. A Cryptographic Evaluation of Ipsec. Technical report, Counterpane Internet Security, February 1999.

[9] M. Guerrero Zapata. Secure ad hoc on-demand distance vector routing. *ACM Mobile Computing and Communications Review (MC2R)*, 6(3): 106–107, July 2002.

[10] M. Guerrero Zapata. Secure ad hoc on-demand distance vector (SAODV) routing. First published in the IETF MANET Mailing List (October 8, 2001), August 2002. INTERNET-DRAFT — work in progress. draft-guerrero-manet-saodv-00.txt.

[11] M. Guerrero Zapata. Secure ad hoc on-demand distance vector (SAODV) routing, Sept. 2006. INTERNET-DRAFT—work in progress. draft-guerrero-manet-saodv-06.txt.

[12] M. Guerrero Zapata and N. Asokan. Securing Ad hoc Routing Protocols. In *Proceedings of the 2002 ACM Workshop on Wireless Security (WiSe 2002)*, pp. 1–10, September 2002.

[13] R. Hauser, A. Przygienda, and G. Tsudik. Reducing the Cost of Security in Link State Routing. In *Symposium on Network and Distributed Systems Security (NDSS '97)*, pp. 93–99, San Diego, February 1997. Internet Society.

[14] Y. C. Hu, D. Johnson, and A. Perrig. SEAD: Secure Efficient Distance Vector Routing for Mobile Wireless Ad hoc Networks. In *4th IEEE Workshop on Mobile Computing Systems and Applications (WMCSA '02)*, pp. 3–13, June 2002.

[15] Y.-C. Hu and A. Perrig. A survey of secure wireless ad hoc routing. *IEEE Security and Privacy*, 2(3): 28–39, 2004.

[16] Y. C. Hu, A. Perrig, and D. Johnson. Ariadne: A Secure On-demand Routing Protocol for Ad hoc Networks. Technical report TR01-383, Rice University, December 2001.

[17] D. B. Johnson et al. The dynamic source routing protocol (DSR) for mobile ad hoc networks for IPv4. IETF Request for Comments, RFC4728, February 2007.

[18] C. Karlof and D. Wagner. Secure routing in wireless sensor networks: Attacks and countermeasures. *Elsevier's Ad Hoc Networks Journal, Special Issue on Sensor Network Applications and Protocols*, 1(2–3): 293–315, September 2003.

[19] S. Kent and R. Atkinson. Security architecture for the internet protocol. IETF Request for Comments, RFC 2401, November 1998.

[20] H. Krawczyk. Simple Forward-Secure Signatures from Any Signature Scheme. In *ACM Conference on Computer and Communications Security*, pp. 108–115, 2000.

[21] C. Madson and R. Glenn. The use of HMAC-MD5-96 within ESP and AH. Internet Request for Comments RFC 2403, November 1998.

[22] C. Madson and R. Glenn. The use of HMAC-SHA-1-96 within ESP and AH. Internet Request for Comments RFC 2404, November 1998.

[23] S. Marti, T. J. Giuli, K. Lai, and M. Baker. Mitigating Routing Misbehavior in Mobile Ad hoc Networks. In *Proceedings of the 6th Annual International Conference on Mobile Computing and Networking*, pp. 255–265, 2000.

[24] P. Papadimitratos and Z. J. Haas. Secure Routing for Mobile Ad hoc Networks. SCS Communication Networks and Distributed Systems Modeling and Simulation Conference (CNDS 2002), January 2002.

[25] C. E. Perkins, E. M. Belding-Royer, and S. R. Das. Ad hoc on-demand distance vector (AODV) routing. Internet Request for Comments RFC 3561, November 2003.

[26] C. E. Perkins and E. M. Royer. Ad hoc On-Demand Distance Vector Routing. In *Proceedings of the 2nd IEEE Workshop on Mobile Computing Systems and Applications*, New Orleans, pp. 90–100, February 1999.

[27] A. Perrig, R. Canetti, D. Song, and D. Tygar. Efficient and Secure Source Authentication for Multicast. In *Network and Distributed System Security Symposium (NDSS'01)*, February 2001.

[28] A. Perrig, R. Szewczyk, V. Wen, D. E. Culler, and J. D. Tygar. SPINS: Security Protocols for Sensor Networks. In *Proceedings of the 7th Annual International Conference on Mobile Computing and Networking*, pp. 189–199, 2001.

[29] S. Ramanathan and M. Steenstrup. A survey of routing techniques for mobile communications networks. *Mobile Networks and Applications*, 1(2): 89–104, 1996.

[30] E. M. Royer and C.-K. Toh. A review of current routing protocols for ad hoc mobile wireless networks. *IEEE Personal Communications*, pp. 46–55, April 1999.

[31] K. Sanzgiri, B. Dahill, B. Levine, and E. Belding-Royer. A secure routing protocol for ad hoc networks. In *International Conference on Network Protocols (ICNP)*, Paris, November 2002.

[32] B. R. Smith, S. Murthy, and J. J. Garcia-Luna-Aceves. Securing distance-vector routing protocols. In *Symposium on Network and Distributed Systems Security (NDSS '97)*, pp. 85–92, San Diego, February 1997. Internet Society.

[33] F. Stajano and R. Anderson. The Resurrecting Duckling: Security Issues for Ad-hoc Wireless Networks. In *Proceedings of the 7th International Workshop on Security Protocols*, number 1796 in Lecture Notes in Computer Science, pp. 172–194. Springer-Verlag, Berlin Germany, April 1999.

[34] H. Yang, H. Luo, F. Ye, S. Lu, and L. Zhang. Security in mobile ad hoc networks: challenges and solutions. *Wireless Communications, IEEE* [see also *IEEE Personal Communications*], 11(1): 38–47, 2004.

[35] Y. Yang, J. Wang, and R. Kravets. Designing Routing Metrics for Mesh Networks. In *Proceedings of the First IEEE Workshop on Wireless Mesh Networks* (WiMesh-2005), September 2005.

[36] K. Zhang. Efficient Protocols for Signing Routing Messages. In *Proceedings of the Symposium on Network and Distributed Systems Security* (NDSS'98), July 2001.

[37] L. Zhou and Z. J. Haas. Securing ad hoc networks. *IEEE Network Magazine*, 13(6): 24–30, November/December 1999.

Chapter 6

Hop Integrity in Wireless Mesh Networks

Chin-Tser Huang

Contents

Message manipulation has become one of the major threats to the security of wireless mesh networks because of the open medium in such networks. An adversary can launch a message insertion attack or a message replay attack such that the next mesh router that receives an inserted or replayed message will unwittingly forward it toward the destination. Even if a message from these attacks fails the authentication at the destination and gets discarded, it has already consumed the communication resources along the forwarding path in the wireless mesh network. Repeated attempts of these types of attack may result in a denial-of-service attack that may paralyze the network. To counter these attacks, it is necessary to provide message authentication and message integrity at every hop. In this chapter,

197

we first address the need of sufficient and efficient authentication and integrity checks at every hop by presenting several attack scenarios and explaining possible constraints on wireless mesh routers. Then, we present a novel protocol suite aimed to provide hop integrity for multi-hop wireless mesh networks. This protocol suite consists of three protocols: (1) an initial authentication protocol for a joining mesh router to use a certificate to achieve mutual authentication and set up an initial shared secret with each of its adjacent mesh routers; (2) a secret exchange protocol used by two adjacent mesh routers to periodically update the secret they share for the purpose of computing message digests; and (3) an integrity check protocol used for computing and verifying message digests and sequence numbers. Together, these three protocols can provide hop integrity for wireless mesh networks to counter message insertion attacks and message replay attacks. Furthermore, these three protocols are specified using a formal notation called Abstract Protocol Notation, and the correctness of these protocols is verified with state transition diagrams.

6.1 Introduction

Wireless mesh networks [1–3] are networks consisting of mesh routers. Some of the mesh routers may be connected to the wired infrastructure of the Internet, but most of them are not. These ad hoc mesh routers are able to dynamically self-organize and self-configure, which is one of the major advantages of wireless mesh networks. By forwarding packets via mesh routers, wireless mesh networks provide communication paths to client nodes that are not within direct radio transmission range with another client node or an Internet attachment point. As the popularity of wireless mesh networks grows, there are more and more attacks directed at wireless mesh networks and the security of them draws increased concern. In particular, message manipulation has become one of the major threats to the security of wireless mesh networks because of the open medium in such networks. The threat of message manipulation can be realized by the following two attacks:

1. Message insertion attack: An adversary impersonates a legitimate mesh router and inserts messages fabricated by itself. Alternatively, the adversary can intercept a message in transit, arbitrarily modify the content of the message, and insert the modified message into the network.
2. Message replay attack: An adversary makes copies of legitimate messages intercepted between one pair of adjacent mesh routers and replays them between the same pair or another pair of adjacent mesh routers in the same wireless mesh network, thanks to the multi-hop nature of such network.

The next mesh router that receives an inserted or replayed message will unwittingly forward it toward its ultimate destination if no appropriate protection is provided. Even if a message originated from one of the above two attacks fails the authentication and integrity check mechanism (such as IPsec [4–6]) at the destination and gets discarded, it has already consumed the communication resources along the forwarding path in the wireless mesh network. If no appropriate protection is provided, repeated attempts of these types of attack may result in a denial-of-service attack that may paralyze the wireless mesh network. To counter these attacks, it is necessary to provide message authentication and integrity check at every hop of the network.

In this chapter, we apply the concept of hop integrity to address the above problems. This chapter consists of two major components. First, we address the need for sufficient and efficient authentication and integrity check at every hop by presenting several attack scenarios and introducing the concept of hop integrity. Second, we present a novel protocol suite aimed to provide hop integrity for multi-hop wireless mesh networks. This protocol suite consists of three protocols. The first protocol is an initial authentication protocol used for a joining mesh router to use a certificate issued by the certificate authority to achieve mutual authentication and set up an initial shared secret with each of its adjacent mesh routers. The second protocol is a secret exchange protocol used by two adjacent mesh routers to periodically update the secret they share for the purpose of computing message digests. The third protocol is an integrity check protocol used for computing and verifying message digests. In the integrity check protocol, a soft sequence number is attached to each message as a freshness identifier. Together, these three protocols can provide hop integrity for wireless mesh networks to counter message insertion attack and message replay attack. Furthermore, these three protocols are specified using a formal notation and the correctness of these protocols is verified with state transition diagrams.

The protocols in this chapter are specified using a version of the Abstract Protocol Notation presented in [7]. We use this notation because it provides a well-defined set of semantics that is suitable for a distributed environment and is not provided by programming languages like C/C++. In this notation, each process in a protocol is defined by a set of inputs, a set of variables, a set of parameters, and a set of actions. For example, in a protocol consisting of two processes x and y, process x can be defined as follows.

process x

inp	⟨name of input⟩	:	⟨type of input⟩
	. . .		
	⟨name of input⟩	:	⟨type of input⟩

var	⟨name of variable⟩	:	⟨type of variable⟩
	. . .		
	⟨name of variable⟩	:	⟨type of variable⟩
par	⟨name of parameter⟩	:	⟨type of parameter⟩
	. . .		
	⟨name of parameter⟩	:	⟨type of parameter⟩

begin

       ⟨action⟩

☐   ⟨action⟩

. . .

☐   ⟨action⟩

end

The inputs of process x have constant values that are assigned by an upper layer process and can be changed, if necessary, only by the assigning process. An input can be read, but not written, by the actions of process x. The variables of process x can be read and updated by the actions of process x. A parameter has a finite number of values and its use will be described next. Comments can be added anywhere in a process definition; each comment is placed between the two brackets { and }.

Each ⟨**action**⟩ of process x is of the form:

$$⟨guard⟩ \rightarrow ⟨statement⟩$$

The guard of an action of x is either a Boolean expression over the constants and variables of x, a receive guard of the form rcv ⟨**message**⟩ from y, or a time-out guard of the form time-out ⟨**time expression**⟩. The ⟨**time expression**⟩ refers to a time period because some action has executed last and a Boolean expression that involves the constants and variables of the process. A parameterized action that refers to one parameter is a shorthand notation for a finite set of actions: each of them refers to a different value in the domain of the parameter.

Executing an action consists of executing the statement of this action. Executing the actions (of different processes) in a protocol proceeds according to the following three rules. First, an action is executed only when its guard is true. Second, the actions in a protocol are executed one at a time. Third, an action whose guard is continuously true is eventually executed.

The ⟨**statement**⟩ of an action of x is a sequence of ⟨**skip**⟩, ⟨**assignment**⟩, ⟨**send**⟩, ⟨**selection**⟩, or ⟨**iteration**⟩ statements of the following forms:

⟨skip⟩   :   **skip**

⟨send⟩   :   **send** ⟨message⟩ **to** y

⟨assignment⟩ :   ⟨list of variables of x⟩ :=
                               ⟨list of expressions⟩

⟨selection⟩   :   **if** ⟨Boolean expression⟩ →
                               ⟨statement⟩

                                . . .

                          ☐ ⟨Boolean expression⟩ →
                                ⟨statement⟩
                          fi

⟨iteration⟩   :   **do** ⟨Boolean expression⟩ →
                                ⟨statement⟩
                          od

6.2   Hop Integrity

Before we present the protocols, we introduce the concept of hop integrity between adjacent wireless mesh routers as discussed in [8–10]. Hop integrity is fundamental to the three protocols in the hop integrity protocol suite that are aimed to counter the aforementioned attacks and strengthen the security of wireless mesh networks. The basic idea of hop integrity is straightforward: whenever a mesh router p receives a message m from an adjacent mesh router q, p should be able to determine whether m was indeed sent by q or it was modified or replayed by an adversary that operates between p and q.

Next, we discuss the requirements of hop integrity. A wireless mesh network is said to provide hop integrity if and only if the following two conditions hold for every pair of adjacent mesh routers p and q in the network:

1. Detection of message modification: Whenever mesh router q receives a message m claimed to be transmitted from mesh router p, q can determine correctly whether message m was modified by an adversary after it was sent by p and before it was received by q.
2. Detection of message replay: Whenever mesh router q receives a message m claimed to be transmitted from mesh router p, and determines that message m was not modified, then q can determine correctly whether message m is another copy of a message that is received earlier by q.

The above two conditions infer receiving integrity, in which whenever a receiver receives a message from a sender, the receiver can verify whether m was indeed sent by the sender or it was modified or replayed by an adversary that operates between the receiver and the sender. Note that the sender and the receiver referred to in our presentation of hop integrity are one hop away from each other, i.e., a message transmitted by the sender can be received directly by the receiver without the forwarding of other nodes.

Next, we present the three protocols that are used to provide hop integrity for wireless mesh networks. These protocols belong to two thin layers, namely, the secret exchange layer and the integrity check layer, that need to be added to the network layer of the protocol stack of each mesh router in a wireless mesh network. The function of the secret exchange layer is to allow adjacent mesh routers to periodically generate and exchange (and so share) new secrets. The exchanged secrets are made available to the integrity check layer, which uses them to compute and verify the integrity check for every data message transmitted between the adjacent mesh routers.

Figure 6.1 shows the protocol stacks in two adjacent mesh routers p and q. The secret exchange layer has two protocols: the initial authentication protocol and the secret exchange protocol. The initial authentication protocol consists of the two processes pa and qa, and the secret exchange protocol consists of the two processes pe and qe in mesh routers p and q, respectively. The integrity check layer has two protocols: the weak integrity check protocol and the strong integrity check protocol. The weak version consists of the two processes pw and qw in mesh routers p and q, respectively. This version can detect message modification, but not message

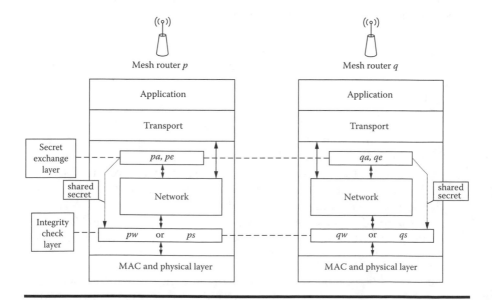

Figure 6.1 Protocol stack for hop integrity protocols.

replay. The strong version of the integrity check layer consists of the two processes ps and qs in mesh routers p and q, respectively. This version can detect both message modification and message replay.

In Section 6.3, we present the initial authentication protocol. In Section 6.4, we present the light-weight secret exchange protocol. In Section 6.5, we present the two versions of the integrity check protocol: weak version and strong version. The combination of these three protocols constitutes a protocol suite that provides hop integrity to wireless mesh networks.

6.3 Initial Authentication Protocol

Before two adjacent mesh routers can forward messages to each other for the first time, they have to use the initial authentication protocol to authenticate each other. When a mesh router moves to a different location in the network or is replaced by another mesh router, the initial authentication protocol also needs to be executed. The initial authentication protocol is designed to achieve three things. First, it assures the two mesh routers that they are communicating with a legitimate mesh router. Second, it allows the two mesh routers to exchange their certified public key. Third, it sets up the initial shared secrets that will later be periodically updated by the secret exchange protocol. There are other upper layer protocols that provide authentication; for example, TLS [11] at the transport layer and Kerberos [12] at the application layer. However, those protocols provide end-to-end authentication and do not fit our needs well. In our case, we want to provide authentication at the network layer for each pair of adjacent mesh routers that are only one hop away.

In many authentication protocols, an online authentication server is commonly used to provide authentication service for clients or other servers. Examples of this design include Kerberos [12] and RADIUS [13]. However, in the context of wireless mesh networks, initial authentication does not occur frequently because most mesh routers are relatively static. Therefore, we choose to use certificates to achieve this purpose. A certificate is simply the binding of a host's identifier and a host's public key, with an expiration time specified, and is signed by a certificate authority using its private key. The most common type of certificate is called X.509, whose format and details can be found at [14,15]. If the recipient of a certificate belongs to the same domain as the sender (namely, the owner) of the certificate, it should know the public key of the certificate authority and can use the certificate authority's signature to verify whether it is a legitimate and valid certificate and whether to accept and use the public key contained in the certificate. In case a certificate is stolen and spoofed by an adversary, a challenge-and-response scheme, as is used in the initial authentication protocol, can be used to counter this attack. (Note that a mesh router can renew its expiring certificate with the certificate authority in an offline manner, but this is beyond the scope of our discussion.)

In the initial authentication protocol, each mesh router has a process responsible for executing the protocol. Before two adjacent mesh routers perform initial authentication, they undergo an association procedure to negotiate necessary parameters for MAC layer and PHY layer. During the association procedure they also exchange the router identifier. The mesh router with a larger identifier will perform active initial authentication; we call this mesh router p and its authentication process pa. The mesh router with a smaller identifier will perform passive initial authentication; we call this mesh router q and its authentication process qa. An authentication request message sent by the mesh router with a smaller identifier will simply be dropped to avoid conflict.

Because the communication between mesh router p and mesh router q is bidirectional, two shared secrets, one for each direction, need to be generated and maintained. (How the two shared secrets are used will be explained in the next section.) Processes pa and qa both have a public key and a private key that they use to encrypt and decrypt the messages that carry the new secrets between them. A public key has to be certified by the certificate authority in the form of a certificate, whereas a private key is known only to its owner process. The public and private keys of process pa are named B_p and R_p, respectively; similarly, the public and private keys of process qa are named B_q and R_q, respectively.

There are five steps in the initial authentication protocol. In the first step, process pa sends a request message rqst($CERT_p$, e) to process qa, where $CERT_p$ is the certificate of mesh router p and e is the encryption of the concatenation of p's identifier and a time stamp. The identifier is used to verify that p is indeed the owner of the certificate, and the time stamp is used both as a freshness identifier to protect against message replay attacks and as a challenge to protect against certificate spoofing attacks. Process pa encrypts the identifier and time stamp using its private key R_p to provide a signature that this message is generated by pa and protect it from arbitrary modification by an adversary.

In the second step, process qa receives the request message from pa, decrypts p's certificate to derive public key B_p, and uses B_p to decrypt the identifier and the time stamp. Process qa verifies that p is the owner of the certificate and that the certificate is still valid. If successful, qa will use a random function to generate a new shared secret sp, and qa sends a reply message rply($CERT_q$, d, e) to pa, where $CERT_q$ is the certificate of mesh router q, d is the encryption of the concatenation of q's identifier and the same time stamp which qa received from pa in the request message, and e is the shared secret sp encrypted using pa's public key B_p. The same time stamp is used here as a response to the challenge. Field d is encrypted using qa's private key R_q to provide a signature that this message is generated by qa and protect it from arbitrary modification by an adversary. Field e is encrypted using p's public key B_p to ensure that only pa can derive the shared secret sp.

In the third step, pa receives the reply message rply(c, d, e) from qa, decrypts q's certificate to derive public key B_q, and uses B_q to decrypt the identifier and the time stamp. Process pa verifies that q is the owner of the certificate and that the certificate is still valid. If successful, pa decrypts e using its private key R_p to derive the shared secret generated by qa, and uses a random function to generate a new shared secret sq. Then, pa sends a first acknowledgment message ack(e) to qa, where e is the encryption of the concatenation of the shared secret sp received from qa and the shared secret sq generated by pa.

In the fourth step, qa receives the first acknowledgment message ack(e) from pa, and uses its private key R_q to decrypt e and verify that the first half of the result is equal to the shared secret sp it generated earlier. This ensures qa that pa has successfully received and installed sp. Then, qa derives the shared secret generated by pa from the second half of the result, uses pa's public key B_p to encrypt this value, and sends the encrypted result in a second acknowledgment to pa.

In the fifth step, pa receives the second acknowledgment message sack(e) from qa, and uses its private key R_p to decrypt e and verify that the result is equal to the shared secret sp it generated earlier. The success of the fifth step ensures pa that qa has successfully received and installed the shared secret sq and concludes the initial authentication between pa and qa.

In addition, if the initial authentication between pa and qa has not completed for an extended period of time (for example four times of the round trip time between pa and qa), then it is an indication that one of the above five messages was lost, and pa times out to resend the rqst message to qa.

Process pa and process qa in the initial authentication protocol can be defined as follows:

process pa

inp	B_a	: **integer**	{public key of authentication authority}
	B_p, R_p	: **integer**	{public key and private key of p}
	$CERT_p$	: **integer**	{certificate's value = NCR(R_a, (B_p; ID_p; exp_p))}
	ID_p	: **integer**	{identifier of p}
	tr	: **integer**	{upper bound on round-trip time}
var	ts	: **integer**	{current value of p's system clock}
	exp	: **integer**	{expiration time of q's certificate}
	sp	: **integer**	
	sq	: **array** [0 .. 1] **of integer** {initially $sq[0] = sq[1] = 0$}
	c, d, e	: **integer**	

```
        t, id        : integer
        B_q          : integer {public key of q}
        ID_q         : integer {identifier of q}

begin
    (process pa and process qa have not performed initial authentication) →
                ts := TMSTP;
                e := NCR(R_p, (ts; ID_p));
                send rqst(CERT_p, e) to qa
    ▯       rcv rply(c, d, e) from qa →
                (B_q, ID_q, exp) := DCR(B_a, c);
                (t, id) := DCR(B_q, d);
                if t ≠ ts ∨ id ≠ ID_q ∨ (current time > exp) →
                    {authentication fails} skip
                ▯ t = ts ∧ id = ID_q ∧ (current time ≤ exp) →
                    {authentication succeeds}
                    sp := DCR(R_p, e);
                    sq[0] := any;
                    sq[1] := sq[0];
                    e := NCR(B_q, (sp; sq[0]));
                    send ack(e) to qa
                fi
    ▯       rcv sack(e) from qa →
                d := DCR(R_p, e);
                if d = sq[0] →
                    {secret exchange succeeds} skip
                ▯ d ≠ sq[0] →
                    {secret exchange fails} skip
                fi
    ▯       timeout ((4*tr seconds passed since rqst message sent last) ∧
                (pa and qa have not completed initial authentication)) →
                ts := TMSTP;
                e := NCR(R_p, (ts; ID_p));
                send rqst(CERT_p, e) to qa
end

process qa

inp  B_a          : integer {public key of authentication authority}
     B_q, R_q     : integer {public key and private key of q}
     CERT_q       : integer {certificate's value = NCR(R_a, (B_q; ID_q; exp_q))}
     ID_q         : integer {identifier of p}
     tr           : integer {upper bound on round-trip time}
```

var *ts*   : **integer** {time stamp received from *p*}
    exp   : **integer** {expiration time of *p*'s certificate}
    sq   : **integer**
    sp   : **array** [0 . . 1] **of integer** {initially $sp[0] = sp[1] = 0$}
    c, d, e   : **integer**
    id   : **integer**
    B_p : **integer** {public key of *p*}
    ID_p : **integer** {identifier of *p*}

begin

        rcv rqst(*d, e*) **from** *pa* →
            $(B_p, ID_p, exp) := \mathrm{DCR}(B_a, d)$;
            $(ts, id) := \mathrm{DCR}(B_p, e)$;
            if $id \neq ID_p$ ∨ (current time > *exp*) →
                {authentication fails} **skip**

            ▯ $id = ID_p$ ∧ (current time ≤ *exp*) →
                {authentication succeeds}
                $d := \mathrm{NCR}(R_q, (ts; ID_q))$;
                $sp[0] :=$ **any**;
                $sp[1] := sp[0]$;
                $e := \mathrm{NCR}(B_p, sp[0])$;
                send rply($Cert_q, d, e$) **to** *pa*
            fi

▯   **rcv** ack(*e*) **from** *pa* →
            $(c, d) := \mathrm{DCR}(R_q, e)$;
            if $c \neq sp[0]$ →
                {secret exchange fails} **skip**

            ▯ $c = sp[0]$ →
                {secret exchange succeeds}
                $sq := d$;
                $e := \mathrm{NCR}(B_p, sq)$;
                send sack(*e*) **to** *pa*
            fi
end

Processes *pa* and *qa* use three functions, namely, TMSTP, NCR, and DCR. Function TMSTP takes no arguments, and when invoked, it returns a time stamp that is according to the system clock and is larger than any time stamp generated by the same process in the past. In other words, the time stamps generated by the same process are monotonic. Function NCR is an encryption function that takes two arguments, a key and a data item, and returns the encryption of the data item using the key. For example, execution of the statement

$$e := \mathrm{NCR}(R_p, (ts; ID_p))$$

causes the concatenation of ts and ID_p to be encrypted using the private key R_p, and the result to be stored in variable e. Function DCR is a decryption function that takes two arguments, a key and an encrypted data item, and returns the decryption of the data item using the key. For example, execution of the statement

$$d := \mathrm{DCR}(R_p, e)$$

causes the (encrypted) data item e to be decrypted using the private key R_p, and the result to be stored in variable d. As another example, consider the statement

$$(d, e) := \mathrm{DCR}(R_p, e)$$

This statement indicates that the value of e is the encryption of the concatenation of two values $(v_0; v_1)$ using key R_p. Thus, executing this statement causes e to be decrypted using key R_p, and the resulting first value v_0 to be stored in variable d, and the resulting second value v_1 to be stored in variable e.

Note in particular that in the specification of the initial authentication protocol, process pa has the following variable declaration:

var sp :   **integer**
  sq :   **array** $[0 .. 1]$ **of integer** {initially $sq[0] = sq[1] = 0$}

In process qa, the array sp is defined in a similar way. Array sq in process pa and array sp in process qa will be used in the secret exchange protocol and will be explained next.

6.4 Secret Exchange Protocol

In the secret exchange protocol, processes pe and qe maintain two shared secrets sp and sq. Secret sp is used by mesh router p to compute the integrity check for each data message sent by p to mesh router q, and it is also used by mesh router q to verify the integrity check for each data message received by q from mesh router p. Similarly, secret sq is used by q to compute the integrity checks for data messages sent to p, and it is used by p to verify the integrity checks for data messages received from q.

Recall that the two initial shared secrets sp and sq have been set up by the initial authentication protocol. However, any shared secret grows more vulnerable to statistical attacks as the usage increases. As part of maintaining the two secrets sp and sq, processes pe and qe need to change these secrets periodically, say every te hours, for some chosen value te. Process pe is to initiate the change of secret sq, and process qe is to initiate the change of secret sp. Processes pe and qe both have a public key and a private key that they use to encrypt and decrypt the messages that carry the new secrets between pe and qe. These keys assume the same names and values as defined in the initial authentication protocol.

For process pe to change secret sq, the following four steps need to be performed. First, pe generates a new sq, and encrypts the concatenation of the old sq and the new sq using qe's public key B_q, and sends the result in a rqst message to qe. Second, when qe receives the rqst message, it decrypts the message contents using its private key R_q and obtains the old sq and the new sq. Then, qe checks that its current sq equals the old sq from the rqst message, and installs the new sq as its current sq, and sends a rply message containing the encryption of the new sq using pe's public key B_p. Third, pe waits until it receives a rply message from qe containing the new sq encrypted using B_p. Receiving this rply message indicates that qe has received the rqst message and has accepted the new sq. Fourth, if pe sends the rqst message to qe, but does not receive the rply message from qe for some tr seconds, indicating that either the rqst message or the rply message was lost before it was received, then pe resends the rqst message to qe. Thus tr is an upper bound on the round-trip time between pe and qe.

Note that the old secret (along with the new secret) is included in each rqst message and the new secret is included in each rply message to ensure that if an adversary modifies or replays rqst or rply messages, then each of these messages is detected and discarded by its receiving process (whether pe or qe).

Process pe has two variables sp and sq declared as follows:

var sp :   **integer**

sq :   **array** $[0 .. 1]$ **of integer**

Similarly, process qe has an integer variable sq and an array variable sp.

In process pe, variable sp is used for storing the secret sp, variable $sq[0]$ is used for storing the old sq, and variable $sq[1]$ is used for storing the new sq. The assertion $sq[0] \neq sq[1]$ indicates that process pe has generated and sent the new secret sq, and that qe may not have received it yet. The assertion $sq[0] = sq[1]$ indicates that qe has already received and accepted the new secret sq. Initially,

$$sq[0] \text{ in } pe = sq[1] \text{ in } pe = sq \text{ in } qe,$$
and
$$sp[0] \text{ in } qe = sp[1] \text{ in } qe = sp \text{ in } pe.$$

Process pe can be defined as follows. (Process qe can be defined in the same way except that each occurrence of R_p in pe is replaced by an occurrence of R_q in qe, each occurrence of B_q in pe is replaced by an occurrence of B_p in qe, each occurrence of sp in pe is replaced by an occurrence of sq in qe, and each occurrence of $sq[0]$ or $sq[1]$ in pe is replaced by an occurrence of $sp[0]$ or $sp[1]$, respectively, in qe.)

process pe

inp R_p : **integer**   {private key of p}
        B_q : **integer**   {public key of q}
        te : **integer**   {time between secret exchanges}
        tr : **integer**   {upper bound on round-trip time}
var sp : **integer**
        sq : **array** $[0 .. 1]$ **of integer** {initially $sq[0] = sq[1] = sq$ in qe}
        d, e : **integer**

begin
timeout $(sq[0] = sq[1] \wedge$ (te hours passed since rqst message sent last))$\rightarrow$
         $sq[1] := \text{NEWSCR};$
         $e := \text{NCR}(B_q, (sq[0]; sq[1]));$
         send rqst(e) **to** qe
$\Box$ **rcv** rqst(e) **from** $qe \rightarrow$
         $(d, e) := \text{DCR}(R_p, e);$
         if $sp = d \vee sp = e \rightarrow$
            $sp := e;$
            $e := \text{NCR}(B_q, sp);$
            send rply(e) **to** qe
         $\Box \, sp \neq d \wedge sp \neq e \rightarrow$
            {detect adversary}
            skip
         fi
$\Box$ **rcv** rply(e) **from** $qe \rightarrow$
         $d := \text{DCR}(R_p, e);$
         if $sq[1] = d \rightarrow$
            $sq[0] := sq[1]$

$\square\ sq[1] \neq d \rightarrow$
    {detect adversary}
    skip
fi

$\square$ **timeout** $(sq[0] \neq sq[1] \wedge$ $(tr$ seconds passed since rqst
message sent last)) $\rightarrow e := \text{NCR}(B_q, (sq[0]; sq[1]))$;
    send rqst(e) **to** qe
end

The four actions of process pe use three functions, namely, NEWSCR, NCR, and DCR. Function NEWSCR takes no arguments, and when invoked, it returns a fresh secret that is different from any secret that was returned in the past. Functions NCR and DCR have been described in the last section.

To verify the correctness of the secret exchange protocol, we can use the state transition diagram of this protocol in Figure 6.2. This diagram has six nodes that represent all possible reachable states of the protocol. Every transition in the diagram stands for either a legitimate action (of process pe or process qe), or an illegitimate action of the adversary.

Initially, the protocol starts at a state S.0, where the two channels between processes pe and qe are empty and the values of variables $sq[0]$ and $sq[1]$ in pe and variable sq in qe are the same. This state can be defined by the following predicate:

S.0 : $ch.pe.qe =<> \wedge ch.qe.pe =<> \wedge$
    $sq[0]$ in $pe = sq[1]$ in $pe = sq$ in qe

At state S.0, exactly one action, namely, the first time-out action in process pe, is enabled for execution. Executing this action at state S.0 leads the protocol to state S.1 defined as follows:

S.1 : $ch.pe.qe =< rqst(e) > \wedge ch.qe.pe =<> \wedge$
    $e = \text{NCR}(B_q, (sq[0]; sq[1])) \wedge$
    $sq[0]$ in $pe \neq sq[1]$ in $pe \wedge sq[0]$ in $pe = sq$ in qe

At state S.1, exactly one legitimate action, namely, the receive action (that receives a rqst message) in process qe, is enabled for execution. Executing this action at state S.1 leads the protocol to state S.2 defined as follows:

S.2 : $ch.pe.qe =<> \wedge ch.qe.pe =< rply(e) > \wedge$
    $e = \text{NCR}(B_p, sq) \wedge$
    $sq[0]$ in $pe \neq sq[1]$ in $pe \wedge sq[1]$ in $pe = sq$ in qe

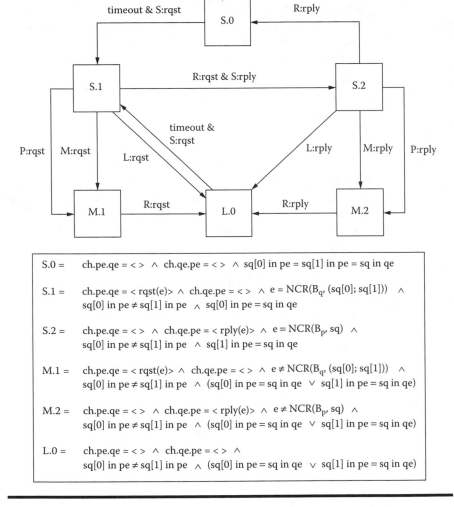

Figure 6.2 State transition diagram of the secret exchange protocol.

At state S.2, exactly one legitimate action, namely, the receive action (that receives a rply message) in process *pe*, is enabled for execution. Executing this action at state S.2 leads the protocol back to state S.0 defined above. States S.0, S.1, and S.2 are called good states because the transitions between these states consist of executing the legitimate actions of the two processes. The sequence of transitions from state S.0 to state S.1, to state S.2, and back to state S.0 constitutes the good cycle of the protocol. If only legitimate actions of processes *pe* and *qe* are executed, the protocol will stay in this good cycle indefinitely. Next, we discuss the bad effects caused by the actions of an adversary, and how the protocol can recover from these effects.

First, the adversary can execute a message loss action at state S.1 or S.2. If the adversary executes a message loss action at state S.1 or S.2, the network moves to a state L.0 defined as follows:

L.0 : $ch.pe.qe =<> \wedge ch.qe.pe =<> \wedge$
$sq[0]$ in $pe \neq sq[1]$ in $pe \wedge$
$(sq[0]$ in $pe = sq$ in $qe \vee sq[1]$ in $pe = sq$ in $qe)$

At state L.0, only the second time-out action in pe is enabled for execution, and executing this action leads the network back to state S.1.

Second, the adversary can execute a message modification action at state S.1 or S.2. If the adversary executes a message modification action at state S.1, the network moves to state M.1 defined as follows:

M.1 : $ch.pe.qe =< rqst(e) > \wedge ch.qe.pe =<> \wedge$
$e \neq$ NCR$(B_q, (sq[0]; sq[1])) \wedge$
$sq[0]$ in $pe \neq sq[1]$ in $pe \wedge$
$(sq[0]$ in $pe = sq$ in $qe \vee sq[1]$ in $pe = sq$ in $qe)$

If the adversary executes a message modification action at state S.2, the network moves to state M.2 defined as follows:

M.2 : $ch.pe.qe =<> \wedge ch.qe.pe =< rply(e) > \wedge$
$e \neq$ NCR$(B_p, sq) \wedge$
$sq[0]$ in $pe \neq sq[1]$ in $pe \wedge$
$(sq[0]$ in $pe = sq$ in $qe \vee sq[1]$ in $pe = sq$ in $qe)$

In either case, the protocol moves next to state L.0 and eventually returns to state S.1.

Third, the adversary can execute a message replay action at state S.1 or S.2. If the adversary executes a message replay action at state S.1, the network moves to state M.1. If the adversary executes a message replay action at state S.2, the network moves to state M.2. As shown above, the protocol eventually returns to state S.1.

From the state transition diagram in Figure 6.2, it is clear that each illegitimate action by the adversary will eventually lead the network back to state S.1, which is a good state. Once the network is in a good state, the network can progress in the good cycle. Hence the following two theorems about secret exchange protocol are proved:

Theorem 1
In the absence of an adversary, a network that executes the secret exchange protocol will follow the good cycle, consisting of the transitions from state S.0 to state S.1, from state S.1 to state S.2, and from state S.2 to state S.0, and will stay in this good cycle indefinitely.

Theorem 2
In the presence of an adversary, a network that executes the secret exchange protocol will converge to the good cycle in a finite number of steps after the adversary finishes executing the message loss, message modification, and message replay actions.

6.5 Integrity Check Protocol

This section introduces the integrity check protocol, starting with a weak version of the protocol, which detects message insertion only, and moving on to a strong version of the protocol, which detects both message insertion and message replay.

6.5.1 Weak Integrity Check Protocol

The main idea of the weak integrity check protocol is simple. Consider the case where a data(t) message, with t being the message text, is generated at a source src, then transmitted through a sequence of adjacent mesh routers $r.1, r.2, \ldots, r.n$ to a destination dst. When data(t) reaches the first mesh router $r.1$, $r.1$ computes a digest d for the message as follows:

$$d := \mathrm{MD}(t; scr)$$

where MD is the message digest function, $(t; scr)$ is the concatenation of the message text t and the shared secret scr between $r.1$ and $r.2$ (provided by the secret exchange protocol in $r.1$). Note that MD can be any common message digest function, such as MD5 [16], SHA [17], or HMAC [18]. Then, $r.1$ adds d to the message before transmitting the resulting data(t, d) message to mesh router $r.2$.

When $r.2$ receives the data(t, d) message, it computes the message digest using the secret shared between $r.1$ and $r.2$ (provided by the secret exchange process in $r.2$), and checks whether the result equals d. If they are unequal, then $r.2$ concludes that the received message has been modified, discards it, and reports an adversary. If they are equal, then $r.2$ concludes that the received message has not been modified and proceeds to prepare the message for transmission to the next mesh router $r.3$. Preparing the message for transmission to $r.3$ consists of computing d using the shared secret between $r.2$ and $r.3$ and storing the result in field d of the data(t, d) message. When the last mesh router $r.n$ receives the data(t, d) message, it computes the message digest using the shared secret between $r.(n-1)$ and $r.n$ and checks whether the result equals d. If they are unequal, $r.n$ discards the message and reports an adversary. Otherwise, $r.n$ sends the data(t) message to its destination dst.

Note that this protocol detects and discards every modified message. More importantly, it also determines the location where each message modification has occurred.

Process pw in the weak integrity protocol has two constants sp and sq that pw reads, but never updates. These two constants in process pw are also variables in process pe, and pe updates them periodically, as discussed in the previous section. Process pw can be defined as follows. (Process qw is defined in the same way except that each occurrence of p, q, pw, qw, sp, and sq is replaced by an occurrence of q, p, qw, pw, sq, and sp, respectively.)

process pw

inp	sp	:	**integer**
	sq	:	**array** $[0 .. 1]$ **of integer**
var	t, d	:	**integer**

begin

    rcv data(t, d) **from** qw $\rightarrow$

        if MD($t; sq[0]$) $= d \lor$ MD($t; sq[1]$) $= d \rightarrow$

            {accept message}

            RTMSG

        ▯ MD($t; sq[0]$) $\neq d \land$ MD($t; sq[1]$) $\neq d \rightarrow$

            {report an adversary}

            skip

        fi

▯   **true** $\rightarrow$

        {p receives data(t, d) from mesh router other than q}

        {and checks that its message digest is correct}

        RTMSG

▯   **true** $\rightarrow$

        {either p receives data(t) from an adjacent host or}

        {p generates the text t for the next data message}

        RTMSG

end

In the first action of process pw, if pw receives a data(t, d) message from qw while $sq[0] \neq sq[1]$, then pw cannot determine beforehand whether qw computed d using $sq[0]$ or using $sq[1]$. In this case, pw needs to compute two message digests using both $sq[0]$ and $sq[1]$, respectively, and compare the two digests with d. If either digest equals d, then pw accepts the message. Otherwise, pw discards the message and reports the detection of an adversary.

The three actions of process pw use two functions named MD and NXT and one statement named RTMSG. Function MD takes one argument, namely, the concatenation of the text of a message and the appropriate secret, and computes a digest for that argument. Function NXT takes one argument, namely, the text of a message (which we assume includes the message header), and determines the next mesh router to which the message should be forwarded. Statement RTMSG is defined as follows:

if NXT$(t) = p \rightarrow$
    {accept message}
    skip
⫿ NXT$(t) = q \rightarrow$
    $d := \text{MD}(t; sp)$;
    send data(t, d) **to** qw
⫿ NXT$(t) \neq p \wedge$ NXT$(t) \neq q \rightarrow$
    {compute d as the message digest of the concatenation of t and
    the secret}
    {for sending data to NXT(t); forward data(t, d) to mesh router NXT(t)}
    skip
fi

To verify the correctness of the weak integrity check protocol, we can use the state transition diagram of this protocol in Figure 6.3, which considers the channel from process qw to process pw. (The channel from pw to qw and the channels from pw to any other weak integrity process in an adjacent mesh router of p can be verified in the same way.) This diagram has two nodes that represent all possible reachable states of the protocol. Every transition in the diagram stands for either a legitimate action (of process pw or process qw), or an illegitimate action of the adversary.

Note that because the weak integrity check protocol operates below the secret exchange protocol in the protocol stack, we can assert that $(sq \text{ in } qw = sq[0] \text{ in } pw \vee sq \text{ in } qw = sq[1] \text{ in } pw)$ is an invariant in every state of the weak integrity protocol. We denote this invariant as I in the specification in Figure 6.3. Also note that the notation Head(data(t, d)) in

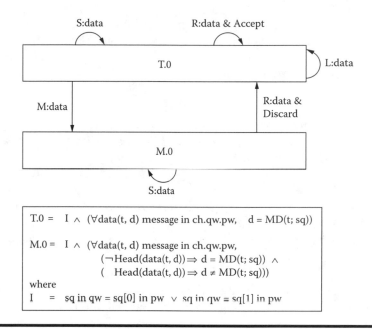

S:data   R:data & Accept

T.0   L:data

M:data   R:data & Discard

M.0

S:data

T.0 = I ∧ (∀data(t, d) message in ch.qw.pw,   d = MD(t; sq))

M.0 = I ∧ (∀data(t, d) message in ch.qw.pw,
           (¬Head(data(t, d)) ⟹ d = MD(t; sq)) ∧
           (Head(data(t, d)) ⟹ d ≠ MD(t; sq)))
where
I   =   sq in qw = sq[0] in pw ∨ sq in qw = sq[1] in pw

Figure 6.3   State transition diagram of the weak integrity check protocol.

the specification in Figure 6.3 is a predicate whose value is true if data(*t, d*) is the head message of the specified channel.

Initially, the protocol starts at state T.0. At state T.0, two legitimate actions, namely, the send action in *qw* that sends a data message and the receive action in *pw* that receives a data message, can be executed. Executing either one of the two actions at state T.0 keeps the protocol in state T.0.

State T.0 is the only good state in the weak integrity protocol. The sequence of the transitions from state T.0 to state T.0 constitutes the good cycle of the protocol. If only legitimate actions of processes *pw* and *qw* are executed, the protocol will stay in this good cycle indefinitely. Next, we discuss the bad effects caused by the actions of an adversary, and how the protocol can recover from these effects.

First, the adversary can execute a message loss action at state T.0. In this case, the predicate that for every data message data(*t, d*) in the channel from *qw* to *pw*, $d = MD(t; sq)$, still holds. Therefore, the protocol stays at state T.0.

Second, the adversary can execute a message modification action at state T.0. In this case, the protocol moves to state M.0. The receive and discard action executed by *pw* at state M.0 leads the protocol back to state T.0. From the state transition diagram, it is clear that each illegitimate action by the adversary will eventually lead the protocol back to T.0, which is a good state. Once the protocol is in a good state, the protocol can progress in the good cycle. Hence the following two theorems about the weak integrity check protocol are proved:

Theorem 3
In the absence of an adversary, a network that executes the weak integrity check protocol follows the good cycle, consisting of the single transition from state T.0 to state T.0, and will stay in this good cycle indefinitely.

Theorem 4
In the presence of an adversary, a network that executes the weak integrity check protocol will converge to the good cycle in a finite number of steps after the adversary finishes executing the message loss and message modification actions.

However, the weak integrity check protocol, while being able to detect and discard all modified messages, cannot detect some replayed messages. The next section introduces the strong integrity protocol that is capable of detecting and discarding all modified and replayed messages.

6.5.2 Strong Integrity Check Protocol

The weak hop integrity protocol can detect message modification, but not message replay. This section discusses how to strengthen this protocol to make it detect message replay as well. The strong hop integrity protocol is presented in two steps: (1) using "soft sequence numbers" to detect and discard replayed data messages, and (2) integrating this soft sequence number protocol into the weak integrity check protocol to construct the strong integrity check protocol.

Before introducing the soft sequence number protocol, a simple protocol is used to illustrate the need for sequence numbers in detecting message replay. Consider a protocol that consists of two processes u and v executing on two adjacent mesh routers. Process u continuously sends data messages to process v. Because process u and process v are only one hop away, the data messages sent by u will be received by v in the same order they were sent. Assume that there is an adversary that attempts to disrupt the communication between u and v by inserting (i.e., replaying) old messages in the message stream from u to v. To overcome this adversary, process u attaches an integer sequence number s to every data message sent to process v. To keep track of the sequence numbers, process u maintains a variable nxt that stores the sequence number of the next data message to be sent by u and process v maintains a variable exp that stores the sequence number of the expected data message to be received by v. (Note that a single variable *exp* at process v is sufficient because there is no reorder.) This is called a "hard sequence number protocol," because process u always remembers the next sequence number to be sent, and process v always remembers the next sequence number it expects to receive.

To send the next data(s) message, process u assigns s the current value of variable *nxt*, then increments *nxt* by one. When process v receives a data(s) message, v compares its variable *exp* with s. If $exp \leq s$, then v accepts the received data(s) message and assigns *exp* the value $s + 1$; otherwise, v discards the data(s) message. Processes u and v of this protocol can be specified as follows:

process u

var   *nxt* : **integer** {sequence number of next sent message}

begin
      true $\rightarrow$
         send data(*nxt*) **to** v;
         nxt := *nxt* + 1
end

process v

var s : **integer** {sequence number of received message}
           exp : **integer** {sequence number expected next}

begin
      rcv data(s) **from** u $\rightarrow$
        if $s < exp$ $\rightarrow$
            {reject message; report an adversary}
            skip
      ☐ $exp \leq s$ $\rightarrow$
            {accept message}
            exp := $s + 1$
        fi
end

Correctness of this protocol is based on the observation that the predicate $exp \leq nxt$ holds at each (reachable) state of the protocol. However, if due to some fault (for example, an accidental resetting of the values of variable *nxt*) the value of *exp* becomes larger than value of *nxt*, then all the data messages that u sends from this point and until the value of *nxt* becomes equal to the value of *exp* will be wrongly discarded by v. Next is a description of how to modify this protocol such that the number of messages, which can be wrongly discarded when the synchronization between u and v is lost due to some fault, is at most N, for some chosen integer N that is larger than one.

The modification consists of adding to process v two variables c and $cmax$, whose values are in the range 0..N-1. When process v receives a data(s) message, v compares the values of c and $cmax$. If $c \neq cmax$, then process v increments c by one (mod N) and proceeds as before, namely, either accepts the data(s) message if $exp \leq s$ or discards the message if $exp > s$. Otherwise, if $c = cmax$, then v accepts the message, assigns c the value 0, and assigns $cmax$ a random integer in the range 0..N-1. We call this modified protocol "soft sequence number protocol" because process v at some instants "forgets" the sequence number it expects to receive next, and accepts the next received sequence number without question.

There are two considerations behind this modification. First, it guarantees that process v never discards more than N data messages when the synchronization between u and v is lost due to some fault. Second, it ensures that the adversary cannot predict the instant when process v is willing to accept any received data message, and so cannot exploit any such predictions by sending replayed data messages at the predicted instant.

Formally, processes u and v in this protocol can be defined as follows:

process u

var nxt : **integer**   {sequence number of next sent message}

begin
          true $\rightarrow$
                    send data(nxt) **to** v;
                    $nxt := nxt + 1$
end

process v

inp N : **integer**
var s : **integer**   {sequence number of received message}
          exp : **integer**   {sequence number expected next}
          $c, cmax$: $0 .. $N-1

begin
          rcv data(s) **from** $u \rightarrow$
                    if $s < exp \land c \neq cmax \rightarrow$
                              {reject message; report an adversary}
                              $c := (c + 1) \mathrm{mod} N$
                    $\square\, exp \leq s \lor c = cmax \rightarrow$

```
                {accept message}
                exp := s + 1
                if c ≠ cmax →
                        c := (c + 1)mod N
                [] c = cmax →
                        c := 0;
                        cmax := RANDOM(0, N − 1)
                fi
        fi
end
```

Processes u and v of the soft sequence number protocol presented above can be combined with process pw of the weak integrity check protocol to construct process ps of the strong integrity check protocol. A main difference between processes pw and ps is that pw exchanges messages of the form data(t, d), whereas ps exchanges messages of the form data(s, t, d), where s is the message sequence number computed according to the soft sequence number protocol, t is the message text, and d is the message digest computed over the concatenation ($s; t; scr$) of s, t, and the shared secret scr. Process ps in the strong integrity check protocol can be defined as follows. (Process qs can be defined in the same way.)

process pw

```
inp     sp      :       integer
        sq      :       array [0 .. 1] of integer
        N       :       integer
var     s, t, d :       integer
        exp, nxt :      integer
        c, cmax :       0 .. N-1

begin
        rcv data(s, t, d) from qw →
                if MD(s; t; sq[0]) = d ∨ MD(s; t; sq[1]) = d →
                        if s < exp ∧ c ≠ cmax →
                                {reject message; report an adversary}
                                c := (c + 1)mod N
                        [] exp ≤ s ∨ c = cmax →
                                {accept message}
                                exp := s + 1
                                if c ≠ cmax →
                                        c := (c + 1)mod N
```

$$\square \, c = cmax \rightarrow$$
$$c := 0;$$
$$cmax := \text{RANDOM}(0, \, N - 1)$$

fi

fi

$$\square \, \text{MD}(s; t; sq[0]) \neq d \wedge \text{MD}(s; t; sq[1]) \neq d \rightarrow$$
{report an adversary}
skip

fi

$\square$ **true** →
{p receives data(s, t, d) from mesh router other than q and}
{checks that its message digest is correct and}
{its sequence number is within range}
RTMSG

$\square$ **true** →
{either p receives data(t) from an adjacent host or}
{p generates the text t for the next data message}
RTMSG

end

The first and second actions of process *ps* have a statement RTMSG that is defined as follows:

if $\text{NXT}(t) = p \rightarrow$
{accept message}
skip
$\square \, \text{NXT}(t) = q \rightarrow$
$d := \text{MD}(nxt; t; sp);$
send data(t, d) **to** qs;

$nxt := nxt + 1 \, \square \, \text{NXT}(t) \neq p \wedge \text{NXT}(t) \neq q \rightarrow$
{compute next soft sequence number s for sending data to NXT(t);}
{compute d as message digest of concatenation of s, t}
{and the secret for sending data to NXT(t);}
{forward data(s, t, d) to router NXT(t)}
skip

fi

To verify the correctness of the strong integrity check protocol, use the state transition diagram of this protocol in Figure 6.4, which considers only the channel from process *qs* to process *ps*. (The channel from *ps* to *qs* and the channels from *ps* to any other strong integrity check process in an adjacent router of *p* can be verified in the same way.) This diagram has

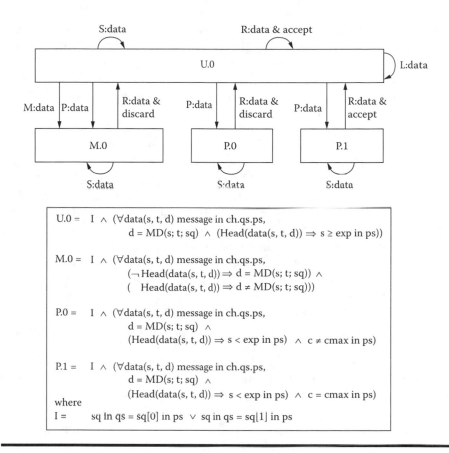

$$U.0 = I \wedge (\forall data(s, t, d) \text{ message in ch.qs.ps,}$$
$$d = MD(s; t; sq) \wedge (Head(data(s, t, d)) \Rightarrow s \geq exp \text{ in ps}))$$

$$M.0 = I \wedge (\forall data(s, t, d) \text{ message in ch.qs.ps,}$$
$$(\neg Head(data(s, t, d)) \Rightarrow d = MD(s; t; sq)) \wedge$$
$$(\quad Head(data(s, t, d)) \Rightarrow d \neq MD(s; t; sq)))$$

$$P.0 = I \wedge (\forall data(s, t, d) \text{ message in ch.qs.ps,}$$
$$d = MD(s; t; sq) \wedge$$
$$(Head(data(s, t, d)) \Rightarrow s < exp \text{ in ps}) \wedge c \neq cmax \text{ in ps})$$

$$P.1 = I \wedge (\forall data(s, t, d) \text{ message in ch.qs.ps,}$$
$$d = MD(s; t; sq) \wedge$$
$$(Head(data(s, t, d)) \Rightarrow s < exp \text{ in ps}) \wedge c = cmax \text{ in ps})$$

where

$$I = sq \text{ in qs} = sq[0] \text{ in ps} \vee sq \text{ in qs} = sq[1] \text{ in ps}$$

Figure 6.4   State transition diagram of the strong integrity check protocol.

four nodes that represent all possible reachable states of the protocol. Every transition in the diagram stands for either a legitimate action (of process *ps* or process *qs*) or an illegitimate action of the adversary.

Note that because the strong integrity check protocol operates below the secret exchange protocol in the protocol stack, the assertion can be made that $(sq \text{ in } qs = sq[0] \text{ in } ps \vee sq \text{ in } qs = sq[1] \text{ in } ps)$ is an invariant in every state of the strong integrity check protocol; this invariant is denoted as I in the specification in Figure 6.4.

Initially, the protocol starts at state U.0. At state U.0, two legitimate actions, namely, the send action in *qs* that sends a data message and the receive action in *ps* that receives a data message, can be executed. Executing either one of the two actions at state U.0 keeps the protocol in state U.0.

State U.0 is the only good state in the strong integrity protocol. The set of transitions that leads the protocol from state U.0 to state U.0 constitutes the good cycle of the protocol. If only legitimate actions of processes *ps* and qs are executed, the protocol will stay in this good cycle indefinitely. Next, the bad effects caused by the actions of an adversary and how the protocol can recover from these effects will be discussed.

First, the adversary can execute a message loss action at state U.0. If the adversary executes a message loss action at state U.0, the predicate that for every data message data(s, t, d) in the channel from qs to ps, $d = \text{MD}(s; t; sq)$, still holds. Therefore, the protocol stays at state U.0.

Second, the adversary can execute a message modification action at state U.0 causing the protocol to move to state M.0. The receive and discard action executed by ps at state M.0 leads the protocol back to state U.0.

Third, the adversary can execute a message replay action at state U.0. There are two cases to consider. First, if the replayed message data(s, t, d) is too old such that the secret used to compute the message digest is different from the current value of constant sq in process qs, then the protocol moves to state M.0, and later returns to state U.0 as discussed above. Second, if the replayed message data(s, t, d) is recent such that the secret used to compute the message digest is equal to the current value of constant sq in process qw, then the protocol moves either to state P.0 or to state P.1. With a high probability of $(cmax - 1)/cmax$, the protocol moves to state P.0, and the replayed message will be received and discarded by ps because the value of field s in the message indicates that the message is replayed. With a probability of $1/cmax$, the protocol moves to state P.1, and the replayed message will be received and accepted. In both cases the protocol returns to state U.0.

From the state transition diagram, it is clear that each illegitimate action by the adversary will eventually lead the protocol back to U.0, which is a good state. Once the protocol is in a good state, the protocol can progress in the good cycle. Moreover, if the adversary replays a recent data message, the replayed message will be detected and discarded with high probability $(cmax - 1)/cmax$. Hence the following two theorems about the strong integrity check protocol are proved:

Theorem 5
In the absence of an adversary, a network that executes the strong integrity check protocol follows the good cycle, consisting of a single transition from state U.0 to state U.0, and will stay in this good cycle indefinitely.

Theorem 6
In the presence of an adversary, a network that executes the strong integrity check protocol will converge to the good cycle in a finite number of steps after the adversary finishes executing any number of message loss or message modification actions. This network will also converge to the good cycle in a finite number of steps after the adversary finishes executing any number of message replay actions.

The protocols used by the weak hop integrity protocol and the strong hop integrity protocol have several novel features that make them correct and efficient. First, whenever the secret exchange protocol attempts to change a secret, it keeps both the old secret and the new secret until it is certain that the integrity check of any future message will not be computed using the old secret. Second, the integrity check protocol computes a digest at every router along the message route so that the location of any occurrence of message modification can be determined. Third, the soft sequence number protocol makes the strong hop integrity protocol tolerate any loss of synchronization between any two adjacent routers.

6.6 Conclusion and Open Issues

This chapter has presented scenarios of message insertion attacks and message replay attacks that may result in denial-of-service attack to wireless mesh networks, and introduces the hop integrity concept, which aims to provide protection against these attacks. Then, the chapter presented the three components of the hop integrity protocol suite for wireless mesh networks, namely, the initial authentication protocol, the secret exchange protocol, and the integrity check protocol. Together, they provide hop integrity to wireless mesh networks and their correctness is verified by state transition diagrams.

There are a few open issues that are worth mentioning. The first open issue is on strategic deployment of hop integrity. Hop integrity protocols are open to incremental deployment, and the security they provide increases with the number of pairs of hop integrity-equipped mesh routers because an adversary will have less venues to apply its attacks. However, due to hardware/software compatibility and efficiency consideration, it may be worthwhile to consider a strategic deployment scheme. For example, a few hotspots in the network can be required to install static hop integrity, in which hop integrity is always turned on; other spots in the network can install dynamic hop integrity, in which hop integrity is randomly on and off.

The second open issue is about interoperability between different wireless mesh networks. The initial authentication protocol is designed for mesh routers that belong to the same domain. For mesh routers from different domains to execute these protocols, the certificates of the involved domains need to be integrated.

The third open issue is about integrity in MAC and PHY layers. Wireless mesh networks are vulnerable to security attacks at various layers. Although the protocols presented in this chapter address the integrity problem at network layer, the same issue at the lower MAC and PHY layers is still an open problem.

References

[1] C.-T. Huang and M. G. Gouda, *Hop Integrity in the Internet*, Springer, December 2005.

[2] M. G. Gouda, *Elements of Network Protocol Design*, Wiley, April 1998.

[3] Y. Zhang, J. Luo, and H. Hu, Eds., *Wireless Mesh Networking: Architectures, Protocols, and Standards*, Auerbach Publications, Boca Raton, FL, 2006.

[4] S. Kent and R. Atkinson, Security Architecture for the Internet Protocol, RFC 2401, November 1998.

[5] S. Kent and R. Atkinson, IP Authentication Header, RFC 2402, November 1998.

[6] S. Kent and R. Atkinson, IP Encapsulating Security Payload (ESP), RFC 2406, November 1998.

[7] M. G. Gouda, *Elements of Network Protocol Design*, Wiley, April 1998.

[8] M. G. Gouda, E. N. Elnozahy, C.-T. Huang, and T. M. McGuire, Hop integrity in computer networks, *IEEE/ACM Transactions on Networking*, Vol. 10, No. 3, June 2002.

[9] C.-T. Huang, Hop Integrity: A Defense against Denial-of-Service Attacks, Ph.D. dissertation, Department of Computer Sciences, The University of Texas at Austin, August 2003.

[10] C.-T. Huang and M. G. Gouda, *Hop Integrity in the Internet*, Springer, December 2005.

[11] T. Dierks and C. Allen, The TLS Protocol Version 1.0, RFC 2246, January 1999.

[12] Kerberos: The Network Authentication Protocol, http://web.mit.edu/Kerberos/.

[13] C. Rigney, S. Willens, A. Rubens, and W. Simpson, Remote Authentication Dial In User Service (RADIUS), RFC 2865, June 2000.

[14] IETF Public-Key Infrastructure (X.509) (pkix) Charter, http://www.ietf.org/html.charters/pkix-charter.html.

[15] R. Housley, W. Polk, W. Ford, and D. Solo, Internet X.509 Public Key Infrastructure Certificate and Certificate Revocation List (CRL) Profile, RFC 3280, April 2002.

[16] R. L. Rivest, The MD5 Message-Digest Algorithm, RFC 1321, 1992.

[17] NIST, FIPS PUB 180-1: Secure Hash Standard, April 1995.

[18] H. Krawczyk, M. Bellare, and R. Canetti, HMAC: Keyed-Hashing for Message Authentication, RFC 2104, February 1997.

Chapter 7

Privacy Preservation in Wireless Mesh Networks[1,2]

Taojun Wu, Yuan Xue, and Yi Cui

Contents

[1] This work was supported in part by TRUST (The Team for Research in Ubiquitous Secure Technology), which receives support from the National Science Foundation (NSF award number CCF-0424422) and the following organizations: Cisco, ESCHER, HP, IBM, Intel, Microsoft, ORNL, Pirelli, Qualcomm, Sun, Symantec, Telecom Italia, and United Technologies.

Multi-hop wireless mesh networking (WMN) has attracted increasing attention and deployment as a low-cost approach to provide last-mile broadband Internet access. Privacy is a critical issue in WMN, as traffic of an end user is relayed via multiple wireless mesh routers. Due to the unique characteristics of WMN, the existing solutions for the Internet are either ineffective at preserving privacy of WMN users, or will cause severe performance degradation.

In this chapter, we propose a lightweight privacy preserving solution aimed to achieve well-maintained balance between network performance and traffic privacy preservation. At the center of this solution is an information-theoretic metric called "traffic entropy," which quantifies the amount of information required to describe the traffic pattern and to characterize the performance of traffic privacy preservation. We further present a penalty-based shortest path routing algorithm that maximally preserves traffic privacy by minimizing the mutual information of "traffic entropy" observed at each individual relaying node, meanwhile controlling performance degradation within the acceptable region. Extensive simulation study proves the soundness of our solution and its resilience to cases when two malicious observers collude.

7.1   Introduction

Recently, multi-hop WMN has attracted increasing attention and deployment as a low-cost approach to provide last-mile broadband Internet access [2–5]. In WMN, each client accesses a stationary wireless mesh router. Multiple mesh routers communicate with one another to form a multi-hop wireless backbone that forwards user traffic to a few gateways connected to the Internet. Some perceived benefits of WMN include enhanced resilience against node failures and channel errors, high data rates, and low costs in deployment and maintenance. For such reasons, commercial WMNs are already deployed in some U.S. cities (like Medford and Chaska). Even large cities are planning to deploy citywide WMNs as well [1].

However, to further widen the deployment of WMN and enable it as a competitive player in the market of broadband Internet access, the issue of privacy must be addressed. Privacy has been a major concern of Internet users [12]. It is a particularly critical issue in the context of WMN-based Internet access, where users' traffic is forwarded via multiple mesh routers. In a community mesh network, this means that the traffic of a residence can be observed by the mesh routers residing at its neighbors. Despite the necessity, limited research has been conducted toward privacy preservation in WMN.

This motivates us to investigate the privacy preserving mechanism in WMN. There are mainly two privacy issues: data confidentiality and traffic confidentiality.

- Data confidentiality: It is obvious that data content reveals user privacy on what is communicated. Data confidentiality aims to protect the data content and prevent eavesdropping by intermediate mesh routers. Message encryption is a conventional approach for data confidentiality.
- Traffic confidentiality: Traffic information such as who the users are communicating with, when and how frequently they communicate, and the amount and the pattern of traffic, also reveals critical privacy information. The broadcasting nature of wireless communication makes acquiring such information easy. In a WMN, attackers can conduct traffic analysis at mesh routers by simply listening to the channels to identify the "ups and downs" of the target's traffic. While data confidentiality can be achieved via message encryption, it is much harder to preserve traffic confidentiality. In this chapter we focus on the user traffic confidentiality issue and study the problem of traffic pattern concealment.

We aim at designing a lightweight privacy preserving mechanism for WMN which is able to balance the traffic analysis resistance and the bandwidth cost. Our mechanism makes use of the intrinsic redundancy of WMN, which is able to provide multiple paths for data delivery. By intuition, if the traffic from the source (i.e., gateway) to the destination (i.e., mesh router) is split to many paths, then all the relaying nodes[3] along the paths could

---

[3] In this chapter we use the following terms interchangeably: wireless mesh router, intermediate relaying node, and wireless node.

only observe a portion of the entire traffic. Moreover, if the traffic is split in a random way both spatially and temporally, then an intermediate node has limited knowledge to figure out the overall traffic pattern. Thus the traffic pattern is concealed.

Based on this intuition, we seek a routing scheme which routes data such that the statistical distributions of the traffic observed at intermediate relaying nodes are independent from the actual traffic from the source to the destination. To achieve this goal, we first define an information-theoretic metric, traffic entropy, which quantifies the amount of information required to describe the traffic pattern. Then we present a penalty-based routing algorithm, which aims to minimize the mutual information of traffic entropy observed at each relaying node, meanwhile controling the network performance degradation under the acceptable level.

Considering the possibility of collusion, we evaluate our scheme under a situation when two observers exchange their knowledge about the same destination. We measure this shared knowledge as "colluded traffic mutual information" and our simulation results show that our scheme is still viable in case of two colluding eavesdroppers.

The rest of this chapter is organized as follows. In Section 7.2, we present the overall architecture for privacy preservation in WMN. Section 7.3 and Section 7.4 focus on the traffic privacy issue. In particular, Section 7.3 presents the model to quantify the performance of traffic privacy pre-servation, and Section 7.4 presents the routing algorithm. The proposed privacy preserving solution is evaluated via extensive simulation study in Section 7.5. Section 7.6 discusses the collusion problem possible with mali-cious traffic observers and its impact on our proposed scheme. Section 7.7 summarizes background knowledge and related work. Section 7.8 con-cludes the chapter and points out the future directions.

7.2 Privacy Preserving Architecture

We consider a multi-hop WMN shown in Figure 7.1. In this network, client devices access a stationary wireless mesh router at its residence. Multiple mesh routers communicate with one another to form a multi-hop wireless backbone that forwards user traffic to the gateway which is connected to the Internet.

Two privacy aspects are considered in this architecture. Data confiden-tiality aims to protect the data content from eavesdropping by the interme-diate mesh routers. Traffic confidentiality prevents the traffic analysis attack from the mesh routers, which aims at deducing the traffic information such as who the user is communicating with and the amount and the pattern of traffic. Our privacy preserving architecture aims to protect the privacy of

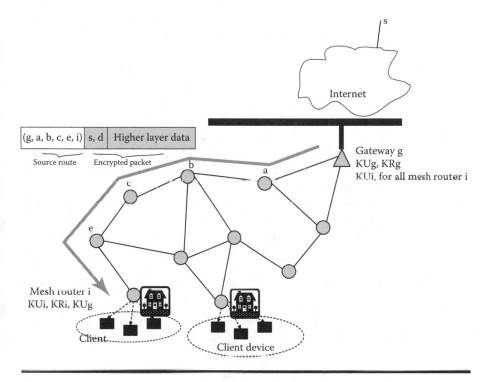

Figure 7.1   Privacy preserving architecture for wireless mesh network.

each wireless mesh router, the basic routing unit in WMN. The architecture consists of the following functional components:

- Key distribution: In this architecture, each mesh node, as well as the gateway, has a pair of public and private keys (KU, KR). The gateway maintains a directory of certified public keys of all mesh nodes, and each mesh node has a copy of the public key of the gateway KU_g. The public key KU_i of mesh node i and KU_g are used to establish the shared secret session key KS_{gi}, which is used to encrypt the messages between them.

- Message encryption: Let M be the IP packet sent from a source s in the Internet to a client d in the mesh network, and let i be the mesh router of client d. The whole IP packet M, which contains the original source and destination address s and d, is encrypted at gateway g via the shared secret key KS_{gi}: $M_e = E(KS_{gi}, M)$. To route the encrypted packet M_e to its destination, the gateway prefixes the source route from the gateway g to the router i to the packet. The encapsulated packet is then forwarded by relaying routers in WMN. Likewise, packets travelling in the reverse direction are treated the same way. As the source address s and other higher-layer header information, such as port, are all encrypted, the relaying routers are unable to obtain the information on who the client of router i is communicating with and what type of application is involved. Because encryption and decryption take place only at the gateway and the destination mesh router, much less computation is required, which is a desired feature in WMN.

■ Routing control: With source route in cleartext in an encapsulated packet, the intermediate mesh routers can still observe the amount and the pattern of the traffic of a particular mesh node i. To address this problem, our privacy preserving mechanism explores the path diversity of WMN and forwards packets between the gateway and the mesh node via different routes. Thus any relaying router can only observe a portion of the whole traffic of this connection. In Section 7.4, we detail the design of a penalty-based routing algorithm, which randomly selects a route for each individual packet such that the observed traffic pattern at each relaying node is independent of the overall traffic. In our design, the gateway maintains a complete topology of the WMN and computes the source routes between the destination mesh nodes and itself.

7.3 Privacy Modeling in WMNs

7.3.1 Network Model

We model the WMN shown in Figure 7.1 as a graph $G = \{V, \mathcal{E}\}$, where V is the set of wireless nodes in WMN, and $\mathcal{E}$ is the set of wireless edges (x, y) between any two nodes x, y. Each node x maintains a logical connection with the gateway node g. Node x receives data from the Internet via g. The source and destination information of a packet is open to the relaying node. The traffic pattern of x can be categorized into two types: incoming traffic pattern and outgoing traffic pattern. In this paper, we mainly consider the first type.

If the traffic between s and x goes through only one route, then any relaying node on this route can easily observe the entire traffic between g and x, thus violating its traffic pattern privacy. To avoid this problem, x must establish multiple paths with g and distribute its traffic along these paths, such that any node can only get a partial picture of x's traffic pattern.

However, the complete traffic pattern information of x could still be obtained by a single node in case of multi-path routing. In the example shown by Figure 7.2, g allocates the traffic to x via three disjoint routes by fixed proportion. Then for any node along any path, although only seeing one third of the flow, the observed traffic shape is isomorphic to the original one. Therefore, the traffic to x must be distributed along multiple

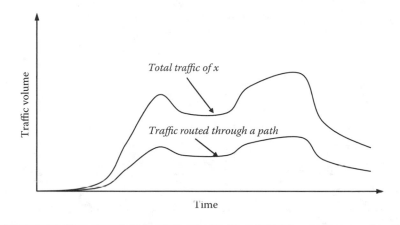

Figure 7.2   An example of isomorphic traffic.

routes in a time-variant fashion, such that the traffic pattern observed at any node is statistically deviant from the original pattern. The notations used in Section 7.3 are listed in Table 7.1.

7.3.2   Traffic Entropy

We propose to use information entropy as the metric to quantify the performance of a solution at preserving the traffic pattern confidentiality. In what follows, we consider two nodes x and y; x is the destination node of the traffic from the gateway g to x, y is the observing node, which relays packets for x and also tries to analyze the traffic of x.

7.3.2.1   Basic Definition

Ideally, we view the traffic of x as a continuous function of time, as shown in Figure 7.3. In practice, the traffic analysis is conducted by dividing time

Table 7.1   Notations Used in Section 7.3

$\mathcal{V}$	Wireless node set
$\mathcal{E}$	Edge set
g	Gateway node
x	Destination node
y	Observing node
X	Random variable describing x's traffic pattern
Y^X	Random variable describing x's traffic pattern observed by y
$H(X)$	Entropy of X
$H(Y^X)$	Entropy of Y^X
$I(Y^X, X)$	Mutual information between X and Y^X

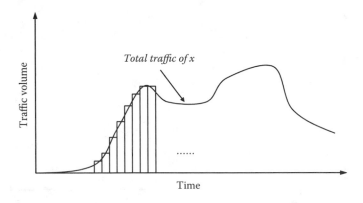

Figure 7.3　Sampling-based traffic analysis.

into equal-sized sampling periods, then measuring the amount of traffic in each period, usually in terms of number of packets, assuming the packet sizes are all equal. Therefore, as the first step, we discretize the continuous traffic curve into piecewise approximation of discrete values, each denoting the number of packets destined to x in a sampling period.

Now, we use X as the random variable of this discrete value. Y^X is the random variable representing the number of packets destined to x observed at node y in a sampling period. We denote $P(X = i)$ as the probability that the random variable X is equal to i ($i \in \mathcal{N}$), i.e., the probability that node x receives i packets in a sampling period. Likewise, $P(Y^X = j)$ is the probability that Y^X is equal to j ($j \in \mathcal{R}$), i.e., j packets destined to x go through node y in a sampling period.

Then the discrete Shannon entropy of the discrete random variable X is

$$H(X) = -\sum_i P(X = i) \log_2 P(X = i) \qquad (7.1)$$

$H(X)$ is a measurement of the uncertainty about the outcome of X. In other words, it measures the information of node x's traffic, i.e., the number of bits required to code the values of X. $H(X)$ takes its maximum value when the value of X is uniformly distributed. On the other hand, if the traffic pattern is CBR, then $H(X) = 0$ because the number of packets at any sampling period is fixed.[4]

---

[4] This offers the information-theoretic interpretation for traffic padding: by flattening the traffic curve with blank packets, the entropy of observable traffic is reduced to 0, which perfectly hides the information of the original traffic pattern.

Similarly, we have the entropy for Y^X as follows:

$$H(Y^X) = - \sum_j P(Y^X = j) \log_2 P(Y^X = j) \qquad (7.2)$$

7.3.2.2 Mutual Information

We then define the conditional entropy of random variable Y^X with respect to X as

$$H(X|Y^X) = - \sum_j P(Y^X = j) \sum_i p_{ij} \log_2 p_{ij} \qquad (7.3)$$

where $p_{ij} = P(X = i | Y^X = j)$ is the probability that $X = i$ given the condition that $Y^X = j$. $H(X|Y^X)$ can be thought of as the uncertainty remaining about X after Y^X is known. The joint entropy of X and Y^X can be shown as

$$H(X, Y^X) = H(Y^X) + H(X|Y^X) \qquad (7.4)$$

Finally, we define the mutual information between X and Y^X as

$$I(Y^X, X) = H(X) + H(Y^X) - H(X, Y^X)$$

$$= H(X) - H(X|Y^X) \qquad (7.5)$$

which represents the information we gain about X from Y^X.

Back to the example in Figure 7.2, let us assume that the observing node y is located on one route destined to x. Because the traffic shape observed at y is the same as x, at any sampling period, if $Y^X = j$, then X must equal to a fixed value i, making $P(X = i|Y^X = j) = 1$. According to Equation (7.3), this makes the conditional entropy $H(X|Y^X) = 0$. According to Equation (7.5), we have $I(Y^X, X) = H(X)$, implying that from Y^X, we gain the complete information about X.

On the contrary, if Y^X is independent from X, then the conditional probability $P(X = i|Y^X = j) = P(X = i)$, which maximizes the conditional entropy $H(X|Y^X)$ to $H(X)$. According to Equation (7.5), we have $I(Y^X, X) = 0,$[5] i.e., we gain no information about X from Y^X.

In reality, because Y^X records the number of a subset of packets destined to node x, it cannot be totally independent from the random variable X. Therefore, the mutual information should be valued between the

---

[5] By the definition of mutual information, $I(Y^X, X) \geq 0$, with equality if and only if X and Y are independent.

two extremes discussed above, i.e., $0 < I(Y^X, X) < H(X)$. This means that node y can still obtain partial information of X's traffic pattern. However, a good routing solution should minimize such mutual information as much as possible for any potential observing node. More formally, we should minimize

$$\max_{Y \in \mathcal{V}-X} I(Y^X, X) \tag{7.6}$$

the maximum mutual information that any node can obtain about X.

7.4   Penalty-Based Routing Algorithm

In this section, we propose a penalty-based routing algorithm to achieve our goal of hiding the traffic pattern by exploiting the richness of available paths between two nodes in WMN. Specifically, we choose to adopt the source routing scheme. Such a choice is enabled by the fact that one node can easily acquire the topology of the WMN it belongs to, which is mid-sized (within 100 nodes) and static.

When designing the algorithm, we also keep in mind the need to compromise between sufficient security assurance and acceptable system overhead. We would show in our algorithm that system performance is satisfactory and security assurance is adequate.

Shown in Table 7.2, the algorithm operates in three phases: path pool generation, candidate path selection, and individual packet routing. The notations used in this section are listed in Table 7.3.

First, in the path pool generation phase, we try to generate a large set of diversified routing paths connecting the gateway g and the destination node x, denoted as S_{paths}. The path generation algorithm is an iterated process of applying a modified version of Dijkstra's algorithm. Here, each node is assigned a penalty weight, and the weight of an edge is defined as the weighted average of penalty weights of its two end nodes. The weight (or cost) of a path is defined as the sum of penalty weights of all edges consisting this path. The algorithm runs in iterations. Initially, we set the penalty weight of each node as 1, then run Dijkstra's algorithm to find the first shortest path from the gateway g to x. Next, we increase the penalty weight for each node on this found path. This will make these appeared nodes less competitive to other nodes in becoming components of next path. After this, the algorithm proceeds to the next iteration, generating the second path, and all nodes appearing on the second path are penalized through increasing their weights. This process goes on until enough numbers of paths are found.

Second, in the candidate path selection phase, we try to choose a combination of diversified routing paths, a subset of paths from the set

Table 7.2   Penalty-Based Routing Algorithm

/*Penalty-Based Shortest Path*/
$PBSP(Snode, Dnode)$
    For each node $v \in V$
        $d[v] \leftarrow \infty$
    For each node $v \in V$
        $prev[v] \leftarrow \infty$
    For each node $v \in V$
        $visited[v] \leftarrow 0$
    $d[SNode] \leftarrow 0$
    Repeat
        Get unvisited vertex v with the least $d[v]$
        If $d[v] \geq \infty$, **Then** v unreachable
        Else $visited[v] \leftarrow 1$
        For all v's neighbors w
            $EdgePenalty = \alpha[pow(\gamma, (w.tag))] + \beta(v.tag)$
            If $d[w] > d[v] + EdgePenalty$
                $d[w] \leftarrow d[v] + EdgePenalty$
                $prev[w] \leftarrow v$
    Until $visited[v] = 1, \forall v \in V$

/*Generate S_{paths} For Each $g - x$ Pair*/
$GenPath()$
For All Non-Gateway Nodes x
    For each node $v \in V$
        $v.tag \leftarrow 1$
    Repeat
        PBSP(g, x)
        Get new $g - x$ path P_{new} from vector $prev[]$
        Store P_{new} in S_{paths}
        For all nodes v on P_{new}
            $v.tag \leftarrow v.tag + 1$
    Until *PathPoolSize* paths found.

/*Select $S_{selected}$ For Each $g - x$ Pair*/
$SelPath()$
Repeat
    $rnd = rand() \bmod PathPoolSize$
    select rndth path from S_{paths}
Until *SelPathNum* paths selected

/*Decide path for arriving packet*/
$RoutePkt(Snode, Dnode)$
    $Packets[Dnode] \leftarrow Packets[Dnode] + 1$
    $rndpath = rand() \bmod SelPathNum$
    route packet along the $rndpath$th path from $S_{selected}$
    If $Packets[Dnode] > ReSelPathCnt$
        $Packets[Dnode] \leftarrow 0$
        SelPath()

Table 7.3 Notations Used in Section 7.4

v, w	Node
$v.tag$	Number of times v is included by a path
α	Factor to slow down penalty rate
β	Factor to avoid many identical paths in beginning stages of path generation
γ	Base of exponential penalty function
$d[]$	Penalty vector for every node
$prev[]$	Vector to store P_{new} reversely
$Packets[]$	Vector to store number of arrived packets for every node

S_{paths}, denoted as $S_{selected}$. The paths in $S_{selected}$ are selected randomly from S_{paths}. After each choice of a path into $S_{selected}$, the probability factor of that path is decreased to lower the chance of multiple identical paths existing in $S_{selected}$. $S_{selected}$ is changed and renewed corresponding to network activities.

Third, in the packet routing phase, we choose randomly from $S_{selected}$ one path for each packet and increase the counter for the selected path subset $S_{selected}$. This $S_{selected}$ path subset expires after counter reaches its predetermined threshold. Then $S_{selected}$ is renewed by calling the second phase again.

Because packets are assigned a randomly chosen path, and all these candidate paths are designed to be disjoint, the chance that packets are routed in similar paths is small. Our experiment results further confirm this intuition.

This algorithm is designed to balance the needs of routing performance (finding paths with smallest hop count) and preserving traffic pattern privacy (finding disjoint paths). The penalty weight update function serves as the tuning knob to maneuver the algorithm between these two contradictory goals. During the initialization, when the penalties of all nodes are equal, the path found by the algorithm is indeed shortest in terms of hop count. As a node is chosen by more routes, its penalty weight monotonically increases, making it less likely to be chosen again. Thus, as the algorithm proceeds, the newly chosen paths (shortest in terms of its aggregate penalty weight) become more disjoint from existing paths, but longer in terms of hop count. The pace of such shift from "smallest hop-count path" to "disjoint path" is controlled by how fast the penalty weight update function grows. Our experiment results confirm this reasoning. Finally, by randomly assigning packets along different paths, the algorithm maximally disturbs the traffic pattern of any $g - x$ pair.

Although penalty-based routing has been used in existing literature [8], we are using it for different objects. Their links were penalized for losses

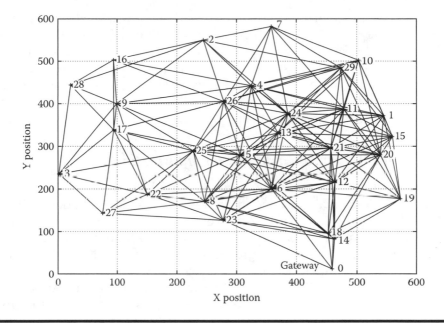

Figure 7.4   Experimental topology.

or malicious behavior while our approach applies it to avoid using links repeatedly to get better path diversity.

7.5   Experimental Results

7.5.1   Simulation Setup

We base our simulation on a randomly generated topology (Figure 7.4) (600 × 600) with 30 nodes. The effective distance between two nodes is set to be 250. The whole process of simulation consists of 400,000 logical ticks. In each single tick, a packet is generated at gateway node 0 and its destination is randomly decided to be one of the other 29 nodes. To better simulate real network traffic, we set the probability of 0.05 that, at one tick, no packet is generated, i.e., idle probability. The distance delay factor is chosen to be 0.003 tick and the hop delay factor is decided as 0.05 tick. We approximate hop delay at any node by multiplying the hop delay factor with its usage count by all paths chosen initially.

With a relatively small node set, we choose 50 as our *PathPoolSize* and 5 as *SelPathNum*. The selected path subset $S_{selected}$ for any destination node is renewed after sending 50 packets to that node. To obtain multiple diversified paths with Dijkstra's algorithm more quickly, we introduce the exponential penalty function on the *tag* of one node and use γ as the base of exponential function when deciding on which edge to include the candidate path. To slow down the growing rate of exponential penalty function, we multiply the exponential function with a factor α when calculating $EdgePenalty$. To avoid getting too many identical paths in the beginning stages, we amplify the influence of another node by multiplying the *tag* of another node with β. The penalty parameters α, β, γ are chosen to be 0.5, 15, and 1.85, respectively.

7.5.2 Traffic Entropy and Mutual Information

The total 400,000 ticks is divided into 20 periods. Each period is then divided into 50 intervals and one interval is 400 ticks long. Within each interval, for each destination node x, we count the number of packets that all other nodes y have relayed for x. Then for each period, we independently calculate the traffic entropies $H(X)$, $H(Y^X)$ and mutual information $I(Y^X, X)$ based on their definitions in Section 7.3.2.

Due to the space limit, we only show part of our results. Among all nodes in the network, we choose two sets of nodes. Nodes in the first set $\{1, 6, 11, 15, 23, 24, 25, 29\}$ are close to (two to three hops) the gateway node 0. Nodes in the second set $\{2, 3, 7, 16, 17, 28\}$ are at the edge of the network, four to five hops away from the gateway. We choose two representative nodes, 1 and 16, out of each set.

Figure 7.5 shows the variance of traffic entropy and mutual information along the time. In Figure 7.5 (a), $H(1-1)$ denotes the traffic entropy of node 1. $H(23-1)$ denotes the traffic entropy of node 23 based on its observation on node 1. $MI(23-1, 1-1)$ denotes the mutual information node 23 shares with node 1. The same notation rules apply for Figure 7.5 (b), where node 16 is the destination and 9 is the observer. In both pictures, the observing node only shares 40 percent or less of information about the observed destination node at any sampling period.

This observation is further confirmed in Figure 7.6, where we plot the time-variant mutual information that destinations 1 and 16 share with other randomly chosen observing nodes. These results show that with our algorithm, the destination node is able to consistently limit the proportion of mutual information it shares with the observing nodes.

7.5.3 Which Nodes Have More Mutual Information?

In Figure 7.7(a), we calculate the time-averaged mutual information for all observing nodes with respect to the destination node 1, and sort them in the ascending order. Here, we observe an almost linearly-growing curve except at its head and tail. For nodes at the head of the curve, their mutual information is 0 because they lie at the outer rim of the network, hence are not chosen by our routing algorithm to relay traffic for node 1. At the

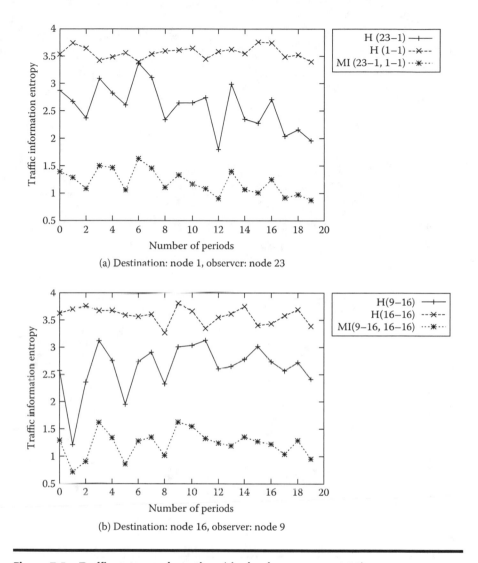

(a) Destination: node 1, observer: node 23

(b) Destination: node 16, observer: node 9

Figure 7.5   Traffic entropy along time (single observer, $\gamma = 1.85$).

tail of the curve is destination node 1, whose mutual information is actually the traffic entropy of its own. In Figure 7.7 (b), we observe the same phenomenon for destination 16, except at the head of the curve. This is because its network location is at the opposite end of the gateway, making every node of the network to be its candidate relaying node.

   This leads us to investigate if such distribution of mutual information is related with any other factors. We tried to connect mutual information of each node with certain metrics, such as its distance to the destination, but failed to find any causal relationship. We then sort observing nodes

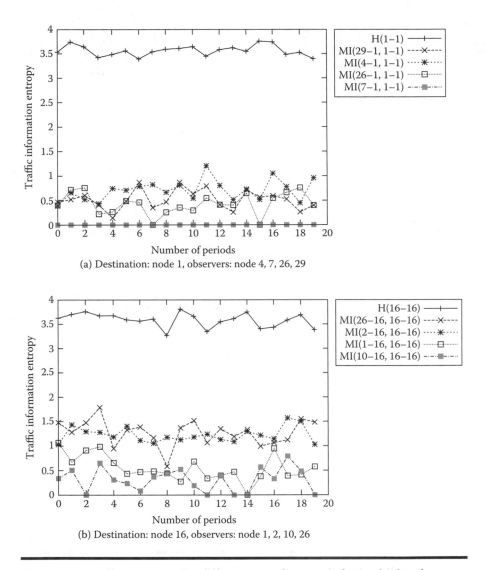

(a) Destination: node 1, observers: node 4, 7, 26, 29

(b) Destination: node 16, observers: node 1, 2, 10, 26

Figure 7.6 Traffic entropy in different sampling periods (multiple observers, $\gamma = 1.85$).

based on the averaged relayed traffic (average number of packets each node relays in a sampling period) on a log-log scale, and find the linear distribution as shown in Figure 7.8.

Obviously, such a power-law correlation tells us that the more traffic an observing node relays for a destination node, the more mutual information can be obtained about its traffic entropy. Furthermore, it gives us one way to experimentally quantify the relationship of these two metrics. Let T be the amount of traffic relayed and I be the mutual information; then their

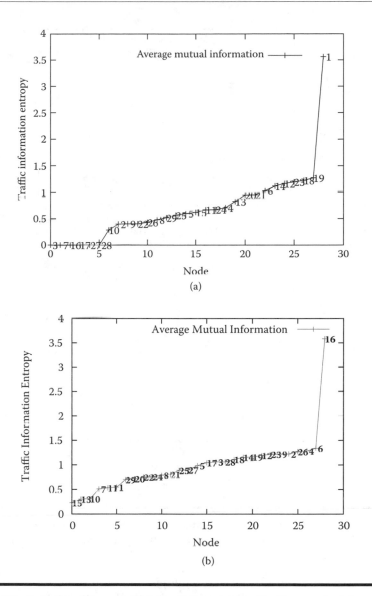

Figure 7.7 **Sorted traffic mutual information. (a) Destination: node 1 ($\gamma = 1.85$); (b) Destination: node 16 ($\gamma = 1.85$).**

power-law relationship can be written as

$$I = aT^k \tag{7.7}$$

where a is the constant of proportionality and k is the exponent of the power law, both of which can be measured from Figure 7.8. If $k < 1$, then the mutual information of an observing node grows in a sub-linear fashion as the amount of its relayed traffic increases, and in a super-linear fashion

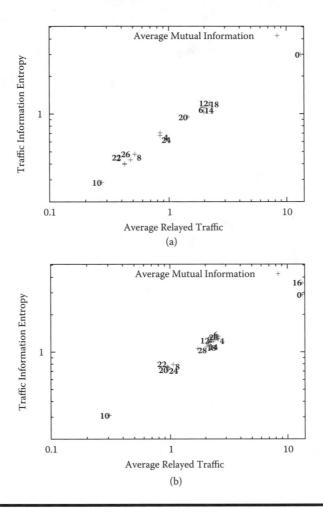

Figure 7.8 Power law correlation of mutual information and amount of traffic re-layed. (a) Destination: node 1 ($\gamma = 1.85$); (b) Destination: node 16 ($\gamma = 1.85$).

otherwise. From what we have in Figure 7.8 and the same results for other destination nodes, $k < 1$. This means that each time to make its mutual information further grows with the same increment, an observing node has to relay more and more traffic.

7.5.4 Trade-Off between Performance Degradation and Traffic Privacy

Finally, we study the performance trade-off of our algorithm by tuning its exponential penalty function base γ. The performance degradation introduced by our algorithm is captured by the average hop ratio. For each

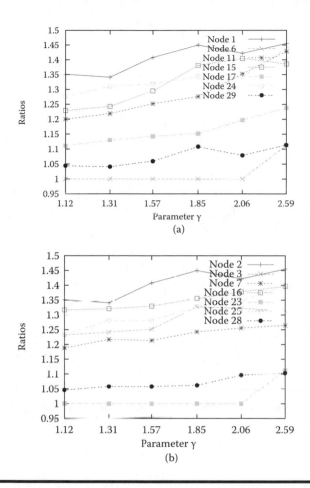

Figure 7.9 Average hop ratio. (a) Hop ratio of nodes of first set; (b) Hop ratio of nodes in the second set.

gateway-destination pair $g - x$, this metric is defined as the ratio between the average number of hops a packet goes through using our algorithm and the number of hops of the shortest path between g and s. From Figure 7.9, we can see that the average hop ratio increases as γ increases. The direct neighbors of the gateway are less sensitive to the change of γ, like node 6 in Figure 7.9(a) and node 23 in Figure 7.9(b).

In Figure 7.10 and Figure 7.11 we find that under shortest path routing, the mutual information of a node is 0 if it is not on the path to destination node. Otherwise, the mutual information node is much higher than the case of our algorithm. Also worth noting is that increasing of γ has a different impact on different nodes, depending on distance to gateway, destination, and location in the WMN. Take nodes 12 (Figure 7.10) and 6 (Figure 7.11) for example, because they lie near to the gateway node and

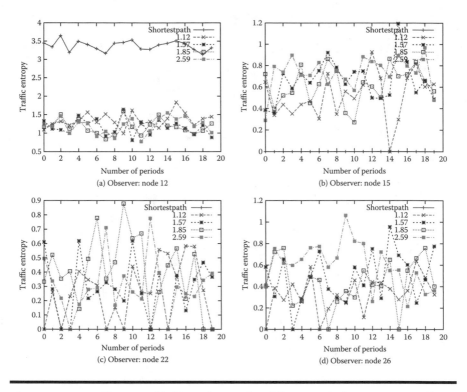

Figure 7.10   Traffic mutual information under different penalty parameters (destination: node 1).

are relatively centrally situated, their observed mutual information varies little with respect to the change of γ. Whereas for node 22 (Figure 7.10), which is far away from destination node 1 and on the edge of the WMN, mutual information shared with node 1 increases with the growth of γ, indicating more traffic is routed through farther nodes. This tendency of routing packets from farther nodes leads to a higher average number of hops, which is confirmed by our analysis about average hop ratio. However, traffic mutual information tends to decrease once the γ parameter gets too high (2.59 in this figure). This is due to the fact that when penalty values of many possible edges get large quickly, their relative differences become less. Consequently, candidate paths become less. The great fluctuation of node 26 (Figure 7.10) is due to its position in the center of the topology and equal distance to both gateway and destination. Similar observations can be made about mutual information values of destination node 16 (Figure 7.11).

We also observe from Figure 7.12 that our algorithm achieves our goal of preserving traffic pattern. In the first place, it is easy to conclude that in

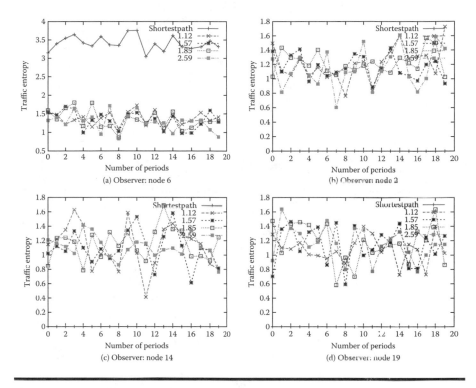

Figure 7.11 Traffic mutual information under different penalty parameters (destination: node 16).

normal shortest path routing, all relaying nodes share the same traffic information with the destination node, as shown by the tail of the ShortestPath curve in Figure 7.12. However, for our algorithm, the mutual information shared between relaying nodes and the destination node varies much less among all relaying nodes, and the higher γ is, the more leveled off the curve becomes and the closer we are to the goal of minimizing the greatest mutual information, formulated in Equation 7.6. It is also interesting to observe that mutual information is 0 for some nodes far away from both gateway and destination; for example, in Figure 7.12(a), when destination is 1, while all nodes participate in relaying packets for destination 16, because destination and gateway nodes are in opposite directions with respect to WMN topology.

7.6 Collusion Analysis

The relative small size of a typical WMN makes it easy for spatially close eavesdroppers to find each other. This alerts us to the high possibility of collusion of two malicious observers by exchanging their observed traffic

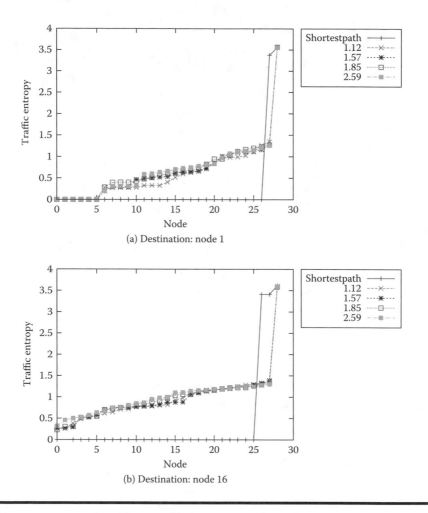

Figure 7.12 Sorted traffic mutual information under different penalty parameters.

pattern, and motivates us to make our proposed solution resilient to such collusion threats.

To analyze the extent to which collusion reveals original traffic pattern, we study the fluctuation of the observed traffic information. In this way, we can know how much in addition the colluders can observe about the original traffic.

7.6.1 Problem Description

Previously, we focused on traffic confidentiality and studied the problem of traffic pattern concealment via routing control. However, the relative small size of a WMN, aided by the stationary adjacent routers, invites a

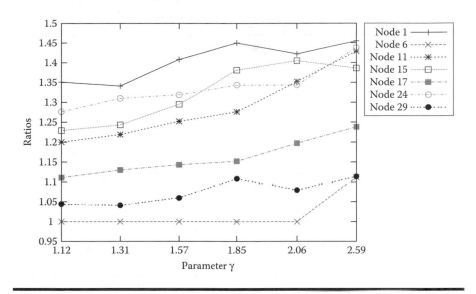

Figure 7.13 Collusion reveals significant portion of original traffic pattern.

high possibility of collusion of several observing relaying routers in the community. Because it is highly possible that different observers will know about various "ups and downs" of target's traffic, if malicious observers interchange their observed traffic information of target users, the combined observation could reveal a significant portion of the original traffic pattern. This is illustrated in Figure 7.13.

Given the size of the community network (less than 100 neighbor nodes), we have a reasonable estimation that three or more malicious observers are unlikely to exist simultaneously, and hence we will focus on analyzing the collusion problem of two observers in this work.

The parameters that affect significantly our collusion analysis include the choice of cooperating observers and destination target node. Because any routing algorithm will largely depend on topology of the network, the relative positions of observers and source and destination nodes can affect portions of revealed traffic pattern greatly. Another important parameter is the base of the exponential penalty function explained in Section 7.4.

7.6.2 Colluded Traffic Mutual Information

Our modeling of colluded traffic analysis tries to study the influence of collusion to observed traffic patterns of every period. This can help us to evaluate the resilience of our proposed A (PBSP) routing algorithm against collusion attack. The notations used in this section are listed in Table 7.4.

Table 7.4   Notations Used in Section 7.6.2

$\mathcal{V}$	Wireless node set
$\mathcal{E}$	Edge set
g	Gateway node
x	Destination node
y, z	Observing nodes
X	Random variable describing x's traffic pattern
Y^X, Z^X	Random variables describing x's traffic pattern observed by y, z, separately
(Y^X, Z^X)	Random variable describing x's traffic pattern observed by y, z together
$H(X)$	Entropy of X
$H(Y^X)$	Entropy of Y^X
$H(Y^X, Z^X, X)$	Joint entropy of Y^X, Z^X, X
$I(Y^X; X)$	Mutual information between X and Y^X
$I(Y^X, Z^X; X)$	Colluded mutual information between X and (Y^X, Z^X)

In what follows, we consider three nodes x and y, z. x is the destination node of the traffic from the gateway g to x. Nodes y, z are the observing nodes, which relay packets for x and also try to analyze the traffic of x. Due to the uncertainty of routing, y, z may or may not be on the same path over time.

To begin with, we need to identify a measurement for colluded observations. Based on the definition of traffic mutual information given in Section 7.3.2, we can measure the colluded observation about destination x with mutual information between x and (y, z). The traffic observations by y and z together can be deemed as the joint distribution of variable Y^X and Z^X. The colluded traffic mutual information $I(Y^X, Z^X; X)$ of random variable (Y^X, Z^X) with respect to X can then be defined as

$$I(Y^X, Z^X; X) = H(Y^X, Z^X) + H(X) - H(Y^X, Z^X, X) \qquad (7.8)$$

where $H(Y^X, Z^X, X)$ is the joint entropy of Y^X, Z^X, and X. $I(Y^X, Z^X; X)$ can represent the information we could gain about X from (Y^X, Z^X), i.e., from y, z together. Their relationship is shown in Figure 7.14.

7.6.3   Simulation Results

For ease of notation, in the following discussion, we would use $H(Y, X)$ to denote $H(Y^X, X)$, i.e., the entropy of traffic that y observes about x. Similarly, we simplify the joint traffic entropy $H(Y^X, Z^X)$ as $H(y, z, x)$, where Y^X, Z^X denote the portions of traffic that Y, Z observes about X. In a subtly different way, we denote $I(Y^X; X)$ as $I(Y; X)$ and $I(Y^X, Z^X; X)$ as $I(Y, Z; X)$.

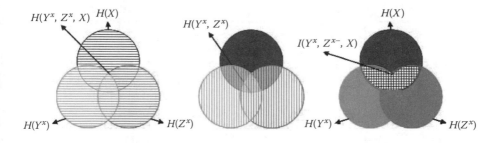

Figure 7.14 Vein graph representation of $I(Y^X, Z^X; X)$, $H(Y^X, Z^X)$, **and** $H(Y^X, Z^X, X)$.

7.6.3.1 Traffic Curves

In the first place, we will present the measured traffic curves along a time line. In Figure 7.15, node 1 is the destination and we can easily conclude that its traffic (node 1 observing itself) is always the largest in amount. This is because any node can observe the whole traffic of itself while other nodes can only observe a portion of it.

Another observation we can make is the fact that the colluded knowledge about traffic activity of node 1 (in squares), as expected, is higher than any single observer, either 15 or 28. Moreover, we are confirmed by this traffic curve figure that, although generally speaking, node 15 observes much more traffic of node 1; during some intervals, node 28 outperforms 15 and elevates the aggregated knowledge about traffic activity of node 1. Example intervals are those near intervals 100 and 150.

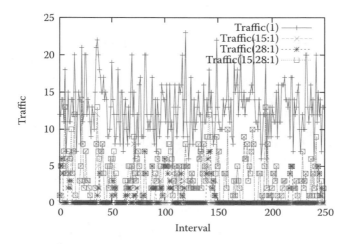

Figure 7.15 Sampled traffic curves from experiment.

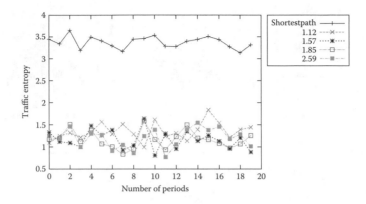

Figure 7.16   Colluded traffic mutual information (destination: 1, $\gamma = 1.85$).

7.6.3.2   Colluded Traffic Mutual Information: Single Pair of Observers

Our next results are the comparisons of colluded traffic mutual information ($I(y, z; x)$), single observer mutual information ($I(y; x)$ and $I(z; x)$), original traffic entropy ($H(x)$), separately observed traffic entropy ($H(y, x)$ and $H(z, x)$), and joint entropy ($H(y, z, x)$).[6] From our analysis in Section 7.6.2, we can conclude the following relations among these values:

1. $H(y, x), H(z, x) \leq H(y, z, x) \leq H(x)$;
2. $I(y, x), I(z, x) \leq I(y, z, x) \leq H(x)$;
3. $I(y, x) \leq H(y, x) \leq H(x)$;
4. $I(z, x) \leq H(z, x) \leq H(x)$.

Now we can verify if the simulation results shown in Figure 7.16 satisfy these relations. This means our modeling of traffic activity not only characterizes the traffic pattern fluctuation along the time, but also stands with the test of collusion problem. The simulation results of our model conform with our conjecture.

The overlapping curves in Figure 7.16(b) indicate node 23 does not observe any traffic of node 1. This could be true because 23 and 1 are on the opposite side of the network.

On the other hand, Figure 7.17 shows similar results, except this time node 16 is the destination.

---

[6] Please note that $H(y, z, x)$, according to our notation, means $H(Y^X, Z^X)$.

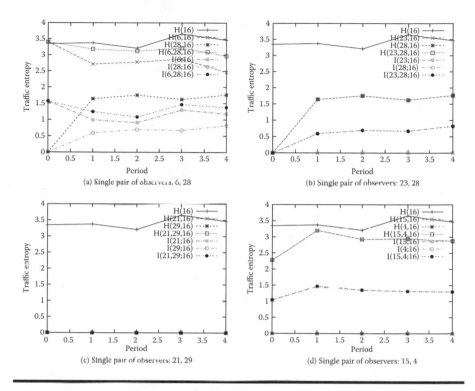

Figure 7.17 Colluded traffic mutual information (destination: 16, $\gamma = 1.85$).

7.6.3.3   Colluded Traffic Mutual Information: Multiple Pairs of Observers

Now that the simulation results have satisfied the necessary relations listed in the previous part, we would like to know how collusion can affect the performance of the PBSP routing algorithm under discussion. To do so, we will study the colluded traffic mutual information of several pairs of observers in one figure. In this way, we can compare the ratio of traffic information revealing of different pairs of observers.

From Figure 7.18 we can observe that the conditions above still hold. Additionally, based on average values of the colluded traffic mutual information curves in both figures, we can guess that the PBSP algorithm still works well when there are two observers colluding to share their knowledge about one destination.

To further confirm this conjecture, we can examine another set of simulation results, as shown in Figure 7.19. The colluded traffic mutual information of all observer pairs in this figure does not exceed half of total traffic information either. In Figure 7.19(b), however, we notice some small error of curves, i.e., the value of $I(15, 6; 16)$ is a little less than that of $I(15; 16)$

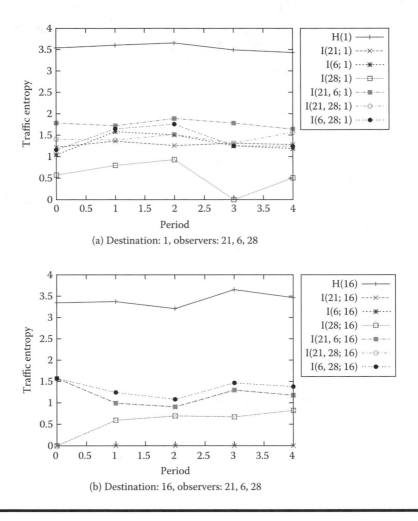

Figure 7.18 Colluded traffic mutual information (multiple pairs of observers, $\gamma = 1.85$).

for period 2. Although this is a small error, it reminds us of an approximation when computing $H(Y^X, Z^X, X)$. Instead of employing three parallel
PacketCounters to get the aggregate traffic information, the simulation program approximates it based on the packet count value dictionary, which
results in a lower $I(Y^X, Z^X; X)$ value.

The same explanation applies for the discrepancy in Figure 7.20(a). In
the meantime, the average value of colluded traffic mutual information of
all observer pairs in Figure 7.20 remains approximately less than half of the
traffic entropy of the target node along the time.

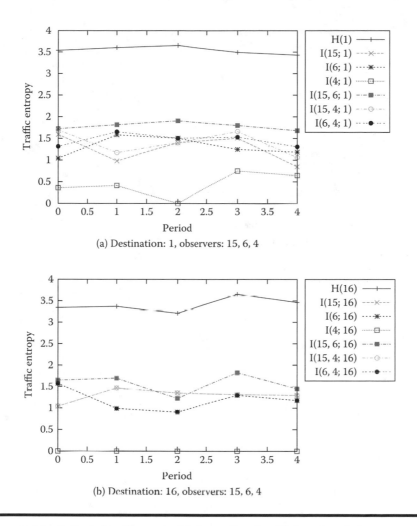

(a) Destination: 1, observers: 15, 6, 4

(b) Destination: 16, observers: 15, 6, 4

Figure 7.19   Colluded traffic mutual information (multiple pairs of observers, $\gamma =$ 1.85).

7.7   Related Work

Currently, multi-hop WMN is gaining more popularity, as deployments of WMN either serve as a substitute of traditional WLAN Internet connection, or aim at providing infrastructural large-scale network access [24].

Existing research [3,7,10,19] on WMN has focused on how to better utilize the wireless channel resource and enhance its performance. For example, some researchers [18] try to derive the optimal node density following capacity analysis, while others strive to devise more efficient protocols [13]. A survey paper by Akyildiz et al. [6] provides a good source for existing

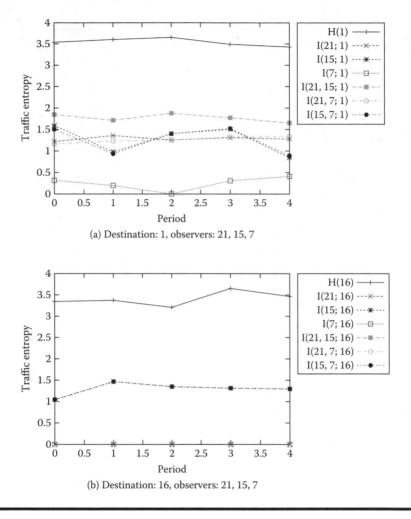

Figure 7.20 Colluded traffic mutual information (multiple pairs of observers, $\gamma =$ 1.85).

and ongoing research about wireless mesh networks. Some of the proposed solutions include equipping mesh routers with multiple radios and distributing the wireless backbone traffic over different wireless channels, routing the traffic through different paths [15,33], or a joint solution of these two [25,26]. Theoretical study shows that these approaches can significantly increase the capacity of WMN [21,22]. These results make a significant step toward enabling WMN as an attractive alternative for broadband Internet access.

Information theory is widely used and proves to be a useful tool. It works in situations where variations are frequent and unpredictable and helps to identify pattern and extent of variation. Serjantov and Danezis [29] define an information theoretic anonymity metric and suggest developing more sophisticated probabilistic anonymity metrics. Existing research [20] in the Internet setting employs information theoretical coding, which is too complex and impractical for WMNs. The book by David Mackay [23] provides a good source for background knowledge in information theory.

Privacy has been a major concern of Internet users [12,31]. In the existing literature of traffic pattern concealment, anonymous overlay routing [9,14,16,17,28,34] and traffic padding [30] have been proposed to preserve user traffic privacy and increase the difficulty for traffic analysis [9,27]. The former approach provides user anonymity in an end-to-end connection through layered encryption and multi-hop overlay routing. The latter one conceals the traffic shape by generating a continuous random data stream at the link level. However neither of them can be applied to WMN directly. First, the number of nodes in a WMN is limited. Second, the traffic forwarding relationship among nodes is strongly dependent on their locations and the network topology. To better utilize the wireless channel resource and enhance the data delivery performance, a short path is usually selected or a load-balanced routing scheme is employed. Such observations show that the anonymity systems, which rely on relaying traffic among nodes (randomly selected out of thousands) to gain anonymity, cannot effectively preserve users' privacy in WMN, or do so at the cost of significant performance degradation. On the other hand, the traffic padding mechanism consumes a considerable amount of network bandwidth, which makes it impractical in resource-constrained WMNs.

The schemes designed in wireless ad hoc networks [11,32] are more focused on location and identity privacy. While these are still issues in WMN, the traffic rates and temporal variations are more meaningful and consequential.

To the best of our knowledge, no existing works have studied collusion problems about traffic privacy in the scenario of wireless mesh networks.

7.8 Conclusion

This chapter identifies the problem of traffic privacy preservation in wireless mesh networks (WMN). To address this problem, we start by introducing a lightweight architecture for WMN, then propose "traffic entropy," an information theoretic metric to quantify how well a solution performs at preserving the traffic pattern confidentiality, all of which pave the way to our penalty-based shortest path routing algorithm. Furthermore, we evaluate our scheme against collusion of two malicious nodes. Simulation results show that our algorithm is able to maximally preserve the traffic privacy, meanwhile managing the network performance degradation within the acceptable region. Our simulation analysis also proves the resilience of our solution against two colluding observers.

For the future work, we will focus on the following problems. First, although our algorithm is evaluated in a single-radio, single-channel WMN setting, it can be easily enhanced to exploit the advantage of multiple radios and multiple channels available in WMNs. Performance evaluation of the enhanced algorithm in such settings will be interesting. It is also beneficial to research the possibility of devising a distributed routing that achieves the same goal, but supports better scalability.

References

[1] Chaska wireless solutions. http://www.chaska.net/.

[2] Mesh Networks Inc. http://www.meshnetworks.com.

[3] MIT Roofnet. http://www.pdos.lcs.mit.edu/roofnet/.

[4] Radiant Networks. http://www.radiantnetworks.com.

[5] Seattle Wireless. http://www.seattlewireless.net.

[6] Ian F. Akyildiz, Xudong Wang, and Weilin Wang. Wireless mesh networks: A survey. *Comput. Netw. ISDN Syst.*, 47(4): 445–487, 2005.

[7] Mansoor Alicherry, Randeep Bhatia, and Li Li. Joint channel assignment and routing for throughput optimization in multi-radio wireless mesh networks. In *Proceedings of ACM MOBICOM*, 2005.

[8] B. Awerbuch, D., Holmer, C. Nita-Rotaru, and H. Rubens. An on-demand secure routing protocol resilient to byzantine failures. In *ACM Workshop on Wireless Security*, 2002.

[9] Adam Back, Ulf Möller, and Anton Stiglic. Traffic analysis attacks and trade-offs in anonymity providing systems. In *Information Hiding Workshop (IH)*, 2001.

[10] John Bicket, Daniel Aguayo, Sanjit Biswas, and Robert Morris. Architecture and evaluation of an unplanned 802.11b mesh network. In *Proceedings of ACM MOBICOM*, pp. 31–42, 2005.

[11] S. Capkun, J.P. Hubaux, and M. Jakobsson. Secure and privacy-preserving communication in hybrid ad hoc networks. Technical report IC/2004/104, EPFL-DI-ICA, 2004.

[12] Roger Clarke. Internet privacy concerns confirm the case for intervention. *Communications of the ACM*, 42(2): 60–67, 1999.

[13] Douglas S.J. De Couto, Daniel Aguayo, John Bicket, and Robert Morris. A high-throughput path metric for multi-hop wireless routing. In *Proceedings of ACM MobiCom*, pp. 134–146, New York, ACM Press, 2003.

[14] Roger Dingledine, Nick Mathewson, and Paul Syverson. Tor: The second-generation onion router. In *USENIX Security Symposium*, 2004.

[15] R. Draves, J. Padhye, and B. Zill. Routing in multi-radio, multi-hop wireless mesh networks. In *Proceedings of ACM MOBICOM*, pages 114–128. ACM Press, 2004.

[16] Michael J. Freedman and Robert Morris. Tarzan: A peer-to-peer anony-mizing network layer. In *Proceedings of ACM CCS*, 2002.

[17] D. Goldschlag, M. Reed, and P. Syverson. Onion routing for anonymous and private internet connections. *Communications of the ACM*, 42(2): 39–41, 1999.

[18] P. Gupta and P. R. Kumar. The capacity of wireless networks. *IEEE Transactions on Information Theory*, 46(2): 388–404, 2000.

[19] R. Karrer, A. Sabharwal, and E. Knightly. Enabling large-scale wireless broadband: The case for taps. In *HotNets*, 2003.

[20] Sachin Katti, Dina Katabi, and Katarzyna Puchala. Slicing the onion: Anonymous routing without PKI. MIT CSAIL Technical report 1000, 2005.

[21] Murali Kodialam and Thyaga Nandagopal. Characterizing the capacity region in multi-radio multi-channel wireless mesh networks. In *Proceedings of ACM MOBICOM*, 2005.

[22] Pradeep Kyasanur and Nitin H. Vaidya. Capacity of multi-channel wireless networks: Impact of number of channels and interfaces. In *Proceedings of ACM MOBICOM*, pp. 43–57, New York, 2005.

[23] David J.C. Mackay. *Information theory, inference, and learning algorithms*. Cambridge, 2003.

[24] Krishna Ramachandran, Milind M. Buddhikot, Scott Miller, Kevin Almeroth, and Elizabeth Belding-Royer. On the design and implementation of infrastructure mesh networks. In *Proceedings of IEEE WiMesh*, 2005.

[25] A. Raniwala and T. Chiueh. Architecture and algorithms for an IEEE 802.11-based multi-channel wireless mesh network. In *Proceedings of IEEE INFOCOM*, 2005.

[26] A. Raniwala, K. Gopalan, and T. Chiueh. Centralized channel assignment and routing algorithms for multi-channel wireless mesh networks. *Mobile Computing and Communications Review*, 8(2): 50–65, 2004.

[27] Jean-François Raymond. Traffic analysis: Protocols, attacks, design issues, and open problems. In *International Workshop on Design Issues in Anonymity and Unobservability*, 2000.

[28] Michael G. Reed, Paul F. Syverson, and David Goldschlag. Anonymous connections and onion routing. *IEEE Journal on Selected Areas in Communications*, 16(4): 482–494, 1998.

[29] A. Serjantov and G. Danezis. Towards an information theoretic metric for anonymity. In *Proceedings of ACM MOBICOM*, 2002.

[30] W. Stallings. *Cryptography and network security*. Prentice Hall, 2003.

[31] Huaiqing Wang, Matthew K.O. Lee, and Chen Wang. Consumer privacy concerns about Internet marketing. *Communications of the ACM*, 41(3): 63–70, 1998.

[32] Xiaoxin Wu and Bharat Bhargava. Ao2p: Ad hoc on-demand position-based private routing protocol. *IEEE Transactions on Mobile Computing*, 4(4): 335–348, 2005.

[33] Yuan Yuan, Hao Yang, Starsky H.Y. Wong, Songwu Lu, and William Arbaugh. Romer: Resilient opportunistic mesh routing for wireless mesh networks. In *Proceedings of IEEE WiMesh*, 2005.

[34] Li Zhuang, Feng Zhou, Ben Y. Zhao, and Antony Rowstron. Cashmere: Resilient anonymous routing. In *Proceedings of USENIX NSDI*, 2005.

Chapter 8

Providing Authentication, Trust, and Privacy in Wireless Mesh Networks

Hassnaa Moustafa

Contents

Security is a big concern in wireless mesh networks (WMNs), where providing a robust secure system is considered one of the most critical challenges promoting the commercial deployment of WMNs and influencing their usage. The security requirements in WMNs will determine what type of link level security protection is needed, at what protocol level intrusion detection and prevention must be performed, and what amount of overhead due to security can be tolerated in the network. This will be a constant battle requiring continuous security enhancements, continuous monitoring, and rapid responses to intrusions. This chapter starts by discussing the security challenges in WMNs, showing the possible types of attacks in these networks, and stating the different security requirements. Then the problem of authentication is presented, showing some authentication mechanisms that are useful in WMNs. The different contributions, employing the emerging standards for authentication and secure links setup with a mobility management support are presented, and the role of authentication, authorization, and accounting (AAA) in such environment is illustrated. The importance of trust provision is shown, where security mechanisms will have to leverage special capabilities to detect untrusted elements and to protect the mesh's integrity. The chapter ends by discussing privacy provision in WMNs considering traffic privacy and confidential transfer.

8.1 Introduction

Wireless mesh networks (WMNs) have emerged as a key technology for next-generation wireless networks, showing rapid progress and inspiring numerous applications. WMNs, however, are not yet ready for wide-scale deployment due to two main reasons: the interference caused by the wireless communication and the non-security guarantees. The fact that all wireless communications are prone to interference causes delay constraints in WMNs. Nevertheless, it is believed that technological solutions would be able to overcome this problem, for example, using multi-radio and multi-channel Terminal Access Points (TAPs) [1]. The lack of security guarantees is another factor slowing down the deployment of WMNs. In fact, security in WMNs is still in its infancy and very little attention has been devoted thus far to this topic by the research community. As these networks continue to grow and as access to the mesh is available for any wireless-enabled device, it should be ensured that only authorized users are granted network access. There is still a strong need for efficient solutions adapted for different security requirements and for different usage scenarios. These solutions have to counter attacks in all protocol layers, guaranteeing collaborative behaviors between mobile nodes. Trust relationships should exist among stakeholders for authentication, authorization, and accounting of end users. Well-performing tools need to be developed for mesh design, maintenance, and management such that future mesh networks should be self-managed rather than unmanaged ones [2]. A number of challenges have to be considered during the design of security mechanisms and solutions, and appropriate security requirements should be defined considering the different existing threats.

8.2 Security Challenges in Wireless Mesh Networks

WMNs have special characteristics distinguishing them from other network technologies and consequently imposing a broad range of design challenges to be solved. This section gives an overview on various security challenges and requirements in WMNs. Security in WMNs is one of the widely discussed topics and one of the major inherent caveats of wireless ad hoc networking. Classical security approaches suffer from the inadequate usage of redundant paths, and hence could not be directly applied in WMNs. The mobility of nodes, the hybrid wireless environment created by the different wireless mesh architectures, the density of connections in these networks, and the unpredictable behavior of nodes are critical factors influencing the security requirements of WMNs and posing new security challenges. One possible approach in providing practically feasible solutions is to deploy, combine, and adopt existing security approaches and protocols in wireless networks in general, and in ad hoc networks in particular. However, specific security mechanisms must be developed allowing intense load sharing while taking into account local capacity limitations and dynamic load changes.

8.2.1 Mobility of Nodes

An attractive point in commercial WMN deployment is the seamless access of mobile clients to services offered by these networks, in a completely transparent manner to clients' mobility. However, clients' mobility itself is a challenge which poses some constraints on WMN security. A part of this challenge lies in the mobile devices themselves. First, mobile devices are susceptible to thefts and thus can be misused by attackers in either an unauthorized access or a communication corruption. Second, the fact that most mobile devices are "thin clients" of limited power, CPU, and storage capacity, leads to difficulty in running some security mechanisms in WMN (as for example, encryption algorithms requiring special resources).

Another part of this challenge arises from the mobility itself, where mobile clients are susceptible to roaming across different administrative domains that may have different security policies. Thus, efficient security mechanisms are needed for handling clients' roaming in a secure manner. Finally, the mobility of mobile clients can facilitate tracing the mobile clients' existence at different places. Privacy protection mechanisms are thus important so that an attacker could not hack client privacy by tracing its mobility.

8.2.2 Hybrid Wireless Environment

WMNs are expected to offer seamless wireless network access for mobile users within a hybrid wireless environment. In such an environment, hybrid wireless communication allows multi-hop access mode combining peer-to-peer communication between mobile nodes as well as mobile nodes' communication with a fixed infrastructure. Peer-to-peer communication can be considered as pure ad hoc networks' communication. In addition, each mobile node may access a fixed infrastructure either directly or via other nodes (mesh routers) in a multi-hop fashion. In spite of the seamless access feature provided by WMNs, there are no mechanisms in place implementing security services when a mobile terminal roams between disparate networks. Consequently, some essential features like secure roaming, authentication, and authorization should be highly considered in that type of environment. The security mechanisms must guarantee that only authorized users can use the network resources and access the services offered by the provider. Furthermore, eavesdropping as well as the modification of the transmitted data during the multi-hop communication, must be prevented. There is a lack of efficient security mechanisms that offer secure links setup and confidential data transfer among mobile clients in hybrid wireless environments. This is in part because the security and mobility management solutions, in wireless networks in general, are often implemented at different protocol layers with limited amount of interaction between these layers.

8.2.3 Capacity and Density of Connections

The capacity of WMNs is an important issue that is worth consideration during the development of security mechanisms in these networks. Many factors can affect the capacity of WMNs such as network architecture, mobile nodes' density, number of channels used for each mobile node, transmission power level, and nodes' mobility [3]. Hence, a clear understanding of the relationship between network capacity and the above factors provides guidelines for protocols' development as well as architecture design in WMNs. Nevertheless, the current security mechanisms and protocols in WMNs do not take this fact into account, although some security issues are related to radio resource's management or can arise due to the nature of the radio medium and the resource constrained devices. Consequently, a number of problems take place due to non-coherence between the security mechanisms and the capacity and density of connections in WMNs. Some examples:

- The possibility of spoofing power control messages among nodes, which can result in an unstable situation within a group of mesh cells, causing loss of services and increasing the load in the neighboring mesh cells.
- The power resources constraint poses an obstacle to running key management protocols in high-density mesh cells, requiring a lot of messages and keys exchanges.
- The difficulty in managing cryptography over all the mesh connections, especially in mesh cells of high connection density.
- The decentralized authentication process, which is a significant requirement in WMNs, becomes more complex in high-density mesh cells, and adequate authenticators' delegation should take place.

8.2.4   Individual Behavior of Nodes

Cooperation among nodes is a primary requirement for WMN functioning. Node cooperation in WMNs is critical for multi-hop transmission, collective data processing, and cooperative security functions. However, providing service to each other consumes resources, which are generally scarce in mobile nodes. Thus, cooperation cannot be taken for granted, especially in opened mesh networks scenarios, because each user would prefer to maximize his own benefit while minimizing his contribution. Mobile nodes in WMNs are supposed to be rational in the sense that they try to maximize their own utilities in a self-interested way. The cooperation issue concerns different layers of the node's protocol stack, with different aims and ways of acting, where a self-interested node can misbehave by:

1. Non-adherence to the protocols specification
2. Optimization of a particular utility function, possibly at the expense of other nodes

Consequently, selfishness and greediness are two misbehaviors that are likely to take place in WMNs. Nodes may behave selfishly by not forwarding packets for others to save power, bandwidth, or just because of security and privacy concerns. Watchdog [4], Confidant [5], and Catch [6] are three approaches developed to detect selfishness and enforce distributed cooperation and are suitable for WMNs. Watchdog is based on monitoring neighbors to identify a misbehaving node that does not cooperate during data transmission. However, Confidant and Catch incorporate an additional punishment mechanism making misbehavior unattractive through isolating misbehaving nodes. On the other hand, a node may behave greedily in consuming channel and bandwidth for its own benefits at the expense of the other users. A mechanism that modifies 802.11 for facilitating the detection of greedy nodes is proposed in [7]; also the DOMINO mechanism [8] solves the greedy sender problem in 802.11 WLANs with a possible extension to multi-hop wireless networks and WMNs.

To provide secure cooperation mechanisms that are suitable for WMNs, the following factors are important to be considered:

■ The vulnerability of wireless links, compared to wired ones, in terms of eavesdropping and jamming
■ The weak connection of each node with the network authority
■ The fact that devices are becoming more and more programmable

Because the above mechanisms require maintaining a great deal of state information at each node while monitoring its neighbors, adaptive schemes are needed for right functioning in WMNs. Two other important issues to be considered are the distributed detection of selfishness and greediness misbehaviors and providing incentives to mobile nodes to stimulate cooperation.

8.3   Threats and Security Requirements in Wireless Mesh Networks (WMNs)

Because WMNs are based on the concept of wireless distribution system (WDS), they are vulnerable to a variety of threats. Security measures should be taken to avoid these threats and allow reliable communication. Also, the notion of WDS requires end-to-end security assurance for each end user.

8.3.1   Threats in Wireless Mesh Network Environment

Threats in WMNs are mainly due to the nature of the radio links, the ubiquity of wireless communications, and the multi-hop communication. The main target of security solutions in WMNs is to encounter the following types of security threats [9]:

- Eavesdropping and data modification: The nature of radio environment can cause eavesdropping and modification of the data sent by mobile nodes. The presence of wireless links and intermediate mobile nodes in WMNs requires the existence of encryption and integrity protection mechanisms to prevent eavesdropping and data modification, allowing confidentiality and integrity of the transmitted data.
- Unauthorized access: The possibility of setting up wireless connections to any mesh network can result in unauthorized nodes access to WMNs, posing a critical threat to these networks. In closed WMNs, which have a centralized administration, successful authentication should be a requirement for joining a mesh network. However, in open mesh networks with no central control, alternative solutions must be in place to allow authentication between mobile nodes in a distributed manner.
- Denial of service (DoS): A traditional DoS may take place during multi-hop transmission by an intermediate mobile node selectively dropping traffic frames. The DoS characteristic of WMNs is generally caused by routing misbehavior of a mobile node. The black hole attack [10] is an example of the DoS, where the malicious mobile node can tamper with the routing messages in a network, or spoof the MAC address of a mobile node into claiming a fake shortest path so as to get all the packets routed to itself, without any intention to route the packets to destination. Indeed, any mobile node that is correctly authenticated when joining the mesh network may suddenly start misbehaving causing DoS. Thus, it is very difficult to discover and prevent the DoS in WMNs.

Countermeasures need to be devised for WMNs using the security options according to the size of risks. An intrusion detection system may be used in such case to address some of the threats. A useful approach to counter security threats is to study the threats with respect to their likelihood of occurrence, their possible impact on individual users and on the whole system, and the expected risk from these threats [11]. The likelihood evaluates the possibility of conducting attacks related with the threat, taking into account the motivation for an attacker and the technical difficulties that he needs to resolve. The impact can evaluate the consequences of an attack related to the threat. This depends on whether the attack is directed to an individual user or to the whole system. It also depends on the possibility of service loss caused by the attack. Consequently, the risk can be defined as a function of the likelihood and the impact values.

8.3.2   Different Types of Attacks to WMNs

Attacks can exist at different layers in WMNs causing network failure. At the physical layer, an attacker may jam the transmission of wireless antennas or simply destroy the hardware of a certain node. At the MAC layer, an attacker may abuse the fairness of medium access by sending MAC control and data packets or impersonating a legal node. Attacks may occur in routing protocols such as advertising wrong routing updates. At the application layer, an attacker could inject false fake information, thus undermining the integrity of the application. Attackers may also sneak into the network by misusing the cryptographic primitives. Consequently, the exchange of cryptographic information should take place through special schemes, for example, the rational exchange scheme [12], ensuring that a misbehaving party cannot gain anything from misbehavior. Furthermore, the absence of a central authority, a trusted third party, or a server to manage security keys necessitates distributed key management.

Two classes of attacks are likely to occur in WMNs:

1. External attacks, in which attackers not belonging to the network jam the communication or inject erroneous information, mostly take place at open mesh networks that are not controlled by a central authority.
2. Internal attacks, in which attackers are internal, compromised nodes that are difficult to be detected.

Both types of attacks may be either passive (intending to steal information and to eavesdrop on the communication within the network) or active (modifying and injecting packets to the network).

Generally, there are two approaches to dealing with security attacks: prevention and detection. Prevention aims at thwarting security breaches from occurring in the first place, whereas detection and reaction are necessary in case of prevention failure. On the other hand, detection aims at discovering malicious nodes that carry out attacks to the network. Special mechanisms can be in place to detect attackers, for example, intrusion detection mechanisms. However, it is difficult to detect internal attackers even in the presence of detection mechanisms. The ideal method is integrating the two approaches; however, the cost of a security system in this case may be too expensive for mobile nodes in this environment. We notice that most of the security mechanisms and protocols follow the prevention approach.

8.3.3   Requirements for Security Architectures and Mechanisms in WMNs

The existence of robust authentication mechanisms is an important security requirement in WMNs to prevent unauthorized user access. Mutual authentication of mesh nodes is a critical issue that should be satisfied. It is important to distinguish between the nodes' authentication at the initialization phase and the nodes' authentication during the session while sending and receiving packets. For authenticating mobile nodes at the initialization phase, public key cryptography can be useful in closed mesh network scenarios. Mutual authentication can take place in this case through using certified public/private key pairs assigned to the mobile nodes by the operator that is managing them. However, the use of public key cryptography to authenticate mobile nodes during the session is a heavy process causing important delay constraints. Instead, the nodes can rely on symmetric key cryptography, using session keys which they establish during the initialization phase or long-term shared keys that can be originally loaded in the devices. In open mesh network scenarios, using per-session per-connection keys seems a feasible solution while considering the knowledge of the key as a stepping stone for authentication.

Once the nodes are authenticated, it is necessary to ensure the integrity of the exchanged messages and prevent messages modification. A possible way to do so is through using symmetric keys that are derived during the session establishment. Consequently, employing encryption mechanisms in WMNs can assure the integrity and confidentiality of transmissions, where reliable encryption solutions are needed while minimizing complexity and overhead. These solutions should allow hop-by-hop encryption and should avoid the possibility of eavesdropping on or tampering with the data by intermediate mobile nodes.

Hybrid security architectures are mostly suitable in WMNs, comprising two phases. The first phase concerns mutual authentication and encryption [13]. In mutual authentication phase, a public key infrastructure (PKI) is generally applied. However, this step requires the deployment of a central node functioning as a trust center and running a database against which key verification can take place. The authenticity of central nodes can also be verified by public/private keying. Based on this authentication, the second step is the exchange of symmetric keys per connection to encrypt all data transfer. This second step can be optional, because the mutual authentication enables a security level that can be sufficient for many systems. On the other hand, the encryption can pose relatively high requirements on the node's resources.

Considering the characteristics of WMNs, security mechanisms and protocols should satisfy most of the following requirements:

- Scalability: The performance of protocols and mechanisms, in terms of computational and communication cost, should not degrade with the network size. To achieve this, every node should not be required to have the global knowledge of the network, for example, sharing a pairwise key with every other node in the network.

- Efficiency: Mechanisms and protocols must be resource efficient. Although security should have a cost, the protocols should incur as little overhead as possible. Security mechanisms and protocols should not require large bandwidth overhead and operations that require high computations such as those based on public key techniques should be minimized.
- Routing protocol independence: One important point in designing security mechanisms and protocols is the independence of the routing protocol. Although it is possible to design mechanisms that work with specific routing protocols, this would require the design of a new customized protocol for every routing protocol, which is clearly undesirable.
- Transparency: It is undesirable that the deployment of security mechanisms requires modification or redesign of other protocols in the protocol stack. Security mechanisms and protocols should work transparently with other protocols and without affecting the functionality of other protocols such as routing protocols or application layer protocols.
- Fast authentication: There should be no high delay for authentication. Otherwise, the authentication latency would be unacceptably high in such a multi-hop communication environment, especially when authentication is needed between different administrative domains.

8.4 Authentication

Authentication of mobile nodes in WMNs can assure authorized clients participation. The simplest solution is to employ an authentication key shared by all nodes in the network. Although this mechanism is simple, it has the following disadvantages:

- An attacker only needs to compromise one node to break the security of the system and paralyze the entire network.
- If the global key is divulged, it is not possible to identify the compromised node.
- It is expensive to recover from a compromise as it usually involves a group key update process.
- Mobile nodes do not usually belong to the same community, which leads to a difficulty in installing/pre-configuring the shared keys.

Another well-known approach that can provide strong source authentication is attaching digital signature to packets. However, signing every packet can be prohibitively expensive because the computational capacity and battery power of mobile nodes are quite constrained. Therefore, the challenge is to design authentication mechanisms for the more vulnerable yet more resource-constrained environment of WMNs.

Authentication and authorization are important counter-attack measures in WMN deployment, allowing only authorized users to get connections via the mesh network and preventing adversaries to sneak into the network disrupting the normal operation or service provision. Authentication, authorization, and accounting (AAA) are provided in most of the WLANs applications and commercial services through a centralized server such as RADIUS or DIAMETER. However, the centralized scheme is not appropriate in the case of multi-hop WMNs and secure key management is much more difficult. Thus, distributed authentication and authorization schemes with secure key management are required in such an environment. Because WMNs can be managed by more than one operator/provider, authentication should be performed during mobile nodes' roaming across different wireless mesh routers and across different administrative domains. This allows users' mobility with seamless and secure access to the offered services in the mesh network. A possible approach for distributed authentication is the continuous discovery and mutual authentication between neighbors, whether they are mobile clients or fixed/mobile mesh nodes. Nevertheless, if mobile nodes move back to the range of previous authenticated neighbors or mesh nodes, it is necessary to perform re-authentication to prevent an adversary from taking advantage of the gap between the last association and the current association with the old neighbor to launch an impersonation attack. The IEEE 802.11i standard proposed the storage of session keys at authenticators to mitigate the overhead of re-authentication; however, it is vulnerable to impersonation attacks, in which a malicious access point can use previously stored keys to dupe user nodes. Other vendors' specific solutions are proposed by Cisco, Aruba, and Trapeze networks, integrating a switched architecture in the 802.11i authentication aiming to centralize the storage of the authentication keys, therefore to accelerate the re-authentication. These solutions work well in WLAN applications, resolving expensive overhead of re-authentication. However, there are no associated security mechanisms to prevent attacks on stored keys. As well, these solutions are not scalable to WMNs, where decentralized key management is necessary.

The following sub-sections describe some authentication mechanisms and protocols that are useful for application in WMNs. Four approaches are mainly considered:

1.   Adapting the 802.11i authentication to the mesh network environment to authenticate nodes and to allow secure links setup at layer 2.
2.   Authenticating data packets transmitted or received aiming to prevent non-authorized nodes from injecting erroneous packets in the network.

3. Using new AAA infrastructures adapted to the dynamic and decentralized WMN environment.
4. Extending some existing authentication protocols to the WMN environment.

8.4.1 802.11i Authentication Model

In most commercial deployments of WLANs, IEEE 802.11i [14] is the most common approach for assuring authentication and secure links setup at layer 2. However, the IEEE 802.11i authentication does not fully address the problem of WLAN vulnerability. In IEEE 802.11i authentication, as depicted in Figure 8.1, the mobile station and the authentication server (AS) apply the 802.1X [15] authentication model carrying out some negotiation to agree on Pairwise Master Key (PMK) by using some upper layer authentication schemes or using a pre-shared secret. This key is generated by both the mobile client and the AS, assuring the mutual authentication between them. The access point (AP) then receives a PMK copy from the AS, authenticating the mobile client and authorizing its communication. Afterward, a four-way handshake starts between the AP and the mobile station to generate encryption keys from the generated PMK. Encryption keys can assure confidential transfer between the mobile station and the AP. If the mobile station roams to a new AP, this mobile station will perform another full 802.1X authentication with the AS to derive a new PMK. For performance reasons, the PMK of the mobile station can be cached by the mobile station and the AP to be used for later re-association without another full

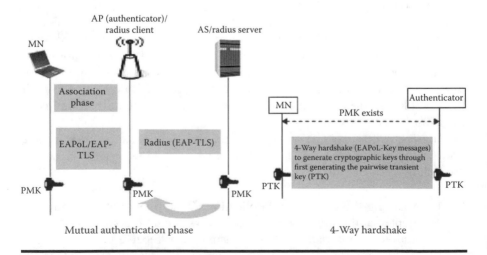

Figure 8.1   IEEE 802.11i authentication model.

authentication. The features of 802.11i exhibit a potential vulnerability because a compromised AP can still authenticate itself to a mobile station and gain control over the connection. Furthermore, IEEE 802.11i authentication does not provide a solution for multi-hop communication. Consequently, new mechanisms are needed for authentication and secure layer 2 links setup in WMNs.

Wireless Dual Authentication Protocol (WDAP) [16] is proposed for 802.11 WLAN and can be extended to WMNs. WDAP provides authentication for both mobile stations and access points and overcomes the shortcomings of other proposed mutual authentication protocols. The name "dual" returns to the fact that the AS authenticates both the mobile station and access points. As in the four-way handshake in IEEE 802.11i, this protocol also generates a session key for confidentiality of communications between the mobile station and the AP after a successful authentication. WDAP provides authentication during the initial connection state and while roaming including three sub-protocols: an authentication protocol, a de-authentication protocol, and a roaming authentication protocol. Figure 8.2 illustrates the WDAP authentication process. In the authentication protocol, the AP that receives the mobile station authentication request, creates also an authentication request for itself concatenating this request to the received request from the mobile station and sending the concatenated request to the AS. The dual part of WDAP lies in this phase, because both the mobile station and the AP do not trust each other until the AS authenticates both of them. In case of successful authentication, a session

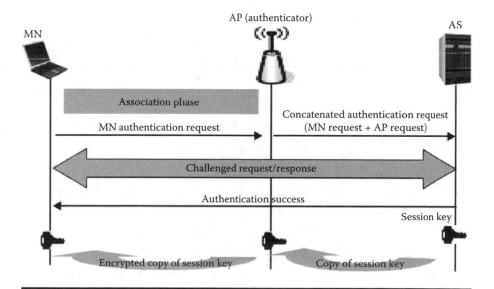

Figure 8.2 Authentication in WDAP.

key authenticating both the AP and the mobile station is generated by the AS and sent to the AP. The AP then sends this key to the mobile station encrypting it with the mobile station secret key. This key is thus shared between the AP and the mobile station for their secure communication and secure de-authentication when the session is finished. When a mobile station finishes a session with an AP, secure de-authentication takes place to prevent the connection from being exploited by an adversary. In case of a mobile station roaming to a new AP, it sends out a roaming authentication request message to the new AP, where the new AP concatenates its authentication request to this message and then sends the concatenated request to the AS. After the AS verification of the previous authentication of the mobile station and the successful authentication of the new AP, it sends a session key revoke message to the old AP and a new generated session key to the new AP to be shared with the mobile station. Applying WDAP in WMN environments allows the mutual authentication between mobile nodes and WMRs. Also, WDAP can be used to assure the authentication between the WMRs themselves through authentication requests concatenation. In case of multi-hop communication in WMNs, each pair of nodes can mutually authenticate through the session key generated by the AS. However, a solution is needed in case of open mesh networks scenarios, where the AS is not always in place. Another problem comes from the roaming authentication approach in WDAP which is not quite suitable for WMN environments, as it restricts the roaming to only new APs and does not consider the case of "back roaming" where the mobile node might need to re-connect with another mobile node or an AP with whom it was authenticated before. Consequently, the WDAP session key revoke mechanism brings some disadvantages to WMNs and another mechanism is required.

An approach that adapts IEEE 802.11i to the multi-hop communication is presented in [17]. An extended forwarding capability in 802.11i is proposed without compromising its secure features, to set up authenticated links on layer 2 and achieve secure wireless access as well as confidential data transfer in ad hoc multi-hop environments. The general objective of this approach is supporting mobile clients' secure and seamless access to the Internet, near public WLAN hotspots, even when they move beyond WLAN communication ranges. To accomplish the AAA process for a mobile client existing in the WLAN communication range, classical 802.11i authentication and messages' exchange takes place. On the other hand, as illustrated in Figure 8.3, for accomplishing the AAA process for mobile clients that do not exist in the WLAN communication range and are consequently belonging to ad hoc clusters, 802.11i is extended to support forwarding capabilities. In this case, the notion of friend nodes is introduced allowing each mobile client to initiate the authentication process through a selected node in its proximity. The friend node plays the role of an auxiliary authenticator and forwards the authentication request of the

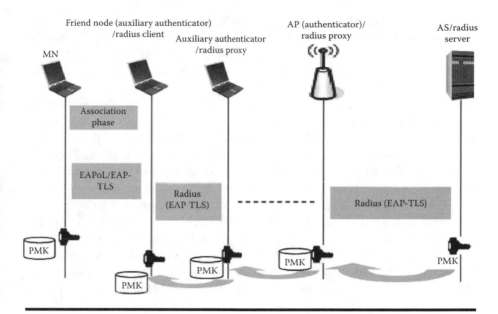

Figure 8.3 Adapted 802.11i with EAP-TLS for multi-hop communication.

mobile node to the actual authenticator (which is the AP in this case). If the friend node does not fall in the communication range of the AP, it invokes other friend nodes in a recursive manner until reaching the AP. The concept of proxy RADIUS [18] is used for forwarding compatibility and secure multi-hop messages' exchange, where proxy chaining [19] takes place if the friend node is not directly connected to an AP. To obtain increased security on each authenticated link between each communicating parties, 802.11i encryption phase takes place through employing the four-way handshake between each mobile node and its authenticator (AP or friend node). This approach is useful in open mesh network scenarios to allow authentication by delegation among mesh nodes. In addition, this approach allows authentication keys storage among intermediate nodes, which optimizes the re-authentication process in case of mobile nodes' roaming. However, an adaptation is needed in terms of allowing multiple connections to authenticators whether APs or auxiliary authenticators (friend nodes) in case of a dense mesh topology. Also, a solution is needed to support fast and secure roaming across multiple WMRs. A possible solution is through sharing session keys of authenticated clients among WMRs.

8.4.2 Data Packets Authentication

Authenticating transmitted data packets is another approach preventing unauthorized nodes' connection to the WMNs. A Lightweight Hop-by-hop

Access Protocol (LHAP) [20, 21] is proposed for authenticating mobile clients in wireless dynamic environments, preventing resource consumption attacks through employing packet authentication. LHAP implements lightweight hop-by-hop authentication, where intermediate nodes authenticate all the packets they receive before forwarding them. This protocol allows a mobile node to first perform some inexpensive authentication operations to bootstrap a trust relationship with its neighbors, then to apply a lightweight protocol for subsequent traffic authentication. LHAP is mainly proposed for ad hoc networks, where it resides between the data link layer and the network layer and can be seamlessly integrated with secure routing protocols to provide a more secure ad hoc network.

LHAP employs a packet authentication technique based on the use of one-way hash chains [22]. Also, LHAP uses Tesla [23] to reduce the number of public key operations for bootstrapping and maintaining trust between nodes. For every traffic packet received from the network layer, LHAP adds its own header, which includes its node ID, a packet type field indicating a traffic packet, and an authentication tag. Afterward, LHAP passes the packet to the data link layer and generates its own control packets for establishing and maintaining trust relationships with neighbor nodes. For a received traffic packet, LHAP verifies its authenticity based on the authentication tag in the packet header. If the packet is valid, LHAP removes the LHAP header and passes the packet to the network layer; otherwise, it discards the packet. LHAP control packets are not passed to the network layer with the goal to allow LHAP execution without affecting the operation of other protocols' layers.

This protocol is quite adaptable to WMN environments, especially open mesh scenarios when the AS is not in place, preventing unauthorized clients' participation in the communication and allowing hop-by-hop authentication. For secure roaming, LHAP can be useful in distributing session keys among mobile clients employing a special type of packet designated for this issue. However, the focus of this protocol on resource consumption attacks' prevention restricts its application to a number of scenarios. Also, the fact that LHAP does not prevent insider attackers from carrying out malicious actions necessitates complementary solutions with such protocol.

8.4.3 AAA Architectures for WMNs

WMN deployment requires appropriate architectures for the different types of scenarios. An important step toward the wide commercial deployments of WMNs is the trust relationship between stakeholders of different access networks, each having its own security mechanisms. To provide seamless service across heterogeneous access networks, there must be a trust relationship among the stakeholders for authentication, authorization, and accounting, and billing of end users.

A lightweight AAA infrastructure is proposed in [24] providing continuous, on-demand, end-to-end security in heterogeneous networks including WMN scenarios. This infrastructure presents an AAA model for supporting secure global mobility in access networks that are managed by different administrators. The notion of a security manager is used through employing an AAA broker. The broker acts as a settlement agent, providing security and a central point of contact for many service providers (stakeholders). This architecture dynamically provides AAA through forming a virtual layer on top of the underlying mesh of network domains, thus supporting user as well as service mobility across multiple access networks. Through using the DIAMETER protocol [25] in this architecture, the number of security association required by each mobile node is reduced to only one. Each mobile node is just required to have a security association with its home AAA server. In addition, by using the roaming capabilities of the DIAMETER protocol, the home DIAMETER server (AAAH) can communicate with foreign DIAMETER servers (AAAFs) in other administrative domains. This architecture is illustrated in Figure 8.4. Through the required security association between the AAAH and the mobile node, keys can be created for each security association. The keys destined for the foreign and home agent are propagated to their nodes via the Diameter protocol, while the key destined for the mobile node is sent via the MIP protocol [26] resulting in an

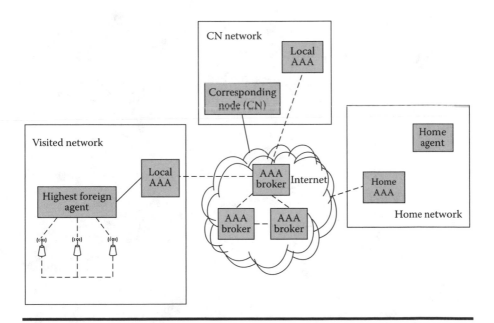

Figure 8.4 Lightweight AAA infrastructure for mobility support across multiple domains.

integrated MIP/DIAMETER architecture. This AAA infrastructure is useful in commercial WMNs deployment allowing dynamic AAA, providing some useful improvements compared to the basic mobility protocol: authentication for signaling messages, accounting of network usage, minimal use of cryptographic keys, and the non-use of digital signatures.

The concept of advanced wireless network architecture is introduced in [27] for efficient communications in complex environments, where diffraction, attenuation, multi-path, scattering, and fading phenomena are frequent. A hybrid network architecture using WLAN is proposed that can be used for high bandwidth applications such as voice and video snapshots. This architecture is depicted in Figure 8.5. The WLAN APs are connected using a mesh topology while the mobile nodes are to be connected to one of the APs using a star topology. The mesh connections between APs allow redundant routes that are desirable in dynamic wireless environments. It is proposed to use the 802.11f [28] Inter Access Point Protocol (IAPP) to handle mobile nodes hand-offs from one AP to another without losing the IP connectivity. Thus, APs need to be connected to a centralized server such as RADIUS server. The inter-network handoffs is proposed to be handled using MIP. Applying this architecture in WMNs has two advantages: (1) allowing better performance of the AAA process, and (2) providing fast secure roaming. The fact that APs are connected through a mesh topology

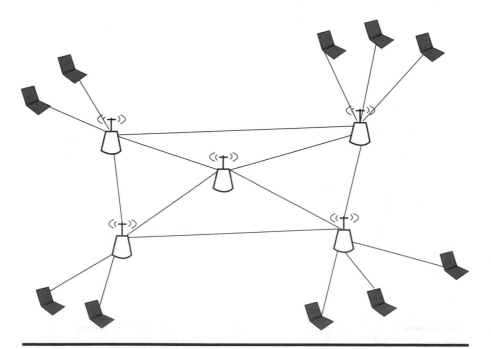

Figure 8.5   WLAN mesh topology.

facilitates the exchange of authentication messages between the APs during the authentication or re-authentication of each mobile node. In addition, in case of roaming of a previously authenticated mobile node to a new AP, the authentication process is optimized thanks to the possible communication between APs. However, the main limitation of this architecture lies in the non-support of multi-hop communication between mobile clients. One way to overcome this limitation is by allowing extended mesh topology among the mobile clients. Furthermore, employing the IAPP limits the application of this architecture to a specific type of mobile devices. Consequently, an alternative solution is needed for general applicability in WMNs; for example, broadcasting between APs in ad hoc mode can be a simple means of communication between APs.

8.4.4 Extensible Authentication Protocol Variants

The mesh network model with no structure and no trust between the nodes makes the security problem more complex, especially that attackers do not need physical access and they can access layer 2 informations. Also, the attacker's job is easier in terms of finding multiple points of attachments to the network. IEEE 802.1X has been applied to resolve some of the security problems introduced in the 802.11 standard, where the mobile station and the AS authenticate each other through applying an upper layer authentication protocol like EAP-TLS (Extensible Authentication Protocol encapsulating Transport Layer Security) protocol [29] in most of the cases. Although EAP-TLS offers mutual authentication, it introduces high latency in WMNs because each terminal behaves as an authenticator for its neighbor to reach the AS, which can result in longer paths to the AS. Furthermore, in case of high mobility of terminals frequent re-authentications due to frequent hand-offs can make the network unusable with real-time traffic. Consequently, variants of EAP are proposed as individual research contributions to adapt the 802.1X authentication model to the multi-hop communication as well as the WMN environment. This section discusses some recent related contributions.

8.4.4.1 EAP with Token-Based Re-Authentication

The dynamic environment together with the multiple possible connectivities in WMNs raise the need for secure fast hand-off protocols. Because each node requiring access to the mesh network initially performs a full and costly authentication, then re-using the information of this initial authentication can speed up the following re-authentications and enhance protocol performance. In this context, a fast secure hand-off protocol is presented in [30], which allows mutual authentication and provides access control protection through limiting the possibility of insider attackers during the re-authentication process. To achieve this, old authentication keys are removed from one host to the other. Thus, any host on the network should not receive keys it does not need, but should rather ask for keys from its neighbors or from the AS when they are needed.

The present solution proposes a token-based re-authentication scheme based on a two-way handshake between the host that performs the hand-off and the AS. It is chosen to involve the AS in every hand-off to have a centralized entity for monitoring the network. An authentication token, in the form of keying material, is provided by the authenticator of the network (whether an AP or a host in the mesh network) to the AS to obtain the PMK key. Initially, the mobile client performs a full EAP-TLS authentication, generating a PMK key that is then shared between the mobile client and its authenticator. Whenever the mobile client performs a hand-off to another authenticator, the new authenticator should receive the PMK key to avoid a full re-authentication. The new authenticator must issue a request to the AS to receive the PMK, adding to the request a token in the form of cryptographic material to prove that it is in contact with the mobile client who owns the requested PMK. Actually, this token is generated by the mobile client while performing the hand-off and is transmitted to the new authenticator. If the AS verifies the token, it then issues the PMK to the new authenticator.

The fast re-authentication presented in this approach permits centralized and hence secure management of the network. However, the need to involve the AS with each re-authentication may cause some constraints in WMNs in which mobile nodes have random and mostly high dynamic behavior. A distributed-based token verification will be more suitable to WMNs, especially for open and multi-hop communication scenarios. Furthermore, the presented solution does not explain the authentication/ re-authentication in case of multi-hop communication, which is a liable scenario in WMNs. Delegation or distribution of the authenticator's role among mobile clients is a useful solution in such a context.

8.4.4.2   EAP-TLS over PANA

A security architecture suitable for multi-hop mesh network is presented in [31], employing EAP-TLS over PANA (Protocol for carrying Authentication and Network Access) [32]. This work proposes an authentication solution for wireless mesh networks growing in an ad hoc manner and using ad hoc network capabilities. An authentication architecture is developed, and data confidentiality is assured. IEEE 802.1X is adapted so that mobile nodes can be authenticated by mesh access routers that can be APs as well as mobile hosts. The authentication between mobile nodes and mesh access routers depending on MAC addresses, according to the 802.1X authentication model, requires mobile clients to be directly attached to mesh routers. Because PANA enables clients to authenticate to the access network using

IP protocol, it is used in this work to overcome the problem of association between mobile clients and mesh access routers that can be attached through more than one intermediate node. Because PANA is an EAP lower layer, any EAP method is suitable for clients' authentication.

When a new mobile node joins the network, it first gets an IP address (pre-PANA address) from a local DHCP server. Then, PANA protocol is initiated so that the mobile node discovers the PANA Access Router (PAA) to authenticate. After successful authentication, the mobile client initiates the Internet Key Exchange (IKE) protocol with the mesh router for establishing a security association. Finally, IPSec tunnel ensures data protection over the radio link and a data access control by the mesh router. During the authentication and authorization phases, PANA uses EAP message exchange between the client and the PAA, where the PAA relays EAP messages to the AS using EAP over RADIUS. EAP-TLS message is used in this approach; however, any other application suitable EAP method can be used.

Because this solution proposes an architecture which is independent of the wireless media, it is appropriate for heterogeneous WMNs' future appli cations and in WMNs that are managed by different operators/administrative domains employing similar or different technologies. However, employing PANA necessitates the existence of IP addresses among mesh nodes, which is still an unsolved problem in the WMN environment.

8.4.4.3   EAP-TLS Using Proxy Chaining

The contributions of [17] and [33] propose adaptive EAP solutions for authentication and access control in the multi-hop wireless environment. In [17], an adapted EAP-TLS approach is used to allow authentication of mobile nodes that do not exist in any AP communication range. A delegation process is used among mobile nodes, through selecting auxiliary authenticators in a recursive manner until reaching the AS. To allow extended forwarding and exchange of EAP-TLS authentication messages, proxy RADIUS is involved using proxy chaining among the intermediate nodes between the mobile client requesting the authentication and the AS. This approach permits the storage of mobile clients' authentication keys among auxiliary authenticators, which speeds up the re-authentication process and enhances the performance of this adaptive EAP-TLS mechanism. This solution is applicable in the WMN environment, especially in scenarios of multi-hop communication. However, a sort of communication is required between auxiliary authenticators to exchange the authentication information concerning the roaming clients. To support secure roaming across different wireless mesh routers (WMRs), communication is required between old and new WMRs during mobile clients' roaming. This can take place through installing central elements/switches linking WMRs and allowing information centralization and distribution between them.

Another adaptive EAP-TLS solution is presented in [33], which is mainly proposed for vehicular networks environment; however, it can be useful in WMNs. This solution employs a Kerberos authentication server as a central server for all mobile nodes. At a first step, each mobile node should authenticate to the Kerberos server prior to connection to the network. As a result of this initial authentication, each mobile node obtains a public key certificate for later use in the network. During communication between the nodes, each two communicating parties can mutually authenticate using EAP-TLS in an ad hoc mode following a client/server model without involving the AS, but rather the previously obtained public key certificates are used. Employing the Kerberos authentication model in WMNs is useful in managing authorization to different services, especially in case of several communicating mesh clusters managed by more than one operator. WMRs can mutually authenticate through the distributed authentication approach proposed; also, this approach is useful for mobile clients authentication during multi-hop communication that can take place in open WMN scenarios. To manage roaming of mobile clients between different WMRs, communication between WMRs is required. Because mutual authentication is possible between WMRs, they can communicate in ad hoc mode to share the authentication information of roaming clients in a secure manner.

8.4.5 AAA in Multi-Operator WMNs

A major objective in WMNs future deployment is services commercialization, which will observe a cooperation between different operators and service providers belonging to different administrative domains. However, some challenges need to be resolved to allow ubiquitous services provision to mobile clients in such a heterogeneous environment. An important challenge concerns the AAA process. Appropriate AAA operation is needed to permit wide and scalable WMNs commercial deployment. This necessitates a trust relationship between operators and providers allowing the continuous authentication of mobile clients during their roaming across different authentication domains. Roaming of clients between WMRs managed by different operators requires authentication of clients each time they connect to a new operator in a rapid manner with no impact on the continuity or the quality of the provided services, especially for real-time applications that are so sensitive to hand-offs delay. Thus trust should exist between the operator of the home network to which the clients belongs and the new operator which is visited by the mobile client. Trust establishment between operators/service providers can take place by signing roaming agreements or by using long-term keys shared between the different operators/service providers.

The charging and accounting of mobile clients across multiple administrative domains should be achieved in a transparent means to services provision. Special accounting mechanisms and tailored billing systems should be in place, with appropriate business models considering the benefits of both mobile clients and service providers. In this context, inter-domain accounting is important in assuring service availability and continuity. The economic interests require the application of usage-sensitive billing systems based on the gathered accounting information for each client. It is recommended that these systems allow online payment or pre-paid tokens. However, processing delay constraints should be considered as well as the need for authentication and integrity.

Considering WMNs operating in an unlicensed spectrum, another important challenge in multi-operators coexistence concerns the spectrum sharing. Because the same WMN can be managed by different operators or WMNs of different operators can interoperate, the utilization of the same unlicensed frequency band by different operators is possible. In such case, mobile clients attachment to WMRs is based on the received signal strength level. Consequently, each operator can authorize its WMRs to transmit using the maximum authorized level to assure that it is heard by the maximum of its own mobile clients, which results in a bad WMN performance increasing the interference. Policy agreements should take place between operators handling the spectrum sharing without bad performance effects. Mobile clients should freely roam across WMRs of different operators attaching to the one offering the best signal quality irrespective of the operator to which the WMRs belong. This roaming policy is expected to be beneficial for both operators and clients. Operators can decrease the transmission power of their devices while serving an increased set of clients. On the other hand, mobile clients can easily discover the closest WMRs and benefit from different services offered by multiple operators.

8.5 Trust

In commercially deployed WMNs, users do not belong to a common group and they do not necessarily trust each other or the different operators. At the same time, each operator does not trust the different users. Because WMN deployment is essentially driven by business considerations, trust is fundamental in such networks, and any security mechanism requires some level of trust in its underlying components.

Building and maintaining trust is not an easy task in WMNs. Trust can be defined as the belief of a network element that another network element, with which it communicates, is functioning in a way that does not disrupt the network operation/services continuity and according to certain predefined rules. However, a trust relation is not symmetric; i.e., if X and Y are two communicating network elements and X trusts Y, this does not imply that Y trusts X, which complicates the problem of trust building. In addition, trust is difficult to quantify or to measure. Consequently, rules enforcement by organizations or governmental authorities is sometimes necessary to facilitate trust building between the different communicating entities. An example of rules enforcement is the governmental regulation of the radio spectrum utilization by network operators. Another example is the control of the mobile devices usage to the radio spectrum by network operators. Besides rules enforcement, there is a need for technical mechanisms deployment to encourage users to some desired behavior during their participation in the network. These mechanisms can also detect/prevent attacks caused by nodes misbehavior, and are typically based on security and cryptographic techniques.

There is a traditional focus on securing routing protocols via ensuring the authenticity of routing messages, aiming to provide transmission among trusted elements. However, this approach is insufficient as the key characteristics of WMNs make it possible for attackers, including malicious users, to add routers, establish links, and advertise routes. In addition, an attacker could steal the credentials of a legitimate user or a legitimate user could himself turn malicious, and thereby inject authenticated-but-incorrect routing information into the network. Thus, beyond ensuring the security of routing protocols, two important issues worth consideration for trust assurance in WMNs environment are:

1. Creating a trust relationship between each pair of communicating nodes as well as between nodes on the redundant routing paths between any communicating parties: Reputation-based mechanisms can help in providing a sort of trust among different network elements in a distributed manner.

2. Securing the packet forwarding and dealing directly with the packet forwarding misbehavior: A way is needed to securely detect and localize the source of the packet forwarding misbehavior. Consequently, the problem of forwarding misbehavior can be solved by controlling the trouble spot, invalidating the compromised credentials, or taking offline action through a human interface.

8.5.1 Using Reputation for Building Trust

Because future business of WMNs is expected to allow interoperability among different operators/service providers, a possible example is the integration of different mesh clusters that belong to different operators/ service providers including wireless Internet service providers (WISPs). However, one of the major problems in this approach is the lack of trust between the heterogeneous communicating entities that belong to different operators/providers. In this context, reputation-based mechanisms seem useful for building up trust between mobile users and the different operators/providers, and at the same time building trust between mobile users belonging to different administrative domains.

The work in [34] treats the problem of interoperability between service providers. A reputation system is developed, using an appropriate trust model. The trust model considers that the home network of a particular service provider can be the home network for some mobile nodes and a foreign network (provider) for other nodes. Thus, the home network of any provider could not be considered as an always-trusted element for all mobile nodes. Furthermore, a mechanism is presented that can enable service providers to predict the QoS they can offer to mobile nodes according to the level of trust.

This work is basically developed for WiFi networks; however, it can be adapted to WMNs. Applying this approach in the WMN environment is beneficial in terms of having interoperability between multiple providers in a secure manner. The reputation-based system can allow mobile nodes to evaluate the behavior of service providers and at the same time can allow service providers to authorize mobile users services access according to their level of trust.

8.5.2 Detecting Forwarding Misbehavior

Secure packet forwarding is an approach to detect malfunctioning among the network elements and estimate a level of trust for each network element according to its forwarding behavior. Although a tool such as traceroute [35] could be used in detecting forwarding misbehavior and identify the offending mesh routers, an attacker can still treat traceroute packets differently or can tamper with the traceroute responses sent by other nodes. A secure traceroute SecTrace protocol [36] is developed to securely trace the existing traffic paths. SecTrace allows intermediate routers to prove the traffic reception rather than using implicit responses. In addition, SecTrace responses are authenticated to verify their origin and prevent spoofing and tampering. SecTrace is recommended for the community WMN environment to monitor end-to-end connectivity to other mesh nodes and to detect connectivity problems.

The operation of SecTrace, as in normal traceroute, takes place in a hop-by-hop manner to identify the offending routers. Each node on the path is being asked to respond to traceroute traffic, where each responding node provides a next-hop router identity for the packet in addition to its own identity. A shared key is established by the tracing node prior to sending the traceroute packets, where this key is used to encrypt and authenticate the communication to and from the expected next node. In replying to a SecTrace packet, a node sends some agreed-upon identifying marker for the packet to prove to the tracing node that the packet has been received. Also, a strongly secure Message Authentication Code (MAC) is contained in the reply packet, ensuring its authentic origin. After replying to SecTrace, the replying node becomes the next node for the next step of traceroute.

SecTrace is useful in the context of deployable WMNs to detect and localize the cause of packet forwarding misbehavior, because securing routing only is insufficient in such environment. An implementation of SecTrace [36] in a WMN scenario shows that it has a negligible performance overhead, making it suitable for monitoring of end-to-end paths and estimating a trust level for each contributing network element, whether it is a mobile client or a mesh router.

8.5.3 Trusted Routing

Mesh networks rely on participation and cooperation of nodes within the network during the routing process. However, the fact that participating nodes are controlled by different owners, nodes may choose to act in their own interest in a way that can impact the networking functioning. In this context, trusted routing is beneficial in providing additional security in this open environment by allowing each mesh node to prove its identity and integrity.

The work in [37] presents a contribution to trusted routing, which extends the Ad hoc On-demand Distance Vector (AODV) [38] routing protocol to ensure that only trustworthy nodes participate in the network. A system is presented that uses trusted computing to prevent selfish or malicious nodes from participating in the network. A new protocol named Trusted Computing Ad hoc On-demand Distance Vector (TCAODV) has been developed to enhance AODV protocol through preventing network abuse by selfish and malicious nodes. In TCAODV, a public key certificate is used by each node, which is stored within a trusted root used for the purposes of routing. The node broadcasts this certificate with Hello messages, where neighbors receiving this certificate first verify it through the signature of the issuer, then store it as the broadcaster's public key in case of validation. The RREQ packet sent by each node is signed with a sealed signature, using integrity metrics from the routing module of the sender. The node that receives the RREQ verifies the signature through using the previously received key for the requester node, and determines if the provided measurements are trustworthy. When the destination is not directly reachable by the RREQ, the intermediate node strips off the signature, replacing it by its own signature and integrity measurements. In addition, a per-route symmetric encryption key is established to ensure that only trusted nodes along the path can use the route. All traffic sent along the route is encrypted using this symmetric key. The TCAODV approach has less overhead on the network and can be applied in WMN scenarios. A typical scenario example is a community wireless mesh network among houses in residential areas. In this scenario, houses are equipped with wireless nodes that forward traffic toward a wired Internet connection, and in turn may also make use of this connection.

8.6 Privacy

Privacy provision is an important issue worth consideration to widen WMN deployment. Privacy concerns hiding the transferred messages/critical data from unauthorized parties, which is an important means for controlling message transfer in WMN environments. However, privacy is difficult to achieve even if messages are protected, as there are no security solutions or mechanisms which can guarantee that data is not revealed by the authorized parties themselves. Thus, complementary solutions are important to be in place. Also, communication privacy could not be assured with messages protection, as attackers could still observe who is communicating with whom as well as the frequency and duration of the communication sessions. This makes personal information susceptible to disclosure. Furthermore, mobile clients in WMNs can be easily monitored/traced in terms of their presence, which causes the exposure of their personal life. Unauthorized parties can learn the mobile clients' positions/locations through observing their communication. Consequently, there is a need to ensure location privacy in WMNs.

To control the usage of personal information and the disclosure of personal data, different types of information hiding from unauthorized parties appear to be efficient. The following approaches can be useful in information hiding, depending on what is needed to be protected:

- Anonymity: This is concerned with hiding the identity of the message sender or the message receiver or both of them. In fact, hiding the identity of both the sender and the receiver of the message can assure communication privacy. Thus, attackers observing transmissions could not know who is communicating with whom, thus no personal information is disclosed.
- Confidentiality: This is concerned with hiding the transferred messages themselves. Instead of hiding the identity of the sender and the receiver of a message, the message itself is hidden.
- Using pseudonyms: This is concerned with replacing the identity of the sender and the receiver of the message by pseudonyms which function as identifiers. Thus, pseudonyms can be used as a reference to the communicating parties without hurting their privacy, which helps to assure untraceability of clients. However, it is important to assure the unlinkability of pseudonyms and real identifiers.

This section discusses privacy protection in WMNs, highlighting some interesting research contributions.

8.6.1 Efficient Key Distribution for Message Protection

Power-efficient encryption and decryption can achieve message protection in WMNs. There is a need for simple, robust, and lightweight security mechanisms that are suitable to the WMN environment and nodes characteristics. Although the second part of the IEEE 802.11i standard uses Advanced Encryption Standard (AES) protocol to overcome the significant processing on every packet caused by the previously used Temporal Key Integrity Protocol (TKIP), the AES also adds an overhead of eight octets on every packet and can still be very expensive. In this context, the contribution of [39] presents a State-Based Key Hop (SBKH) protocol that provides a strong and lightweight encryption scheme suitable for battery operated devices. It is shown that integrating SBKH with 802.11 allows a power and processing cost that is much lower than 802.11i encryption mechanisms. SBKH is based on the concept of state-based encryption, where it does not reinitialize RC4 state for every packet. Instead, the same RC4 seed is maintained for a duration that is known to the communicating nodes. The initialization of the RC4 state is only carried out when the base key changes. SBKH allows mobile nodes to be state synchronized, where they keep using the same cipher stream to encrypt and decrypt packets exchange between them. In fact, applying this scheme in WMNs is important in terms of providing cheap and robust security without additional encryption overhead together with saving significant processing power, especially for applications of large packet sizes. Furthermore, operating with the existing hardware as well as the existing 802.11 protocols is important to millions of 802.11 cards shipped, where a change in the hardware will not solve the security issues with these existing 802.11 cards.

The messages generated in WMNs are sent using multi-hop communication among WMRs and mobile clients relaying the messages. Consequently, the use of public key cryptography is a heavy process introducing important delays, and thus leading to sub-optimal utilization of network resources. A possible solution consists in establishing or pre-defining secret keys between mesh routers that can be used in encrypting messages transferred through the hop-by-hop communication. However, a major problem in WMNs is the distribution of secret keys. To meet the constraints of high and unpredictable mobility together with limited power and storage resources of mobile nodes, particular key distribution protocols are needed taking into account these constraints and maintaining a strong security level. A new approach for random key pre-distribution is proposed in [40], achieving both efficiency and security objectives. This work replaces the use of a key pool for random keys by a developed key-generation technique. In this developed technique, a large number of random keys can be represented by a small number of key-generation keys. Consequently, instead of storing a large number of random keys, each mobile node stores a small number of key-generation keys while computing the shared secret keys during the bootstrapping phase. This solution is useful in WMN scenarios because it is scalable to large network sizes. The distributed solution for secret sharing is appropriate for WMN multi-hop communications, whether through WMR relays or mobile client relays. Furthermore, applying this scheme in the WMN environment allows a significant reduction in storage requirements, while maintaining the required security strength.

8.6.2 Traffic Privacy

Traffic preservation is a useful approach in providing communication privacy. Despite the necessity of traffic preservation, limited research has been conducted on this issue. Indeed, in a community mesh network, the traffic of mobile users can be observed by the mesh routers residing at its neighbors, which could reveal sensitive personal information. A mesh network privacy preserving architecture is presented in [41]. This work targets traffic confidentiality, aiming at deducing the traffic information, such as who the user is communicating with, and the amount and time of traffic. A lightweight traffic privacy-preserving mechanism for WMNs is developed, based on the concept of traffic pattern concealment via routing control, using the intrinsic WMN redundancy in terms of multi-paths. As illustrated in Figure 8.6, the traffic from the source (gateway) to the destination (mesh router) is split to many paths, thus all the relaying nodes along the paths could only observe a portion of the entire traffic. Furthermore, the traffic can be split in a random way (spatially and temporally) so that an intermediate node can have little knowledge to figure out the overall traffic pattern, allowing the traffic pattern to be concealed.

The present work first defines an information-theoretic metric, then proposes a penalty-based routing algorithm to allow traffic pattern hiding by exploiting the multiple available paths between any two mesh nodes. The source routing scheme is adopted which allows a node to easily learn the topology of the WMN that it belongs to through each received packet, while the source and destination ID are encrypted. This work can assure communication privacy in WMNs, where each destination is able to consistently limit the proportion of mutual information it shares with the observing node. This approach needs more adaptation for the WMN environment and applications. The fact of splitting traffic on multiple paths may impact the transmission delay. This can be harmful to the continuity of service of real-time applications, such as VoIP and streaming, which are delay sensitive. Furthermore, when applying this approach in WMN scenarios with multi-hop communication among mobile clients, multi-path transmission among

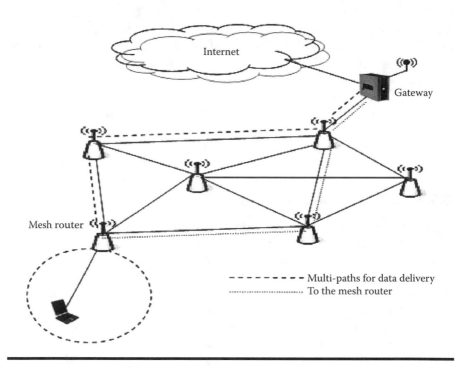

Figure 8.6   Preserving traffic privacy.

mobile nodes (relays) can cause packets loss, which in turn impacts the transmission quality. Consequently, positioning information of the relaying mobile clients is important to be acquired to select the relaying multi-paths according to their mobility behavior and patterns.

8.6.3   Non-Traceability

In fact, the behavior of mesh nodes can be easily traced by adversaries due to the use of wireless channels, multi-hop connections through intermediate nodes, and convergence of traffic to WMRs. Hiding nodes activity is an approach that can prevent nodes traceability, assuring their privacy. Cryptographic approaches are not appropriate to achieve nodes privacy in terms of hiding nodes activities, as they are not efficient in case of internal attackers among the WMRs or the mobile clients. At the same time, redundancy in transmissions through broadcasting at WMRs or gateways can hide the activity of the receiver node; however, an internal attacker can discover the node when it sends a message to a WMR or a gateway. In [42], a solution is proposed with the objective of hiding an active node that connects to a gateway router, where this active mesh node has to be anonymous. A novel communication protocol is designed to protect nodes privacy using both cryptography and redundancy. This protocol uses the concept of onion routing in wired networks [43], adapting it to the WMN environment. In this solution, an end user requiring an anonymous communication sends a request to an onion router (OR). The OR acts as a proxy for the mobile user, and the communication between the end user/mobile client and the OR is protected from adversaries. The proxy constructs a route consisting of other ORs and constructs an onion using the public keys of the routers on the route. The onion is constructed such that the most inner part is the message for the intended destination, and the message is wrapped by being encrypted using the public keys of the ORs in the route with their same order in the route. The ID of the session initiator is not carried in the constructed route, where the initiator is kept anonymous to other mesh nodes. To prevent attackers from monitoring routes from gateways to initiator nodes, the constructed route between the initiator node and the gateway does not end at the initiator; however, it extends for a few extra hops carrying dummy information generated by the initiator node.

This work protects the routing information from insider and outsider attackers, making each node behavior/activity undistinguishable. However, there should be a trade-off between the anonymity and the computing/communication overhead. It should be assured that achieving a higher level of anonymity should not result in higher overhead cost.

8.7   Conclusion and Outlook

To further ensure security of WMNs, some essential strategies need to be considered. Security and privacy mechanisms and architectures for access networks including WMNs have considered the lower layers in the form of security over wireless networks and the upper service layers in the form of application and transport security. However, what is still missing is a general solution which is both adaptable to the network types and also takes into account end-system capabilities as well as enabling inter-domain AAA negotiation.

Security mechanisms need to be embedded into MAC protocols to detect and prevent misbehavior in channel access and into network protocols providing a secure routing. Moreover, new or adaptive upper layer protocols are needed for WMNs, taking into consideration centralized and opened WMN scenarios together with the multi-hop communication principle. Generally, multi-layer security is desired as attacks occur simultaneously in different protocol layers. It might be important to develop cross-layer framework for security monitoring to detect attacks responding quickly to them. Furthermore, it is necessary to provide sufficient authentication for user nodes to authenticate mesh nodes or for a downstream mesh node to authenticate an upstream mesh node. However, it is important to be mindful of the overhead caused by authentication as wireless users or mesh nodes are often constrained by limited battery power, computing power, or memory space. Also, unacceptable authentication delay might impact service continuity. The future deployment of WMNs will observe multi-operators' coexistence, which requires appropriate AAA systems that allow mobile clients authentication and accounting across multiple administrative domains.

Providing trust between mesh nodes is an important aspect; however, the domain of WMNs still lacks appropriate mechanisms capable of introducing trusted elements. In this context, upper layer architectures for trust provision between nodes should be provided, taking into account the special characteristics of WMNs. Consequently, an important issue that should also be considered is the measurement and estimation of the trust levels between nodes. Appropriate metrics should be developed for calculating trust levels in WMNs at lower layers. As well, new architectures for trust infrastructure assurance are needed at the application level.

The open medium property of WMNs makes them vulnerable to privacy attacks. The behavior of mesh nodes can be easily monitored and traced by adversaries due to the use of wireless channel, multi-hop communication, and traffic convergence to mesh routers. Despite the necessity of privacy to protect sensitive personal information and prevent client traceability, limited research contributions have been conducted toward privacy preserving in WMNs. This subject still needs wide investigation and studies, and could impact the type of applications in future WMN deployment.

For future deployment of WMNs, further important open issues are still not covered and need more investigation from the research community as well as the industry. One important issue is the secure auto-configuration of mobile nodes in this environment. Another issue is the fast and secure association between mobile nodes in a totally distributed manner and with high mobility that is mostly taking place in WMN open scenarios. In providing intelligent commercial WMN services, an interesting point to be studied is employing rewarding mechanisms, in terms of providing incentives to mobile nodes to cooperate, as a means of accounting mobile users. Finally, applying the Grid Computing paradigm seems useful in WMNs, in terms of aggregating the mesh nodes resources to carry out heavy security services. A wide take up in Grid Computing is the appropriate security models and the cross-organizational AAA for collaborative business. In this new trend, mobile users with varying context and capabilities act as resource providers and at the same time clients participating in the grid.

References

[1] M. Kodialam and T. Nandagopal, Characterizing the Capacity Region in Multi-Radio Multi-Channel Wireless Mesh Networks, MobiCom, 2004.

[2] H. Moustafa, U. Javaid, T. M. Rasheed, S. M. Senouci, and D. Meddour, A Panorama on Wireless Mesh Networks: Architectures, Applications and Technical Challenges, Wimeshnets 2006, 2006.

[3] P. Krishnamurthy, D. Tipper, and Y. Qian, The Interaction of Security and Survivability in Hybrid Wireless Networks, IEEE International Conference on Performance, Computing, and Communications, 2004.

[4] S. Marti, T. J. Giuli, K. Lai, and M. Baker, Mitigating Routing Misbehavior in Mobile Ad Hoc Networks, MobiCom, 2000.

[5] S. Buchegger and J.-Y. Le Boudec, Performance Analysis of the CONFIDANT Protocol (Cooperation Of Nodes: Fairness In Dynamic Ad-hoc NeTworks), MobiHoc, 2002.

[6] R. Mahajan, M. Rodrig, D. Wetherall, and J. Zahorjan, Sustaining Cooperation in Multi-Hop Wireless Networks, Second Symposium on Networked Systems Design and Implementation (NSDI '05), 2005.

[7] P. Kyasanur and N. H. Vaidya, Detection and Handling of MAC Layer Misbehavior in Wireless Networks, International Conference on Dependable Systems and Networks (DSN '03), 2003.

[8] M. Raya, J. P. Hubaux, and I. Aad, Domino: A System to Detect Greedy Behavior in IEEE 802.11 Hotspots, Second International Conference on Mobile Systems, Applications and Services (MobiSys 2004), 2004.

[9] S. M. Faccin, C. Wijting, J. Kneckt, and A. Damle, Mesh WLAN networks: Concept and system design, *IEEE Wireless Communication*, April 2006.

[10] H. Deng, W. Li, and D. Agrawal, Routing security in wireless ad hoc networks, *IEEE Communication Magazine*, October 2002.

[11] M. Barbeau, WiMax/802.16 Threat Analysis, First ACM International Workshop on Quality of Service and Security in Wireless and Mobile Networks (Q2SWinet '05), 2005.

[12] I. Akylidiz, X. Wang, and W. Wang, Wireless mesh networks: A survey, *Computer Networks—Elsevier Science*, No. 47, January 2005.

[13] A. Sikora, Design challenges for short-range wireless networks, *IEEE WLAN Systems and Internetworking*, Vol. 151, No. 5, October 2004.

[14] IEEE Std 802.11i, Medium Access Control Security Enhancements, 2004.

[15] IEEE Std 802.1X, Local and Metropolitan Area Networks Port-Based Network Access Control, 2001.

[16] X. Zheng, C. Chen, C.-T. Huang, M. Matthews, and N. Santhapuri, A Dual Authentication Protocol for IEEE 802.11 Wireless LANs, IEEE Second International Symposium on Wireless Communication Systems, 2005.

[17] H. Moustafa, G. Bourdon, and Y. Gourhant, Authentication, Authorization and Accounting (AAA) in Hybrid Ad hoc Hotspots Environments, Fourth ACM International Workshop on Wireless Mobile Applications and Services on WLAN Hotspots (WMASH '06), 2006.

[18] C. Rigney, S. Willens, A. Rubins, and W. Simpson, Remote Authentication Dial In User Service (RADIUS), RFC 2865, June 2000.

[19] B. Aboba and J. Vollbrecht, Proxy Chaining and Policy Implementation in Roaming, RFC 2607, June 1999.

[20] S. Zhu, S. Xu, S. Setia, and S. Jajodia, LHAP: A Lightweight Hop-by-Hop Authentication Protocol for Ad-hoc Networks, IEEE 23rd International Conference on Distributed Computing Systems Workshops (ICDCSW '03), 2003.

[21] S. Zhu, S. Xu, S. Setia, and S. Jajodia, LHAP: A lightweight network access control protocol for ad hoc networks, *Elsevier Ad Hoc Networks Journal*, Vol. 4, No. 5, September 2006.

[22] L. Lamport, Password Authentication with Insecure Communication, *Communications of the ACM*, Vol. 24, No. 11, November 1981.

[23] A. Perrig, R. Canetti, D. Song, and J. Tygar, Efficient and Secure Source Authentication for Multicast, Network and Distributed Security System Symposium (NDSS '01), 2001.

[24] N. Prasad, M. Alam, and M. Ruggieri, Light-weight AAA infrastructure for mobility support across heterogeneous networks, *Wireless Personal Communications*, Vol. 29, 2004.

[25] DIAMETER: http://www.diameter.org/

[26] C. Perkins, Mobile networking through Mobile IP, *IEEE Internet Computing*, Vol. 2, Issue 1, January/February 1998.

[27] K. Srinivason, M. Ndoh, and K. Kaluri, Advanced Wireless Networks for Underground Mine Communications, First International Workshop on Wireless Communications in Underground and Confined Areas (IWWCUCA 2005), 2005.

[28] IEEE Std 802.11f, IEEE Trial-use Recommended Practice for Multi-vendor Access Point Interoperability via an Inter-access Point Protocol Across Distribution Systems Supporting 802.11 Operation, 2003.

[29] B. Aboba and D. Simon, PPP EAP TLS Authentication Protocol, RFC 2716, 1999.

[30] R. Fantacci, L. Maccari, T. Pecorella, and F. Frosali, A Secure and Performant Token-based Authentication for Infrastructure and Mesh 802.1X Networks, InfoCom, 2006.

[31] O. Cheikhrouhou, M. Maknavicius, and H. Chaouchi, Security Architecture in a Multi-hop Mesh Network, Fifth Conference on Security and Network Architectures (SAR), 2006.

[32] M. Parthasarathy, Protocol for Carrying Authentication and Network Access (PANA) Threat Analysis and Security Requirements, RFC 4016, March 2005.

[33] H. Moustafa, G. Bourdon, and Y. Gourhant, Providing Authentication and Access Control in Vehicular Network Environment, 21st IFIP TC-11 International Information Security Conference (IFIP/SEC 2006), 2006.

[34] N. B. Salem, J-P. Hubaux, and M. Jakobsson, Reputation-based Wi-Fi deployment, *Mobile Computing and Communication Review*, Vol. 9, No. 3, 2005.

[35] V. Jacobson, The Traceroute Manual Page, Lawrence Berkeley Laboratory, Berkeley, CA, December 1988.

[36] G. Mathur, V. Padmanabhan, and D. Simon, Securing Routing in Open Networks Using Secure Traceroute, Microsoft Research technical report (MSR-TR-2004-66), July 2004.

[37] M. Jarrett and P. Ward, Trusted Computing for Protecting Ad-hoc Routing, IEEE Fourth Annual Communication Networks and Services Research Conference (CNSR 2006), 2006.

[38] C. Perkins, E. Belding-Royer, and S. Das, Ad hoc On-demand Distance Vector (AODV) Routing, RFC 3561, July 2003.

[39] S. Michell and K. Srinivasan, State Based Key Hop Protocol: A Lightweight Security Protocol for Wireless Networks, ACM International Workshop on Performance Evaluation of Wireless Ad hoc, Sensor and Ubiquitous Networks (PE-WASUN '04), 2004.

[40] K. Ren, K. Zeng, and W. Lou, A new approach for random key pre-distribution in large-scale wireless sensor networks, *Wireless Communications and Mobile Computing*, Vol. 6, 2006.

[41] T. Wu, Y. Xue, and Y. Cui, Preserving Traffic Privacy in Wireless Mesh Networks, 2006 International Symposium on a World of Wireless, Mobile and Multimedia Networks (WoWMoM 2006), 2006.

[42] X. Wu and N. Li, Achieving Privacy in Mesh Networks, The Fourth ACM Workshop on Security of Ad Hoc and Sensor Networks (SASN '06), 2006.

[43] M. Reed, P. Syverson, and D. Goldschlag, Anonymous connections and onion routing, *IEEE Journal on Selected Areas in Communication*, special issue on copyright and privacy protection, 1998.

Chapter 9

Non-Interactive Key Establishment in Wireless Mesh Networks[1]

Zhenjiang Li and J.J. Garcia-Luna-Aceves

Contents

[1] This work was supported in part by the Baskin Chair of Computer Engineering at UCSC, the National Science Foundation under Grant CNS-0435522, and the U.S. Army Research Office under Grant no. W911NF-05-1-0246. Any opinions, findings, and conclusions are those of the authors and do not necessarily reflect the views of the funding agencies.

297

Symmetric cryptographic primitives are preferable in designing security protocols for wireless mesh networks (WMNs) because they are computationally affordable for resource-constrained mobile devices forming a WMN. Most proposed key-establishment schemes for symmetric cryptosystems assume services from a centralized authority (either online or offline), or involve interaction between communicating parties. However, requiring access to a centralized authority, or ensuring that correct routing be established before the key agreement is done, is difficult to attain in wireless networks.

We present a new non-interactive key agreement and progression (NIKAP) scheme for wireless networks, which does not require an online centralized authority, can establish and update pairwise shared keys between any two nodes in a non-interactive manner, is configurable to operate synchronously (S-NIKAP) or asynchronously (A-NIKAP), and has the ability to provide differentiated security services wireless routers the given security policies. As the name implies, NIKAP is especially valuable to scenarios in which shared secret keys are desired to be computed without negotiation between mobile nodes over insecure channels, and also need to be updated frequently.

As an application example, we present the Ad hoc On-demand Secure Routing (AOSR) protocol based on NIKAP to secure the signaling of on-demand ad hoc routing, which exploits pairwise keys between pairs of nodes and hash values keyed with them to verify the validity of the path discovered. Analysis and simulation results show that AOSR has low communication overhead caused by the key establishment process due to the use of NIKAP, effectively detects or thwarts a wide range of attacks to on-demand ad hoc routing, and is able to maintain a high packet-delivery ratio, even when a considerable percentage of nodes are compromised.

9.1   Introduction

A wireless mesh network (WMN) is a dynamically self-organized network of wireless nodes that automatically establish and maintain mesh connectivity among themselves (forming, in effect, an ad hoc network). A WMN consists of mesh routers and mesh clients, and each node operates not only as a host, but also as a router that forwards packets for other nodes. This feature enables advantages such as low operation cost, robustness, and extendable service coverage. However, the ad hoc deployment without centralized administration and the highly dynamic nature of wireless networks also bring up new challenges to systems built on them, among which security is a pressing problem.

In general, there are three cryptographic techniques that can be used to devise security mechanisms for WMNs: one-way hash functions, symmetric cryptosystems, and asymmetric (or public key) cryptosystems. An asymmetric cryptosystem is more efficient in key utilization in that the public key of a node can be used by all the other nodes; a symmetric cryptosystem requires the existence of a shared key between two communicating nodes. Hash functions can be implemented quickly, and usually work together with symmetric or asymmetric algorithms to create more useful credentials, such as a digital certificate or a keyed hash value (i.e., a keyed message authentication code).

Portable devices forming a WMN usually have limited battery life and must share a relatively limited transmission bandwidth. Therefore, symmetric cryptosystems are preferable in ad hoc scenarios due to their computational efficiency (conducting an asymmetric algorithm usually is three or four orders of magnitude slower than the symmetric counterpart). For a symmetric cryptosystem to work, a shared key must be established between each pair of communicating entities. The key establishment problem between two network principals is well understood for conventional communication networks, and generally can be resolved by key distribution or key agreement.

The classic key-distribution scheme, such as Kerberos [1], requires an online centralized authority (CA) to generate and distribute keys for nodes. However, this is not suitable for WMNs. In practice, the online CA can be unavailable to some of the nodes, or even the whole network during certain time periods, because of the unpredictable state of wireless links and node mobility. Given that the CA is the single point of failure, compromising the CA jeopardizes the security of the entire system. More importantly, the Kerberos system is designed to provide authentication and key distribution services for networks structured based on the client/server model, which, however, is not the case of WMNs. In WMNs, nodes are assumed to be willing to route packets for other nodes and behave as peers of one another, such that every node has the responsibility of a mobile router in addition to a common network user. Therefore, a WMN is a peer-to-peer communication system for the purpose of routing, into which the conventional client/server model-oriented, centralized key distribution approach does not fit. Recently proposed key distribution protocols [2] for wireless environments replace the functionality of CA by a subset of nodes in the network. However, this approach still relies on a small number of nodes, and it is not clear whether sharing the CA functionality among multiple nodes can perform better than using a single CA, given that applications need to contact multiple nodes that can be multiple hops away, to obtain the desired keys.

Key agreement protocols, such as the Diffie–Hellman key exchange protocol [3] and many variations derived from it, do not need an online CA and compute the shared keys between nodes on-demand. These protocols are interactive schemes in that nodes need to exchange messages between them to establish the desired keys, for which active routes must pre-exist for such approaches to work. The assumption of pre-existing routes between two communicating parties, which may be multiple hops away from each other, contradicts the need to secure the routing discovery process between such nodes in the first place. Even if such an assumption is satisfied, network dynamics can tear routes down in the middle of the key negotiation, and as such no key can be agreed upon. Moreover, interactive key agreement protocols are not scalable in terms of communication overhead, because messages exchanged for key establishment can consume significant CPU cycles and wireless bandwidth in such a highly dynamic environment as WMNs, which can become even worse if the shared keys between nodes need to be updated frequently.

Motivated by the observations above and based on self-certified key (SCK) [4] cryptosystem, we propose new NIKAP protocols to facilitate the key agreement process in WMNs. In NIKAP-oriented protocols, pairwise keys can be computed between two nodes in a non-interactive manner, as well as the subsequent key progression (rekeying) process. NIKAP needs the aid of a CA only at the initial network formation, and the CA can be entirely offline thereafter. Consequently, single-point failures are avoided during the operation of the deployed WMN. Compared with other key distribution and agreement approaches, NIKAP saves scarce energy and bandwidth of wireless nodes in transmitting, receiving, and processing messages. To our knowledge, NIKAP is the first key establishment scheme that supports the non-interactive key agreement and subsequent key progression simultaneously. Though there are a few protocols that can establish shared keys between nodes non-interactively based on either matrix threshold key pre-distribution (MTKP), or polynomial threshold key predistribution (PTKP) [16], none of them supports non-interactive key progression.

The rest of the chapter is organized as follows. For completeness, Section 9.2 reviews the basic idea of the SCK cryptosystem, which was first introduced by Petersen and Horster [4]. Section 9.3 presents S-NIKAP and A-NIKAP, the non-interactive key agreement and progression protocols tailored for WMNs, in which we also discuss scenarios to which NIKAP-based protocols can be applied. Section 9.4, Section 9.5, and Section 9.6 present the results of our recent use of NIKAP to secure the routing process in wireless ad hoc networks. We compare NIKAP with other key distribution and agreement approaches proposed for wireless environments in Section 9.7, and present the concluding remark in Section 9.8.

9.2   Basics of the Self-Certified Key Cryptosystem

In an asymmetric cryptosystem, there are two ways of ensuring the authenticity of a public key: explicit verification and implicit verification. In explicit verification, a trusted centralized authority signs a certificate that binds a public key and the identity (ID) of its owner. Then any user can verify the certificate explicitly provided that the public key of the centralized authority is known. In implicit verification, the authenticity of a public key is verified when it is used for encryption (or decryption), signature verification, key exchanging, or other cryptographic operations. For example, a successful verification of a signature means that the public key matches the private key used to construct this signature. A self-certified key (SCK) system follows the track of implicit verification. In the following, we first summarize the basic primitives used by SCK to establish and update the shared pairwise keys between two communicating parties. In such cases, the authenticity of a public key is verified when the shared keys derived based on it are used, for example, to encrypt and decrypt data, and to generate and check keyed hash values.

- Initialization: A CA Z is assumed to exist before the network formation. Z chooses large primes p, q with $q|(p-1)$ (i.e., q is a prime factor of $p-1$), a random number $k_A \in Z_q^*$, where Z_q^* is a multiplicative sub-group with order q, and generator α; then Z generates its (public, private) key pair (x_Z, y_Z). We assume that the public key y_Z is known to every node that participates in the network. To issue the private key for node A with identifier ID_A, Z computes the signature parameter $r_A = \alpha^{k_A} \ (mod \ p)$ and $s_A = x_Z \cdot h(ID_A, r_A) + k_A \ (mod \ q)$, where $h(\cdot)$ is a collision-free one-way hash function and $(mod \ p)$ means modulo p. Node A publishes the parameter r_A, called the guarantee, together with its identifier ID_A, and keeps $x_A = s_A$ as its private key. The public key of A can be computed by any node that has y_Z, ID_A and r_A using the following equation:

$$y_A = y_Z^{h(ID_A, r_A)} \cdot r_A \ (mod \ p) \qquad (9.1)$$

  We denote this initial key pair as $(x_{A,0}, y_{A,0})$.
- User-controlled key pair progression: Node A can update its (public, private) key pair either synchronously or asynchronously. In the synchronous setting, where A uses the key pair $(x_{A,t}, y_{A,t})$ in time interval $[t \cdot \Delta T, (t+1) \cdot \Delta T)$, node A can choose n random pairs $\{k_{A,t} \in Z_q^*, r_{A,t} = \alpha^{k_{A,t}} \ (mod \ p)\}$, where $1 \le t \le n$, and publishes guarantees $r_{A,t}$. Then the private key of node A progresses as follows:

$$x_{A,t} = x_{A,0} \cdot h(ID_A, r_{A,t}) + k_{A,t} \ (mod \ q) \qquad (9.2)$$

and the corresponding public keys can be computed according to

$$y_{A,t} = y_{A,0}^{h(ID_A, r_{A,t})} \cdot r_{A,t} \ (mod \ p) \tag{9.3}$$

■ Non-interactive pairwise key agreement and progression: Pairwise shared keys between any two nodes A and B can also be computed and updated synchronously or asynchronously based on Algorithm 1.

Algorithm 1   Key agreement between nodes A and B

Node A:

$x_{A,t} = x_{A,0} \cdot h(ID_A, r_{A,t}) + k_{A,t}$

$y_{B,t} = y_{B,0}^{h(ID_B, r_{B,t})} \cdot r_{B,t} \ (mod \ p)$

$K_{A,t} = y_{B,t}^{x_{A,t}} \ (mod \ p)$

$K_t = h(K_{A,t})$

Node B:

$x_{B,t} = x_{B,0} \cdot h(ID_B, r_{B,t}) + k_{B,t}$

$y_{A,t} = y_{A,0}^{h(ID_A, r_{A,t})} \cdot r_{A,t} \ (mod \ p)$

$K_{B,t} = y_{A,t}^{x_{B,t}} \ (mod \ p)$

$K_t = h(K_{B,t})$

The pairwise shared keys obtained by node A and node B are equal because

$$h(K_{A,t}) = h(y_{B,t}^{x_{A,t}} \ (mod \ p))$$

$$= h(\alpha^{x_{A,t} x_{B,t}} \ (mod \ p)) = h(y_{A,t}^{x_{B,t}} \ (mod \ p)) = h(K_{B,t}) \tag{9.4}$$

Two features of SCK are worth pointing out:

1. Given that N nodes participate in the network and their IDs are globally known, N guarantees are advertised to distribute their public keys, instead of N traditional certificates. The advantage is that, unlike a certificate-based approach, such N guarantees can be published and need not be certified (signed) by any centralized authority. This means that the public key of each node can be derived and updated (rekeying) without the aid of an online CA (access to the CA is only required at the initial network formation, as previously described).

2. Given that guarantees are correctly received by each node in the network, then any two nodes can establish and progress the pairwise key shared between them in a non-interactive manner. Consequently, without considering the distribution of guarantees, the communication overhead incurred by key establishment is zero.

9.3 Non-Interactive Key Agreement and Progression

9.3.1 S-NIKAP and A-NIKAP

SCK is particularly attractive to the design of security protocols for wireless networks because it promises a NIKAP scheme. However, the basic primitives of SCK cannot be applied directly to WMNs. In this section, we present two protocols that implement NIKAP to facilitate security mechanisms using symmetric cryptographic primitives, and allow NIKAP to be configurable, depending on whether time synchronization is available to wireless nodes in the network.

For NIKAP to work correctly, we assume that the guarantees of a node are successfully distributed to all nodes participating in the network. To ensure the delivery of nodal guarantees in such an error-prone environment as wireless channel, an efficient and reliable broadcasting scheme (for instance, the reliable broadcasting protocol proposed in [21]) can be used to facilitate the process of guarantee distribution, which tolerates link failures and node mobility.

In S-NIKAP, two nodes negotiate and update the shared keys between them periodically according to the current time instant and the specified security policy. Processes or applications of higher security concern can perform the rekeying (key progression) operation at a high rate and those of lower security concern at a low rate, accordingly. Therefore, communication principals in the network can be distinguished based on different security policies, such as roles, service types, or the sensitivity of data. As a result, differentiated security services can be achieved by specifying high-to-low rekeying rates that correspond to high-to-low security levels. The main limitations of S-NIKAP are the prerequisite of time synchronization and the periodical rekeying at a fixed rate. Though there exist devices or protocols providing time synchronization for wireless networks, it is still not clear if the desired performance can be achieved in such dynamic and unpredictable environments. Another drawback of S-NIKAP is that the pairwise key is independently updated no matter whether there is communication between peer nodes to take place. Therefore, local CPU cycles (and therefore battery life) are wasted if the newly generated keys are not used within its life cycle. Algorithm 2 presents the specification of S-NIKAP.

Algorithm 2 Protocol S-NIKAP (for any node A)

---

1. **Node initialization:** Retrieve the CA's public key y_Z, initial private key $x_{A,0}$, initial guarantee $r_{A,0}$, and key progression interval ΔT

2. **Guarantees distribution**: Advertise ID_A and randomly selected guarantees $r_{A,t}$ where $1 \leq t \leq n$. ($r_{A,t}$ and ID_A can be broadcast over insecure channel)

3. **Pairwise keys agreement and progression**: To communicate with node B within time interval $[T_0 + t \cdot \Delta T, T_0 + (t + 1) \cdot \Delta T)$, first update the key shared with B to K_t, according to the following procedure:

$$x_{A,t} = x_{A,0} \cdot h(ID_A, r_{A,t}) + k_{A,t}$$
$$y_{B,t} = y_{B,0}^{h(ID_B, r_{B,t})} \cdot r_{B,t} (mod \ p)$$
$$K_{A,t} = y_{B,t}^{x_{A,t}} (mod \ p)$$
$$K_t = h(K_{A,t})$$

It follows naturally that an asynchronous version of NIKAP is desired in cases in which time synchronization is not available or portable nodes cannot afford the cost of progressing keys at high rates. A-NIKAP has the same non-interactive rekeying capability as S-NIKAP does, but requires no time synchronization service from the underlying network. Instead, A-NIKAP uses a pseudo-random bit stream to synchronize the rekeying process between nodes, of which "1" invokes new key progression and "0" keeps two nodes using the current key shared between them. According to SCK, an initial shared key can be non-interactively established. Therefore, the pseudo-random bit stream can be generated, encrypted (using the initial key), and securely agreed-upon between nodes sharing the initial key. If the same pseudo-random number generator is used by both ends, to save the bandwidth, only a common seed needs to be exchanged. The progression strategy in A-NIKAP can be specified as per-session based, fixed number of sessions based, or fixed number of packets sent based etc., according to the given security policies. If the bit-synchronization is lost, nodes need to re-establish a new pseudo-random bit stream (by using the last pairwise key working between them, or simply start over). If we count one bit in the random bit stream equal to one time interval used in S-NIKAP, A-NIKAP incurs half of the local CPU cycles than S-NIKAP does, provided that the bit stream is perfectly randomized. Algorithm 3 defines protocol A-NIKAP.

Algorithm 3   Protocol A-NIKAP (for any node A)

1. **Node initialization:** Retrieve CA's public key y_Z, initial private key $x_{A,0}$, and initial guarantee $r_{A,0}$

2. **Guarantees distribution:** Advertise ID_A and randomly selected guarantees $r_{A,t}$, where $1 \leq t \leq n$. ($r_{A,t}$ and ID_A can be broadcast over insecure channel)

3. **Random bits stream generation and exchange:** To communicate with node B, first generate a random bit stream $BITS_A$ and send to B as follows:

$A \Rightarrow B : \{ID_A, ID_B, BITS_A, \ \textbf{hash}(ID_A, ID_B, BITS_A, K_{A,0})\}_{K_{A,0}}$

Where the hashing value **hash**$(\cdot)$ is used by node B to verify the integrity of $BITS_A$

4. **Bit-controlled key progression:**
While *BITS_A is not empty* **do**
   if *new session* **then**   /* Or other triggering events */
     $flag \leftarrow pop(BITS_A)$
     if $flag = 1$ **then**
       update the shared key to K_t
     else
       keep using the current key K_{t-1}

9.3.2 Application Scenarios of NIKAP

The non-interactive progression capability of NIKAP makes it attractive to wireless applications in which shared keys need to be established without negotiation through insecure channels, or need to be updated frequently. Such scenarios include secure ad hoc routing, peer-to-peer communication in combat fields, and surveillance systems.

When mechanisms based on symmetric cryptographic algorithms are used to secure the routing discovery process in wireless ad hoc networks, interactive key agreement protocols are not suitable, because the topology and routes in an ad hoc network are usually unknown when it is first deployed. Consequently, given that there can be no pre-existing routes for nodes to communicate with each other, a common broadcast channel must be used for key establishment, which is easy to be exploited by malicious users. In addition, requiring the collaboration among nodes to establish shared keys while they are establishing routes to one another cannot be done efficiently. The non-interactive nature of NIKAP allows nodes to secure the routing process without incurring undue overhead.

NIKAP can also be used to provide differentiated security services in wireless networks. To achieve better security, the keys shared between nodes can be updated regularly, and the keys used between different nodes can be rekeyed at different rates based on different security policies, such as privilege rankings, roles, and location of the nodes.

Surveillance systems are often used to gather and upload critical data periodically to a command center from monitoring nodes. The topology of a surveillance system is relatively fixed compared with that of a mobile ad hoc network (MANET), which exposes it to high possibility of being identified and attacked. Therefore, keys used between the command center and each monitoring node have to be updated regularly. Moreover, a pairwise key-based scheme is also preferable to a group key-based scheme, to confine the damage caused by key divulgence. In such a case, S-NIKAP can be a good candidate for key establishment because of its periodic, non-interactive key progression capability.

9.4 Ad hoc On-demand Secure Routing Protocol

In this section, we present the secure Ad hoc On-demand Secure Routing protocol (AOSR), which derives pairwise keys using NIKAP and exploits keyed hash values to authenticate the generic on-demand ad hoc routing.

9.4.1 Assumptions

We assume that each pair of nodes (node N_i and node N_j) in the network shares a pairwise secret key $K_{i,j}$, which can be achieved by using the key agreement protocols described in Section 9.3.1. Whether S-NIKAP or A-NIKAP is adopted depends on the availability of time synchronization in the deployed network. We also assume that the MAC address of a node cannot be changed once it joins the network. Even though some vendors of modern wireless cards do allow a user to change the card's MAC address, we will see that this simple assumption can be helpful in detecting some complicated attacks such as wormhole. Moreover, every node must obtain a certificate signed by the CA, which binds its MAC and ID (can be the IP address of this node), before it joins the network. Note that such certificates are used for nodes to verify the authenticity of their neighbors, rather than validating the routes discovered during the process of route discovery. A node presents its certificate to each node that it meets for the first time, and two nodes can communicate with its neighbor nodes only if their certificates have been mutually verified. The approach used to authenticate and maintain neighbor-node information is presented in [5], and as such is omitted here due to space limitations. To be clear, the notation used in the rest of the chapter is summarized in Table 9.1.

9.4.2 Route Discovery

AOSR consists of route request initialization, route request forwarding, route request checking at the destination D, and the symmetric route reply initialization, route reply forwarding, and route reply checking at the source S. The message flow of the route discovery of AOSR is illustrated in Figure 9.1.

9.4.2.1 Route Request Initialization

Source S generates the following route request $RREQ$ and broadcasts to its neighboring nodes, when S wants to communicate with node D, but has no active route maintained for D at that point.

$$RREQ = \{RREQ, S, D, QNum, HC, \{NodeList\}, QMAC_{s,d}\} \quad (9.5)$$

because no node has been traversed by $RREQ$ at the source S, $HC = 0$ and $\{NodeList\} = \{Null\}$. $QMAC_{s,d} = Hash(CORE, HC, \{NodeList\}, K_{s,d})$ is the

Table 9.1   Notation Used in This Chapter

Name	Meaning
S, D, N_i	Node IDs, particularly, $S =$ source, $D =$ Destination
RREQ	The type identifier for a route request *RREQ*
RREP	The type identifier for a route reply *RREP*
RERR	The type identifier for a route error report *RERR*
QNum	The route request ID, a randomly generated number
RNum	The route reply ID, and $RNum = QNum + 1$ for the same round of route discovery
$HC_{i \rightarrow j}$	The hop count from node N_i to N_j
QMAC	The k-MAC[2] used in *RREQ*
RMAC	The k-MAC used in *RREP*
EMAC	The k-MAC used in *RERR*
$K_{i,j}$	The key shared between nodes N_i and N_j, thus $K_{i,j} = K_{j,i}$
{*NodeList*}	Records the accumulated intermediate nodes traversed by messages *RREQ*, *RREP*, or *RERR*. For clarity, they are increasingly numbered from S to D, i.e., $\{S, N_1, N_2 ... N_i ... D\}$
$r\,T_{i \rightarrow j}$	The route from node N_i to node N_j

k-MAC which will be further processed by intermediate nodes and used by the destination D to verify the integrity of *RREQ* and the validity of the path recorded by {*NodeList*}. Parameter

$$CORE = Hash(RREQ, S, D, QNum, K_{s,d}) \tag{9.6}$$

serves as a credential of S to assure D that the *RREQ* is really originated from S and its immutable fields are integral during the propagation.

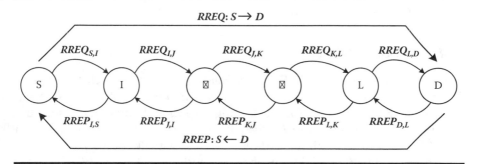

Figure 9.1   Route discovery between source S and destination D.

---

[2] In our discussion, k-MAC refers to keyed-message authentication code (a keyed hash value), while MAC refers to media access control unless specified otherwise.

9.4.2.2 Route Request Forwarding

An *RREQ* received by an intermediate node N_i is processed and further broadcast only if it has never been seen (the ID of node S and the randomly generated *QNum* uniquely identify the current route discovery initialized by S). Because {*NodeList*} records the nodes that have been traversed before the *RREQ* is received at N_i, N_i increases *HC* by one and appends the ID of the upstream node N_{i-1} into {*NodeList*}, and updates QMAC as follows:

$$QMAC_{i,d} = Hash(QMAC_{i-1,d}, HC, \{NodeList\}, K_{i,d}) \qquad (9.7)$$

A reverse forwarding entry is also established at N_i, which is used to relay the corresponding *RREP* back to source S.

9.4.2.3 Checking RREQ at Destination D

Figure 9.2 shows the procedure conducted by destination D to authenticate the validity of the path reported by *RREQ*. Basically, D repeats the computation executed by each intermediate node traversed by *RREQ*, which is recorded in field {*NodeList*}, using the shared keys maintained by D itself. Obviously, the number of hashing that D needs to perform equals *HC*, the number of nodes traversed by the *RREQ*.

If such a verification is successful, D can be assured that the *RREQ* was really originated from S, each node listed in {*NodeList*} actually participated in the forwarding of *RREQ*, and the distance between S and D is equal to $HC_{s \to d}$.

The route reply initialization, reverse forwarding of route reply, and checking *RREP* at the source S are basically symmetric to that of *RREQ*, and as such are omitted for brevity. Note that AOSR forwards traffic on a hop-by-hop basis, and each intermediate node relaying an *RREP* also establishes the forwarding entry for the requested destination D, which is used to route succeeding data packets.

9.4.3 Route Maintenance

A route error message (*RERR*) is generated and unicast back to source S if an intermediate node N_i finds the downstream link of an active route is broken (Figure 9.3). Before accepting an *RERR*, S must make sure that (1) the node generating the *RERR* belongs to the path for the destination, and (2) the node reporting link failure should actually be there when it is reporting the link failure. The process of sending back an *RERR* from node N_i is similar to that of originating a route reply from N_i to the source S.

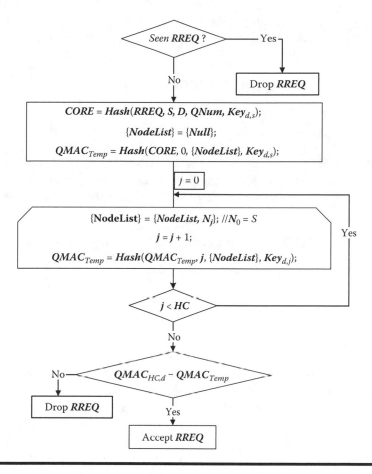

Figure 9.2   Check *RREQ* at destination D.

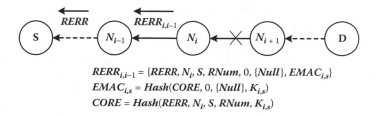

$RERR_{i,i-1} = \{RERR, N_i, S, RNum, 0, \{Null\}, EMAC_{i,s}\}$

$EMAC_{i,s} = Hash(CORE, 0, \{Null\}, K_{i,s})$

$CORE = Hash(RERR, N_i, S, RNum, K_{i,s})$

Figure 9.3 N_i generates *RERR* when the downstream link fails.

Therefore, here we only describe the main differences. An *RERR* has a format similar to that of an *RREP*, except the type identifier *RERR* and the initialization of *CORE*, which is calculated as follows:

$$CORE = Hash(RERR, N_i, S, D, RNum, K_{i,s}) \qquad (9.8)$$

Each intermediate node in the reverse path to the source only processes and back-forwards an *RERR* received from its successor used for destination *D*, which ensures that no node rather than N_i can initialize an *RERR*, and node N_i is still in the path for *D* when reporting the link failure. When source *S* receives the *RERR*, it invokes a verification procedure similar to that of *RREP*. The only difference is the initial value of *CORE*, which is calculated by

$$CORE = Hash(RERR, N_i, S, D, RNum, K_{s,i}) \qquad (9.9)$$

where rather than $K_{i,s}$ of node N_i, the pairwise key $K_{s,i}$ maintained at *S* is used.

9.5 Security Analysis

The attacks to an ad hoc network can be classified into external attacks and internal attacks based on the information acquired by the attackers. External attacks are launched by malicious users who do not have the cryptographic credentials (e.g., the keys required by the cryptographic algorithms being used) that are needed to participate in the route discovery. On the other hand, internal attacks are originated by attackers who have broken into legitimate nodes, and as such have access to cryptographic keys owned by the compromised nodes. As a result, internal attacks are far more difficult to detect and not as defensible as external attacks. For a good description of potential attacks to ad hoc routing, the reader can refer to [6,7]. Figure 9.4 depicts the network topology and notation used for our analysis. In the following, we only consider *RREQ* because the processing of *RREP* is symmetric.

In AOSR, a route request *RREQ* consists of immutable fields *RREQ*, *QNum*, *S*, *D*, and mutable fields *QMAC*, *HC*, and {*NodeList*}. As to immutable parts, they are protected by the one-way hash value *CORE*, which has *RREQ*, *S*, *D*, *QNum*, and $K_{s,d}$ as the input. No node can impersonate the initiator *S* to fabricate *RREQ* due to the lack of key $K_{s,d}$ known only to *S* and *D*. Any modification on such fields can be easily detected by destination *D*, because the *QMAC* carried in the *RREQ* cannot match what *D* recalculates based on {*NodeList*}.

Mutable fields {*HC*, {*NodeList*}, *QMAC*} are modified by intermediate nodes when the *RREQ* propagates to *D*. In AOSR, the authenticity of

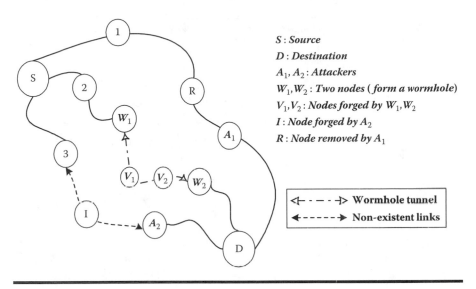

Figure 9.4   Network topology for security analysis.

HC, {*NodeList*}, and *QMAC* is guaranteed by integrating *HC* and {*NodeList*} into the computation of *QMAC*, in such a way that no node can be added into {*NodeList*} by the downstream node, unless it has actually forwarded an *RREQ*; and no node can be maliciously removed from {*NodeList*}, unless it is not used for routing traffic for *D*. For instance, let us assume that attacker A_1 attempts to remove node *R* from {*NodeList*} and decrease *HC* by one. When receiving the *RREQ*, *D* recomputes *QMAC* according to the nodes listed in {*NodeList*}. Because the hashing executed by *R*, i.e., $QMAC_{r,d}$, has been omitted, *D* cannot have a match with the received *QMAC*. The reason is that hashing operation is one-way only, and there is no way for A_1 to reverse the computation of $QMAC_{r,d}$. Another possible attack is for attacker A_2 to insert a non-existent node *I* into {*NodeList*} and increase *HC* by one. To achieve this, A_2 needs to perform one more hashing that requires $K_{i,d}$ as the input, which is impossible because $K_{i,d}$ is only known to *I* and *D*. For the same reason, A_2 cannot impersonate another node (spoofing) and make itself appear on {*NodeList*}.

A wormhole is a special attack that is notoriously difficult to detect and defend against. Wormholes usually consist of two or more nodes working collusively, picking up packets at one point of the network, tunneling them through a special channel, then releasing them at another point far away. The goal is to mislead the nodes near the releasing point to believe that the tunneled packets are transmitted by a nearby node. A demonstrative scenario of wormhole attacks is shown in Figure 9.5. Wormholes are a big

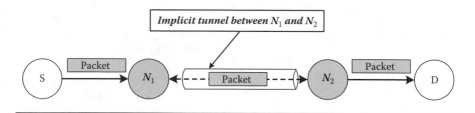

Figure 9.5   Illustration of wormhole attack. Shaded nodes are attackers and white nodes are legal nodes.

threat to ad hoc routing, largely because wrong topology information is learned by the nodes near the releasing point. As a result, data packets are more likely to be diverted into the tunnel, in which attackers can conduct varied malicious operations, such as dropping data packets (black hole attack), modifying packet contents, or performing traffic analysis.

Wormholes can be further classified based on the type of end nodes forming the tunnel. For external attackers (without valid keys or certificates), they need to make themselves invisible due to the lack of required keys to participate in the routing process. Therefore, what they actually perform is passing packets through the tunnel without any modification. On the other hand, internal attackers can "legally" participate in the routing process, and as such manipulate the intercepted packets with many more possibilities.

The chained k-MAC values computed by all intermediate nodes during the route discovery, together with the authenticated neighbor information provided by the neighbor maintenance scheme, enable AOSR to detect a wormhole and varied attacks derived from it. As an example, let us assume that nodes W_1 and W_2 in Figure 9.4 are two adversaries who have formed a tunnel $Tul_{w_1 \leftrightarrow w_2}$. First, they can refuse to forward *RREQ*, but this is not attractive because this actually excludes them from the route discovery. Second, they can attempt to modify *HC* or {*NodeList*}, but this can be detected when destination *D* checks the *QMAC* carried by *RREQ*. They can also insert some non-existent nodes, like V_1, V_2, into {*NodeList*}, but this cannot succeed due to the lack of shared keys $K_{v1,d}$ and $K_{v2,d}$.

Packets tunneled by external attackers can be detected because the MAC address of the outsider cannot match any ID maintained by the neighbor list at the receiving node near the releasing point (or does not exist at all). This can be done because a node's MAC address cannot be changed, any binding of a MAC address and an ID on the neighbor list has been authenticated, and the MAC address of a packet is always in cleartext. For instance, assume again that the nodes W_1 and W_2 in Figure 9.4 are two external attackers and form a tunnel $Tul_{w_1 \leftrightarrow w_2}$, and w_1 or w_2 is tunneling a packet from node 2 to node D. This packet cannot be accepted because the MAC address shown in the packet (the MAC address of W_2) does not match the MAC address of node 2 maintained by node D (or there is no neighbor entry maintained for node 2 at all).

The only variation of wormhole attacks that AOSR cannot detect takes place when the end node at the releasing point is an internal attacker to the network, and owns all the required cryptographic keys or certificates. To date, there is still no effective way to detect this kind of wormhole attack. Though there are other approaches to defending against wormhole attacks [8], time synchronization must be made available to each node for the proposed packet leashes to work. On the other hand, binding on an unalterable MAC address with a nodal identifier is simple to implement and provides almost the same defensive results as packet leashes.

9.6 Performance Evaluation

We implement AOSR in NS2 [9], which can act as the centralized authority at the network formation and provide time synchronization in the course of simulation. Therefore, S-NIKAP is used to serve the purpose of key establishment among mobile nodes. The hash function (used for the computation of k-MAC) and the digital signing function (used by the neighbor maintenance scheme) in our simulation are MD5 (128 bits) and RSA (1024 bits), respectively. In this way, we take into account the cost and delay caused by the cryptographic operations performed by AOSR, in addition to the overhead incurred by processing control messages. The simulation parameters are summarized in Table 9.2, and used throughout the following, unless specified otherwise.

Five metrics are used to evaluate the performance of AOSR:

1. Packet delivery ratio (PDR) is the total number of CBR packets received, over the total number of CBR packets originated, averaged over all nodes in the network.
2. End-to-end packet delay is the average elapsed time between a CBR packet passing to the routing layer and that packet being received at the destination node, averaged over all received packets.
3. Route discovery delay is the average time it takes for the source node to find a route for the requested destination.
4. Normalized routing overhead is the total routing messages originated and forwarded over the total number of CBR packets received, averaged over all nodes.
5. Average route length is the average length (hops) of the routes used to forward data packets, averaged over all routes discovered.

Table 9.2   Simulation Parameters

Parameter	Value
Simulator	NS2 [9]
Topology	30 nodes, 1000 m × 250 m field
Node placement	Uniformly distributed
Propagation model	Two-ray propagation
MAC protocol	802.11 DCF
Transmission range	250 m
Link bandwidth	2×10^6 bps
Traffic pattern	15 constant bit rate (CBR) flows with randomly chosen source and destination, two packets per second, and with a payload size of 512 bytes. Each flow starts randomly within 50 seconds after the simulation is launched, and the lasting time varies between $100 \sim 200$ seconds
Mobility model	Random way-point model with $V_{min} = 0$ and $V_{max} = 15$ mps
Simulation time	300 seconds
# of trials with random seeds	5

Figure 9.6 and Figure 9.7 demonstrate the performance comparison between AOSR using S-NIKAP and the AODV protocol [10]. When there is no attack occurring in the network, the normalized routing overhead of AOSR, as shown in Figure 9.6(b), is almost the same as that of AODV. The reason is intuitive: establishing shared keys using NIKAP does not need the negotiation between nodes or between the nodes and an online CA. In our simulation, the key progression interval is set to five seconds, and in practice, this is adjustable according to the processing power of mobile nodes, or the given security policy. Because shared keys between nodes need to be updated at a fixed rate, we expect that the time it takes for AOSR to discover routes should be longer than that of AODV. Fortunately, as shown in Figure 9.7(a), the average routing delay caused by key progression, measured over all nodes, is only $2 \sim 5$ milliseconds more than that of AODV, which is an acceptable increase of $5 \sim 12$ percent. This indicates that NIKAP efficiently supports the security mechanisms used by the route-discovery process of AOSR without incurring significant routing delay. The average route length of AOSR is a little shorter than that of AODV, as shown in Figure 9.7(c). The reason is that AOSR requires all route requests to reach the destination, while AODV allows intermediate nodes to reply to an RREQ if they cache an active route, which may not be the shortest at that moment. This also explains why the packet delivery delay of AOSR is shorter than that of AODV, as shown in Figure 9.7(b).

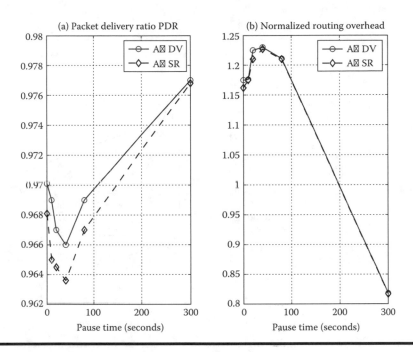

Figure 9.6   PDR and routing overhead comparisons without attackers.

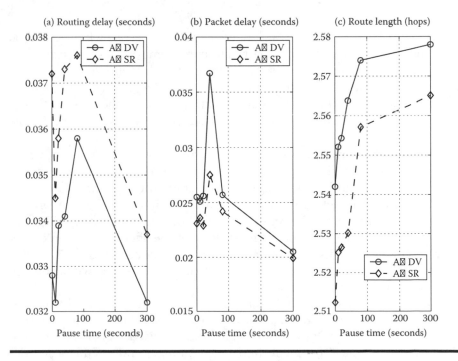

Figure 9.7   Delay and route-length comparisons without attackers.

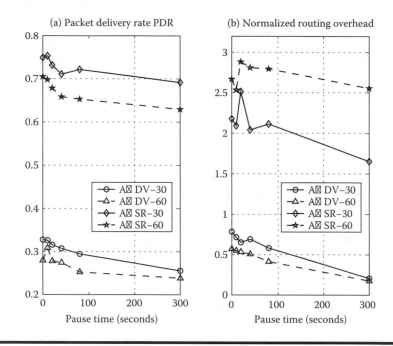

Figure 9.8   PDR and routing overhead comparisons with attackers.

Figure 9.8 and Figure 9.9 present the simulation results when 30 and 60 percent of the nodes in the network are compromised, and fabricate fake route replies to route requests by claiming that they are zero hop away from the specified destination node, in hopes that the querying source node is willing to send its succeeding data packets to them. After that, a compromised node simply drops all the data packets received (black hole attack).

The packet delivery ratio of AODV decreases drastically, as shown in Figure 9.8(a), given that most of the packets are sent to the compromised nodes, which discard them silently. The average route length of AODV is much shorter than when there is no malicious node in the network, as shown in Figure 9.9(c). The reason is that a compromised node is likely to receive and reply to the route requests for the specified destination earlier than the destination itself or other nodes having an active route. This also indicates that most of the successful packets are delivered within one or two hops away from the source.

On the other hand, as shown in Figure 9.8(a), AOSR is still able to sustain over 62 percent packet delivery ratios for all pause time configurations, even when 60 percent of the nodes are compromised. This is achieved at the cost of more routing time to find a route, longer end-to-end packet delay, and higher routing overhead, as shown in Figure 9.9(a), Figure 9.9(b),

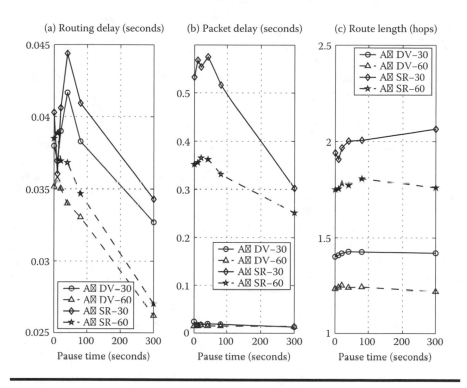

Figure 9.9   Delay and route-length comparisons with attackers.

and Figure 9.8(b), respectively. Lastly, nodes running AOSR cannot be misled by compromised nodes declaring better reachability for the requested destination,[3] and as such are able to find a route to the destination if there is one. Consequently, the average length of routes discovered by AOSR is longer than that of AODV, as shown in Figure 9.9(c).

9.7   Related Work and Open Issues

Existing key distribution protocols for wireless networks generally assume the existence of an online CA. To alleviate the risk caused by the single point of failure, threshold cryptography replaces the CA by a subset of nodes that share and provide the functionality of the CA contributorily [2]. However, this approach cannot completely eliminate the reliance on the functioning of an online CA, which is still of major interest to attackers.

---

[3] AOSR detects the misbehavior of malicious nodes when the verification of *RREQ* or *RREP* fails.

The alternative use of multiple mobile mini-CAs requires nodes to contact up to a certain number of mini-CAs before they can obtain the desired keys. Therefore, we have reason to argue that, in highly dynamic scenarios such as WMNs, the responsiveness of deploying multiple mini-CAs could be worse than schemes based on a single CA. Key distribution protocols using ID-based cryptography [11], or the combination of threshold and ID-based cryptography [12], have the same advantage as SCK because IDs (publishable) are used to obtain the corresponding public keys of nodes, instead of using a certificate to bind the ID and its public key. However, online CA services must exist for such protocols to work, which has the same limitations of protocols based on threshold cryptography.

Another approach to key agreement for wireless networks is to combine threshold secret sharing and probabilistic key sharing [13]. The basic idea is to split the shared secret between a source–target pair into several pieces, and propagate them toward the target in such a way that the target node has a high probability to recover the splitted secret based on the secret pieces it receives. However, the overhead incurred by sending multiple secret pieces toward each target node can be high due to network dynamics. Moreover, if a required number of secret pieces do not reach the target, the original secret cannot be recovered.

Group key agreement protocols [14,15] are very different from S-NIKAP and A-NIKAP. In group key agreement, a shared key needs to be distributed among all possible nodes belonging to a multicast or many-to-many-cast group, while S-NIKAP and A-NIKAP only consider the key agreement between two nodes. The storage complexity of a system using group keys is obviously lower than that of a system using pairwise keys. However, in group communication, the cost of rekeying operation caused by nodes leaving or joining a group, network partition, or merging can be considerably high. The reason is that, whenever the group membership changes, a new group key must be re-established among all group members; otherwise, the subsequent communication within the group becomes insecure due to the possibility of key divulgence. Another drawback of a system using group keys is that the compromise of a group key can jeopardize the communication confidentiality of the entire group, while the compromise of a pairwise key only affects the pair of nodes using the shared key. In practice, whether to use a pairwise key scheme or a group key scheme should be decided according to the application scenario and the security policy.

Future design of key management schemes needs to carefully consider the unique characteristics of wireless networks, i.e., volatile topology, collision-prone transmission channel, and stringent resources of the wireless nodes. Given that no centralized administration exists, a practical key management scheme must also be fully distributed and self-organizing.

Though threshold cryptograph-based approaches [2] divide the centralized authority into a subset of the nodes to improve the service availability and fault tolerance, the inherent idea of central administration limits its applicability to ad hoc networks, and also makes CA-capable nodes the major interests to malicious attackers. A possible modification to threshold cryptograph is to allocate each designated mini-CA more than one share of the CA's secret, such that the probability of successfully issuing a certificate can be increased [19].

The key pre-distribution (KPD) [18] scheme has been demonstrated to be a promising approach for symmetric key establishment for wireless scenarios. Given that N is the set of nodes in the network, each node in N is first pre-loaded a set of keys chosen from a pre-established key pool. Then any sub-group of nodes $N_i \subset N$ can establish a common key shared among them that is unknown to nodes outside N_i. KPD systems have been believed to be the only practical approach for truly ad hoc scenarios. The major limitations of KPD are that (1) the success of key establishment is probabilistic guaranteed and (2) the overhead of key pre distribution can be expensive. An interesting research topic is how to achieve the same key establishment results as that of KPD, but with a deterministic success guarantee.

Signature aggregation [20] is another effective approach to reducing the size of certificate chains by aggregating all certificates in the chains into a single short signature, as such saves the scarce bandwidth of nodes in WMNs. The basic idea of signature aggregation is that, given that N distinct messages are signed by N distinct users, it is possible to aggregate the resulting signatures into a single signature in such a way that a verifier of the aggregated signature can be convinced that each user indeed signed its message. It is an interesting research topic whether such an approach can be utilized for key management for WMNs, especially in the case of group key establishment, to reduce the overhead incurred by group-key creation and rekeying.

9.8 Conclusion

We proposed S-NIKAP and A-NIKAP, two key agreement protocols that achieve non-interactive key establishment and, if needed, the succeeding key progression (rekeying process). NIKAP needs the aid of a centralized authority only at the initial network formation, which is better than other approaches relying on online CA services. Our work using NIKAP for secure ad hoc routing shows that NIKAP bootstraps key establishment in ad hoc networks efficiently, and is promising for other resource-constrained ad hoc scenarios where frequent and non-interactive key rekeying are desired.

References

[1] J. Kohl and B. Neuman, The Kerberos Network Authentication Service (V5), RFC 1510, September 1993.

[2] L. Zhou and Z.J. Haas, Securing ad hoc networks, *IEEE Network*, special issue on network security, Vol. 13, No. 6, pp. 24–30, 1999.

[3] W. Diffie and E. Hellman, New directions in cryptography, *IEEE Tran. Inform. Theory*, Vol. 22, pp. 644–654, November 1976.

[4] H. Petersen and P. Horster, Self-Certified Keys—Concepts and Applications, 3rd Conference of Communications and Multimedia Security, Athens, September 22–23, 1997.

[5] Z. Li and J.J. Garcia-Luna-Aceves, Enhancing the Security of On-demand Routing in Ad hoc Networks, 4th International Conference on Ad-hoc Networks and Wireless (AdhocNow '2005), LNCS 3738, pp. 164–177, Cancun, Mexico, October 6–8, 2005.

[6] K. Sanzgiri, B. Dahill, B.N. Levine, E. Royer, and C. Shields, A Secure Routing Protocol for Ad hoc Networks, 10th Conference on Network Protocols (ICNP), 2002.

[7] Y. Hu, A. Perrig, and D. Johnson, Ariadne: A Secure On-demand Routing Protocol for Ad hoc Networks, 8th ACM International Conference on Mobile Computing and Networking (MobiCom), September 2002.

[8] Y. Hu, A. Perrig, and D. Johnson, Packet Leashes: A Defense against Wormhole Attacks in Wireless Networks, IEEE INFOCOM, San Francisco, March 30–April 3, 2003.

[9] NS2, The Network Simulator, http://www.isi.edu/nsnam/ns/.

[10] C. Perkins, E. Royer, and S. Das, Ad hoc On Demand Distance Vector (AODV) Routing, RFC 3561 (Experimental), July 2003.

[11] D. Boneh and M. Franklin, Identity Based Encryption from the Weil Pairing, Crypto '2001, LNCS 2139, pp 213–229, 2001.

[12] A. Khalili, J. Katz, and W. Arbaugh, Towards Secure Key Distribution in Truly Ad-hoc Networks, IEEE Workshop on Security and Assurance in Ad hoc Networks, Orlando, January 28, 2003.

[13] S. Zhu, S. Xu, S. Setia, and S. Jajodia, Establishing Pairwise Keys for Secure Communication in Ad hoc Networks: A Probabilistic Approach, 11th IEEE International Conference on Network Protocols (ICNP), Washington, DC, 2003.

[14] Y. Amir, Y. Kim, C. Nita-Rotaru, and G. Tzudik, On the Performance of Group Key Agreement Protocols, 22nd IEEE International Conference on Distributed Computing Systems (ICDCS), Vienna, Austria, July 2–5, 2002.

[15] H. Chan, A. Perrig, and D. Song, Random Key Predistribution Schemes for Sensor Network, IEEE Symposium on Research in Security and Privacy, pp. 197–213, 2003.

[16] C. Castelluccia, N. Saxena, and J.H. Yi, Self-Configurable Key Predistribution in Mobile Ad-hoc Networks, IFIP Networking Conference, LNCS 3462, pp. 1083–1095, Waterloo, Canada, May 2005.

[17] S. Capkun, L. Buttyan, and J.P. Hubaux, Self-organized public-key management for mobile ad hoc networks, *IEEE Transactions on Mobile Computing*, Vol. 2, No. 1, pp. 52–64, 2003.

[18] A.C.-F. Chan, Distributed Symmetric Key Management for Mobile Ad hoc Networks, IEEE INFOCOM, Hong Kong, March 7–11, 2004.

[19] D. Joshi, K. Namuduri, and R. Pendse, Secure, redundant and fully distributed key management scheme for mobile ad hoc networks: An analysis, *EURASIP Journal on Wireless Communications and Networking*, pp. 579–589, 2005(4).

[20] D. Boneh, C. Gentry, H. Shacham, and B. Lynn, Aggregate and Verifiably Encrypted Signatures from Bilinear Maps, EuroCrypt '03, LNCS 2656, pp. 416–432, 2003.

[21] E. Pagani, Providing reliable and fault tolerant broadcast delivery in mobile ad-hoc networks, *Mob. Netw. Appl.*, Vol. 4, No. 3, pp. 175–192, 1999.

Chapter 10

Key Management in Wireless Mesh Networks

Manel Guerrero Zapata

Contents

In wireless mesh networks (WMN), nodes use the air to communicate, so a lot of nodes might hear what a node transmits and there are messages that are lost due to collisions. The concept of servers has to be modified: there is no guarantee that a node will be able to reach another node, so things like DNS servers, certification authorities (CAs), and other entities that are assumed to be found in fixed networks cannot be used here.

In a network where the existence of central servers cannot be expected, nodes need to be able to communicate without the risk of malicious nodes impersonating the entities they want to communicate with. In a network where everybody is anonymous, identity and trust need to be redefined.

In addition, if the security protocols that are used in these kind of networks are based in mechanisms that require asymmetric cryptography, the task of having secure routing protocols for such kind of networks will not be completed without an specific key management scheme.

In this chapter, we analyze the problems that arise when designing a key management scheme for WMNs. We will use that analysis to design SAKM (Simple Ad hoc Key Management), a key management system that allows the nodes of an ad hoc network to use asymmetric cryptography with zero configuration, intended to be applied to wireless network routing protocols that provide security features that require the use of asymmetric cryptography (like SAODV). Finally, through simulation results, we will show what kind of cryptographic algorithms are more suitable for SAKM and for key management in WMNs in general.

10.1 Introduction

Currently, there are several secure routing protocols and applications for WMN that use symmetric or asymmetric keys without providing a key management scheme to distribute them. Some of them argue that a CA can be placed as a special fixed node in the WMN. Nevertheless, this is not feasible if some client nodes are not directly connected to the WMN backbone. In addition, that requires that client nodes need to register to that CA. Therefore, there is a need for key management schemes for WMNs that can operate without the help of the WMN backbone, and that allow incorporation of new nodes transparently.

10.2   Related Work

In their paper on securing ad hoc networks [28], Zhou and Haas primarily discuss key management. They devote a section to secure routing, but essentially conclude that "nodes can protect routing information in the same way they protect data traffic." They also observe that denial-of-service attacks against routing will be treated as damage and routed around.

A couple of papers [19,20] have proposed a solution to solve the "address ownership" problem in the context of Mobile IP. It consists in picking a key pair and mapping the public key to a tentative address in some deterministic way. These ideas can be adapted to the context of WMNs to provide an appropriate key management scheme.

The following proposals use symmetric cryptography, and are mainly targeting sensor networks. All of them either assume that there are no malicious nodes, that nodes do not move after deployment, or that no new nodes will be added after deployment.

The paper about secure pebblenets [4] proposes deploying the same secret key on all nodes to provide group authentication. It has a method to select clusterheads to perform the key management. Nevertheless, it assumes that there are no malicious nodes and requires nodes to have a tamper-resistant storage.

Eschenauer and Gligor [8] propose a scheme that uses a random pre-distribution of secret keys. Each sensor node receives a random subset of keys from a large key pool before deployment. Then, to agree on which key they will use to communicate, two nodes try to find one common key within their subsets that they can use as their shared secret key. Clearly, its main drawback is the requirement of pre-distribution that will not allow new nodes to connect to the network in an ad hoc manner.

SPINS [22] is a protocol in which sensor networks are formed around a base station. The base station helps every pair of nodes that need to communicate in a secure manner to do so. Nevertheless, compromising the base station renders the whole network useless. In addition, each sensor node gets a secret shared with the base station and needs to be able to communicate with the base station before establishing a communication.

Du et al. [7] study the problem of random key distribution for networks in which there is the knowledge of how the sensor nodes are going to be deployed, which, of course, simplifies a lot the problems of the key distribution. But, it also limits greatly its applicability.

Another proposal for static networks is presented in [16], where the main idea is that sensor nodes can be deployed with a large amount of keys from the pool of possible keys and, once deployed, decide which keys they keep according to their location and discard the other keys. Nevertheless, that requires that sensor nodes will be aware of their location.

LDK [3] (Location Dependent Key management) uses random key pre-distribution and does not require any knowledge about the deployment of the nodes. Nevertheless, it is only designed for static nodes and the author admits that is vulnerable during an interval after nodes deployment. In addition, it assumes that certain special nodes that are also deployed randomly (called anchors) are tamper proof and that each sensor node is in a transmission range of at least one anchor node.

The next section discusses the convenience of using asymmetric cryptography mechanisms instead of symmetric cryptography ones and the use of solutions that require tampering resistant nodes and misbehavior detection schemes. Related work that is not strictly about key management, but about securing the routing protocol, is discussed in Chapter 9.

10.3   Playing without a Referee

10.3.1   Symmetric versus Asymmetric Cryptography

If in a wireless network all routing messages are encrypted with a symmetric cryptosystem, it means that everybody that we want to be able to participate in the network has to know the key. That is not a big problem if nodes are a "team" that gets to know the "team-key" before they are deployed or try to interconnect, creating an ad hoc wireless network. A member of the team trusts the other members of the team, so they assume that the other members of the team will not act in a malicious or selfish way. They trust the other members and authorize them to change their routing tables.

Maybe this is the best thing to do for military scenarios (besides the problem of the compromised nodes and some others), but it is probably not a good approach for a wireless network where everybody can participate (like in a convention, in a meeting room, on a campus, or in our neighborhood). In this case there is a problem: nodes do not trust each other (and they should not). They are not a team. So what can be done? How can everybody be forced to be honest? A possible approach is to only believe a piece of routing information if the originator of such information is the destination of the route. In this way, if a node lies, the only thing it will achieve is that the other nodes will not be able to communicate with it (because you can only lie about yourself).

In this kind of scenario, the best option is to use an asymmetric cryptosystem (with public and private key pairs) so that the originator of the route messages signs its messages. It would not be needed to encrypt the routing messages because routing messages are not meant to be secret. The only requirement is that the nodes will be able to detect forged routing messages.

10.3.2 Obscurity- and Tamper-Resistant Devices

Because there had not been a clear way to secure ad hoc networks, by the end of the last century, some people decided to dust off the tamper-resistant approaches. There are several papers [1,2,5] which discuss why "trusting tamper resistance is problematic." The attacks against the supposededly tamper-resistant devices range from playing with things like voltage, temperature, fast signals, and clock frequency to affect EEPROM operation to the use of chemicals to remove the covering plastic or the processors.

Those papers show that obscurity is not the way to obtain security. They show that there is no such thing as a tamper-resistant device. Therefore, trying to combine symmetric cryptography solutions with tamper resistant devices to create the same result provided by alternatives that use asymmetric cryptography does not make sense.

In addition, having a secret key stored in so many devices and with the problem that, once the key is known to a malicious entity, the whole security of the network (not only the security of a single node) is compromised, makes the whole approach too risky to be even seriously considered.

10.3.3 Misbehaving Detection Schemes

In the year 2000, a long trail of papers about how to secure ad hoc networks by using misbehavior detection schemes started (e.g., [17]). This kind of approach has two main problems:

1. It is quite likely that it will be not feasible to detect several kinds of misbehavior (especially because it is very hard to distinguish misbehavior from transmission failures and other kind of failures).
2. It has no real means to guarantee the integrity and authentication of the routing messages.

Therefore, unless those problems are addressed, this approach will not be feasible. Any malicious node can generate forged misbehaving reports, making everybody believe that the rest of the nodes are even more evil than itself. Trying to use reputation schemes is just a way of blurring the problem.

10.4 The Concept of Identity

The concept of identity in computer applications is most of the time binded to a person and, on occasion, to a program or to a process. But, in routing protocols it must be binded to the node itself as user and application identification only makes sense at the application level.

10.4.1   Identity in a Place without Authorities

One of the most important consequences of the nature of wireless networks is that one cannot assume that a node that is part of a network will be always reachable by all the other nodes. This implies that there cannot be central servers in the conventional meaning of fixed networks. Therefore, the use of CAs for wireless networks is not feasible.

The approach of distributing the CA functionality among ad hoc nodes (by dividing the private keys into shares) discussed in [28] implies a huge overhead, and it may be ineffective in a network where partitions occur or where there is high mobility. In addition, it will not work at all in trivial scenarios, like when a network partition is composed of only two or three nodes.

The use of key management protocols that require exchange of messages between two nodes that need to forward routing information and that might never see each other again is, most of the time, not a choice. It would be great if the key management scheme would not need to send any additional messages besides the ones used for the routing protocol. Is all this possible?

10.4.2   MAC Addresses Are Not Unique Identifiers

Just in case somebody does not know it yet, MAC addresses are not unique identifiers. Moreover, you can change the MAC address (if you have the proper rights) of your network card under virtually any operating system.

For instance, in most Linux distributions you can just type this as root:

```
/etc/init.d/networking stop
ifconfig eth0 hw ether 01:23:45:67:89:A0
/etc/init.d/networking start
```

If you use Free BSD, you would type:

```
ifconfig fxp0 ether 01:23:45:67:89:A0
```

And, if you use Mac OS X, you would type:

```
sudo ifconfig en0 lladdr 01:23:45:67:89:A0
```

You can also change the MAC address under Windows®, although the method will vary depending on the version you use, and it is not going to be as straightforward as in the UNIX world.

10.4.3 What Identifies Me?

Another characteristic of servers in fixed networks, besides continuous availability, is the fact that clients have to know the server's IP address (or to know its human address and have the IP address of a DNS server). The same thing happens in wireless networks for any node you want to make a request to or initiate an exchange of data.

However, current trends about addressing in ad hoc networks are driving toward dynamic address allocation and auto-configuration [6,25]. In these schemes, typically a node picks a tentative address and checks if it is already in use by broadcasting a query. If no conflict is found, the node is allowed to use that address. If a conflict is found, the node is required to pick another tentative address and repeat the process.

But then, if IP addresses do not identify a node (because they are dynamically allocated), how does a node know the IP address of the node to which it wants to sent data? In fixed networks, if a node wants to send data to another one, it needs to know its address (it cannot send anything to a node that has a dynamic address because it does not know its IP address).

The binding between public keys and other attributes is typically achieved by using public key certificates. In some limited scenarios, a possible approach could be for a certification authority (that would live in a fixed network) to issue such certificates that the nodes could collect before going to the wireless "playground." However, this is not feasible for a large group of the targeted scenarios. An added problem is that the IP address should be one of the attributes binded to the public keys because it is binded to your identity.

In WMNs that are created in an ad hoc manner, node identity must be its private key that can be used to sign messages and be verified by others with the node's public key. We say it must be their key pair because there is nothing else. Another important observation is that, because we are working at the routing layer, those key pairs identify not users, but nodes.

The problem with establishing public pairs as the identity of the nodes is the fact that one can generate as many key pairs as it desires. This, combined with the fact that one can set its own MAC and IP addresses to the values it wants, can lead to a scenario where a malicious node has different sets of key pairs, IP address, and MAC address to use as different personalities. There is no easy way to detect that. But it is feasible to design a key management scheme that prevents one malicious node from impersonating another.

To sum up, what is required is a system that achieves the following: IP addresses will be assigned dynamically, nodes will be identifiable by their IP addresses, and a binding between the public key and the IP address of a node. All this should be achieved without any kind of certification authorities, which is quite a challenge.

10.5 Dynamically Generated IP Addresses

The proposal of SAKM is to generate IP addresses in a similar way [19]. In that paper, they were using what they called SUCV (Statistically Unique and Cryptographically Verifiable) addresses. SUCV addresses were designed to protect binding updates in mobile IPv6. SUCV addresses are generated by hashing an "imprint" and the public key. That imprint (that can be a random value) is used to limit certain attacks related to mobile IP.

For wireless networks, it is only needed to hash the public key. The hash digest (or a sub-string of it) may be formatted in some specific way (to be a valid IP address), and will be a Cryptographically Generated Address (CGA), which will also be statistically unique. When a message that uses the CGA as the source IP address and the public key of a node is signed by its private key, it can be verified by any other node that the node has a certain identity (represented by the knowledge of the secret key).

10.5.1 SAKM IP Address Generation

In SAKM, it is recommended to use IPv6 (instead of IPv4) due to its bigger address length (that would guarantee the statistical uniqueness of the IP addresses). The address can be, then, a network prefix of 64 bits with a 64-bit SAKM_HID (Half IDentifier) or a 128-bit SAKM_FID (Identifier). These two identifiers are generated almost in the same way as the sucvHID and the sucvID in SUCV (with the difference that they hash the public key instead of an imprint):

$$SAKM\_HID = SHA1HMAC\_64(PublicKey, PublicKey)$$

$$SAKM\_FID = SHA1HMAC\_128(PublicKey, PublicKey)$$

There will be a flag in the SAODV (or whatever other protocol that uses SAKM) routing message extensions (the H flag) that will be set to 1 if the IP address is an HID and to 0 if it is an FID.

Finally, if it has to be a real IPv6 address, a couple of things should be done [11]:

■ If HID is used, then the HID behaves as an interface identifier and, therefore, its sixth bit (the universal/local bit) should be set to zero to indicate local scope (because the IP address is not guaranteed to be globally unique).
■ And, if FID is used, then a format prefix corresponding to the wireless network should be overwritten to the FID. Format prefixes 010 through 110 are unassigned and would take only three bits of the FID. Format prefixes 1110 through 1111 1110 0 are also unassigned

and they would take between four and nine bits of the FID. All of these format prefixes need to have 64-bit interface identifiers in EUI-64 format, so universal/local bit should be set to zero.

The length of an IPv4 address is probably too short to provide the statistical uniqueness that this scheme requires when the number of nodes is very big. Nevertheless, if the number of nodes is assumed to be low enough (around 100 nodes or less), it is not very unrealistic to expect that the statistical uniqueness property will hold.

The SAKM IPv4 address will have a network prefix of eight bits and an SAKM_4ID (IPv4 Identifier). The network prefix can be any number between 1 and 126 (both included) with the exception of 14, 24, and 39 [14]. The network prefix 10 can only be used if it is granted that it will not be connected to any other network [23].

The SAKM_4ID will be the first bits of the SAKM_HID and the H flag will be set.

10.5.2 SAKM Message Fields

The public key should be included in the routing messages that are signed, so that the nodes can verify the signature. Because, obviously, the public key should be signed by the signature, it is placed before the signature field.

The identifier of the algorithm that is used to sign the message is specified in the Signature_Method field. The possible values are shown in Table 10.1 (being mandatory to support RSA). Because SAODV (or whatever other protocol uses SAKM) could allow more than one possible signature method, it might happen that a node has to verify a signature with a method it does not know. If this happens, the node will consider that the verification of the signature has failed.

This implies that all the nodes that form part of a wireless network should know all the methods used by all the other nodes to sign their

Table 10.1 Possible Values of the Signature Method Field

Value	Signature Method
0	Reserved
1	RSA [24]
2	DSA [26]
3	Elliptic curve [15]
4–127	Reserved
128–255	Implementation dependent

Table 10.2   Possible Values of the Hash_F_Sign Field

Hash_F_Sign	Hash Length	Value
RESERVED	—	0
MD2	(128 bits)	1
MD5	(128 bits)	2
SHA1	(160 bits)	3
SHA256	(256 bits)	4
SHA384	(384 bits)	5
SHA512	(512 bits)	6
Reserved		7–127
Implementation dependent	—	128–255

messages. This is not a problem because, typically, all nodes of a wireless network will use the same method (or two different methods the most). The fact that there is more than one possible signature method is because different networks may have tighter security requirements than some others and, therefore, use different signature methods.

The same happens with the hash function used to generate the hash that will be signed. The identifier of the hash algorithm is specified in the Hash_F_Sign field. The possible values are shown in Table 10.2 (being mandatory to support SHA1).

The exact codification of the all the fields is shown in Section 10.8.

10.6   Duplicated Address Detection

If a node A receives a routing message that is signed by a node B that has the same IP address as one of the nodes for which A has a route entry (node C), it will not process that routing message normally. Instead, it will inform B that it is using a duplicated IP and it will prove it by adding the public key of C (so B can verify the truthfulness of the claim).

When the node B receives a routing message that indicates that somebody else has the same IP address as itself (or it realizes it by itself), it will have to generate a new pair of public/private keys. After that, it will derive its IP address from its public key and it might inform all the other nodes (through a broadcast) of its new IP address with a special message that contains the two IP addresses (the old and the new ones) and the two public signatures (old and new) signed with the old private key and the new private key. Nevertheless, it is much better if that message is unicast (instead of broadcast) to all the nodes it considers should receive this information (in the case they are just a few). This unicast will be answered with an acknowledge message by the receiver if it verifies that everything is in order.

After this, the node will generate a route error message for the old IP address. Its propagation will delete the route entries for the old IP address and, therefore, eliminate the duplicated addresses. This route error message may have a message extension that tells which is the new address. In this way, the nodes that receive the routing message can already create the route to the new IP address.

This solution allows two nodes to coexist in the same network with the same IP address until one of them realizes it. This can be considered as a good trade-off between the impact of changing address (and having a coexisting period of two nodes with the same IP address) and the extremely low probability of having address collision.

Intermediate nodes could decide to store the IP addresses and public keys of all the nodes they would meet (or of the last N nodes, depending on their capabilities); that would allow an earlier detection of duplicated IP addresses in the network.

An alternative to this solution could be that, when a node detects that another node is using the same IP address, it would keep its public/private key pair and change the used IP address by applying a salt to the algorithm that derives the IP address from the public key. Salt variations of hash algorithms have been used to avoid dictionary attacks of passwords [18]. The "salt" is a random string that is added to the password before being hashed. This idea can be adapted with a very different purpose. If the statistically unique IP address is derived from the public key and a salt (instead of only from the public key), the node that detects or is informed that its IP address is also used by another node can change its IP address without changing its public key by just changing the salt.

Nevertheless, that would imply that the salt used by a node should be included in all the routing messages and stored in all the entries of the routing tables; and still, the node has to inform the others of its change of IP address. Therefore, it will not be used for the purpose of SAKM.

In conclusion, the approach described here handles properly the very unlikely situation of two nodes with the same IP address, without adding any complexity to the typical situation.

The format of the SAKM duplicate address detection messages is shown in Section 10.9.

10.6.1 Duplicated IP Address Detection for SAKM

SAKM can deal with the duplicated IP address problem as described earlier. Duplicate address (DADD) detected message is sent to notify to a node that its address is already being used by another node. New address (NADD) notification message is used to inform that the node has changed key pair and IP address. Finally, new address acknowledgment (NADD-ACK) message is used to confirm the reception of the NADD. In SAKM, NADD is always unicast (never broadcast).

10.6.2 Network Leaders

The original SAODV design established that besides how key distribution is achieved, when distributing a public key, this should be binded to the identity of the node (of course) and also to its netmask (in the case the node is a network leader). This was to prevent an attack in which a malicious node becomes a black hole for a whole subnet by claiming that it is their network leader.

In the new approach presented here, ad hoc nodes will typically never be network leaders. Network leaders will be only fixed nodes that typically give access to the fixed network and the nodes in the wireless network should know their IP addresses, prefix size, and public keys.

Network leaders will not change their IP address in case there is a node that happens to generate the same IP address. A node generating its IP address will check if the resulting IP address corresponds to the network leader or to the subnet corresponding to its prefix size. A node detecting another node using the network leader IP address or any of the ones corresponding to the leader subnet will inform the node and not the network leader.

10.7 Delayed Verification of Signatures

As stated in the Introduction, there has been some concern (e.g., [12, 13, 21]) that using signatures might require a processing power that might be excessive for certain kinds of ad hoc scenarios. Delayed verification addresses this problem by revising one of SAODV's security requirements from the list that was stated in [9].

10.7.1 Revised Security Requirements

The security requirements that will be provided are source authentication and integrity (that combined provide data authentication) and delayed import authorization. Import authorization was defined in [9] as the ultimate authority about routing messages regarding a certain destination node being that node itself. Therefore, a node will only authorize route information in its routing table if that route information concerns the node that is sending the information. In this way, if a malicious node lies about it, the only thing it will cause is that others will not be able to route packets to the malicious node.

Delayed import authorization allows route entries and route entry deletions in the routing table that are pending verification. They will be verified whenever the node has spared processor time or before these entries should be used to forward data packages.

The security requirements will not include confidentiality and non-repudiation because they are not necessarily critical services in the context of routing [10]. They will not include either availability (because an attacker can focus on the physical layer without bothering to study the routing protocol) and they will not address the problem of compromised nodes (because it is arguably not critical in non-military scenarios).

10.7.2 Achieving Delayed Import Authorization

In reactive ad hoc routing protocols, most of the routing messages that circulate in the network are (by far) route requests. This is due to the fact that route requests are broadcast. Route replies are unicast back through the selected path. Route error messages are unicast down through the tree of nodes that had a route to the now-unreachable node that is advertised by the route error message.

When a node receives a routing message, it creates a new entry in its routing table (the so-called reverse route). Therefore, after the broadcast of the route request, all the nodes in the network (or in the broadcast ring) have created reverse routes to the originator of the route request. From all these reverse routes, most of them will expire soon (typically all but the ones that are in the selected path through which the route reply will travel).

Then, the question is why all these route requests should be verified (with the consequent delay in the propagation of the broadcast) when most of them are going to be soon discarded. The answer is that there is no need to verify them until the corresponding route reply comes back and the node knows that it is in the selected path. The other reverse routes will expire without being verified.

Actually, the two signatures (the ones from the route request and route reply) will be verified after the node has forwarded the route reply. In this way transmissions of the route requests and replies occur without any kind of delay due to the verification of the signatures.

Following the same idea, the signature of route error messages (and in general, any routing message that has to be forwarded) can also be verified after forwarding them.

Routes pending verification will not be used to forward any packet. If a packet arrives for a node for which there is a route pending verification, the node will have to verify it before using that route. If the verification fails, it will delete the route and request a new one.

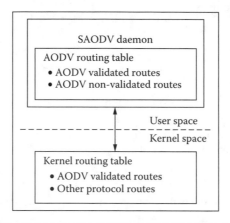

Figure 10.1   SAODV daemon.

10.7.3   SAKM with Delayed Verification

When a node needs to send or to forward a packet to a destination for which it does not have an active route, first it will check if it has a route pending validation. If it does, it will try to validate it and, if it was successfully validated, it will mark it as active and use it. If after all this there is not an active route, the node will start a route discovery process.

As shown in Figure 10.1, only once the validation is done successfully, the route is incorporated in the routing table of the node. That avoids doing dirty hacks into the routing table of the operating system of the node. The packets can be routed normally, and only when there is a route lookup that the routing table cannot resolve, the petition is captured by the SAODV routing daemon.

Figure 10.2 shows that in the case where there is a routing middleware (like Zebra[1] or Quagga[2]), the middleware routing table will contain the validated routes from the SAODV daemon combined with the ones from the other routing daemons, and the routing table in the kernel the ones with lowest "administrative distance" (in case there is a route to the same destination provided by two different routing daemons).

Talking about administrative distances, none of the routing protocols for wireless networks that are being designed or standardized have specified which would be the appropriate administrative distance for them. Let us look to the "standard de facto" (Cisco, Zebra, etc.) default administrative distance values. Probably a good default distance value would be between 160

---

[1] www.zebra.org
[2] www.quagga.net

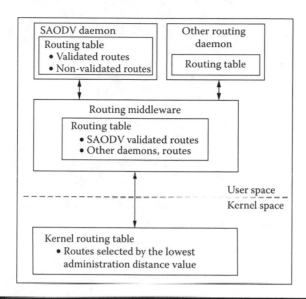

Figure 10.2   SAODV daemon with a routing middleware.

(Cisco's On-Demand Routing) and 170 (external routes in EIGRP). Therefore, a default distance value of 165 for SAODV (and also for AODV in
general) would be appropriate.

10.8   SAKM Encoding of Public Keys and Signatures

This section is provided for completeness, and it shows how public keys
and signatures are encoded under SAKM. When SAODV is used in conjunction with SAKM, it will encode the originator public key for each routing
message before its signature field.

Figure 10.3 and Table 10.3 show the fields of the encoding of the signature. Figure 10.4 and Tables 10.4 and Table 10.5 show the fields of the
encoding of the public key.

10.9   SAKM Duplicate Address Detection Messages

This section serves as a reference of the SAKM duplicate address detection
messages structure. It shows their fields and what they are used for.

Figure 10.5 and Table 10.6 show the fields of the duplicated address
(DADD) detected message. Figure 10.6 and Table 10.7 show the fields of

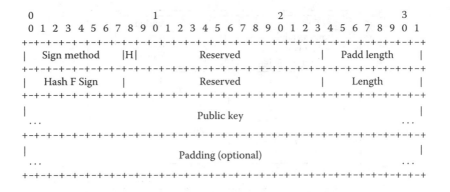

Figure 10.3   Encoding of the signature.

the new address (NADD) notification message. And, finally, Figure 10.7 and Table 10.8 show the fields of the new address acknowledgment (NADD-ACK) message.

Table 10.3   The Fields of the Encoding of the Signature

Field	Value
Signature method	The signature method used to compute the signatures. (RSA is encoded as 1)
H	Half Identifier flag. Set to 1 indicates the use of HID; set to 0, the use of FID
Reserved	Sent as 0; ignored on reception
Padding length	Specifies the length of the padding field in 32-bit units. If the padding length field is set to zero, there will be no padding
Hash_F_Sign	The hash function used to compute the hash that will be signed. Because, typically you do not want to sign the whole message, you sign a hash of the message. (MD5 is encoded as 2 and SHA1 is encoded as 3)
Reserved	Sent as 0; ignored on reception
Length	The length of the Value field (not including the Length and Reserved fields) in 32-bit units
Public key	The public key of the originator of the message. This field has variable length, but it must be 32-bits aligned
Padding	Random padding. The size of this field is set in the Padding Length field

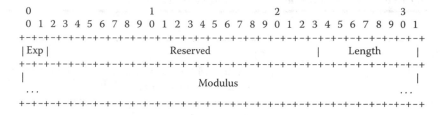

Figure 10.4   Encoding of the public key.

Table 10.4   The Encoding of an RSA Public Key

Field	Value
Reserved	Sent as 0; ignored on reception
Length	The length of the Modulus field (not including the Length and Reserved fields) in 32-bit units
Exp	The Exponent (e) encoded as specified in the next table

Table 10.5   The Encoding of the RSA Exponent

00	The components are encoded in the standard way. The Exponent (e) will be specified after the Modulus (n)
01	Specifies that Exponent (e) is 65537
10	Specifies that Exponent (e) is 17
11	Specifies that Exponent (e) is 3

Note: A message that uses any of these "smartly chosen" exponents must include random padding (in the Padding field). There is no security problem with everybody using the same exponent

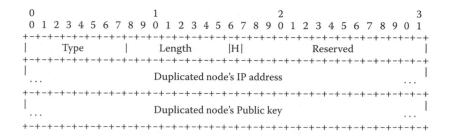

Figure 10.5   Duplicated address (DADD) detected message.

Table 10.6   Duplicated Address Detected Message Fields

Field	Value
Type	64
Length	The length of the type-specific data, not including the Type and Length fields of the message
H	Half Identifier flag; set to 1 indicates the use of HID, set to 0, the use of FID
Reserved	Sent as 0; ignored on reception
Duplicated node's IP address	The IP address of the node that uses a duplicated IP address
Duplicated node's public key	The public key of the node that uses a duplicated IP address

10.10   Simulation Results

The purpose of using SAODV with delayed verification is to obtain the same level of security as with the original SAODV, but without its main drawbacks. These drawbacks are a quite bigger average end-to-end delay and a higher power consumption by the nodes (when compared with AODV).

```
 0                   1                   2                   3
 0 1 2 3 4 5 6 7 8 9 0 1 2 3 4 5 6 7 8 9 0 1 2 3 4 5 6 7 8 9 0 1
+-+-+-+-+-+-+-+-+-+-+-+-+-+-+-+-+-+-+-+-+-+-+-+-+-+-+-+-+-+-+-+-+
|      Type       |     Length      |            Reserved       |
+-+-+-+-+-+-+-+-+-+-+-+-+-+-+-+-+-+-+-+-+-+-+-+-+-+-+-+-+-+-+-+-+
|   Sign method   |H|         Reserved         |  Padd length   |
+-+-+-+-+-+-+-+-+-+-+-+-+-+-+-+-+-+-+-+-+-+-+-+-+-+-+-+-+-+-+-+-+
|                                                               |
| ...                   Old public key                      ... |
+-+-+-+-+-+-+-+-+-+-+-+-+-+-+-+-+-+-+-+-+-+-+-+-+-+-+-+-+-+-+-+-+
|                                                               |
| ...                 Padding (optional)                    ... |
+-+-+-+-+-+-+-+-+-+-+-+-+-+-+-+-+-+-+-+-+-+-+-+-+-+-+-+-+-+-+-+-+
|   Sign method 2  |H|        Reserved         | Padd length 2  |
+-+-+-+-+-+-+-+-+-+-+-+-+-+-+-+-+-+-+-+-+-+-+-+-+-+-+-+-+-+-+-+-+
|                                                               |
| ...                  New public key                       ... |
+-+-+-+-+-+-+-+-+-+-+-+-+-+-+-+-+-+-+-+-+-+-+-+-+-+-+-+-+-+-+-+-+
|                                                               |
| ...                Padding 2 (optional)                   ... |
+-+-+-+-+-+-+-+-+-+-+-+-+-+-+-+-+-+-+-+-+-+-+-+-+-+-+-+-+-+-+-+-+
|                                                               |
| ...               Signature with old key                  ... |
+-+-+-+-+-+-+-+-+-+-+-+-+-+-+-+-+-+-+-+-+-+-+-+-+-+-+-+-+-+-+-+-+
|                                                               |
| ...               Signature with new key                  ... |
+-+-+-+-+-+-+-+-+-+-+-+-+-+-+-+-+-+-+-+-+-+-+-+-+-+-+-+-+-+-+-+-+
```

Figure 10.6   New address (NADD) notification message.

Table 10.7   New Address Notification Message Fields

Field	Value
Type	65
Length	The length of the type-specific data, not including the Type and Length fields of the message
Reserved	Sent as 0; ignored on reception
Signature method	
… padding	The same in the message extensions. Corresponds to the Signature with Old Public Key signature
Signature method 2	
… padding 2	The whole block of fields is repeated. Corresponds to the Signature of the New Public Key signature
Signature with old key	The signature (with the old key) of all the fields in the AODV packet that are before this field
Signature with new key	The signature (with the new key) of all the fields in the AODV packet that are before this field

These drawbacks are due to the computation of asymmetric cryptography primitives (message signature and verification). Through the use of simulations, it was shown that delayed verification actually achieves this.

The simulations were done with 30 nodes moving at a maximum speed of 10 meters per second in a square of 1000 × 1000 meters. They simulated the establishment of ten connections that started between second 0

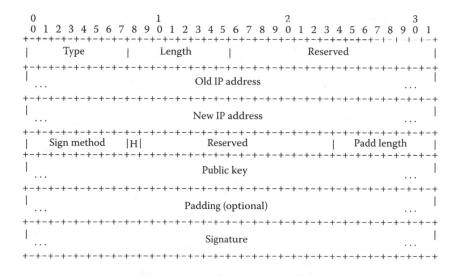

Figure 10.7   New address acknowledgment (NADD-ACK) message.

Table 10.8   New Address Acknowledgment Message Fields

Field	Value
Type	66
Length	The length of the type-specific data, not including the Type and Length fields of the message
Reserved	Sent as 0; ignored on reception
Old IP address	The old IP address
New IP address	The new IP address
Signature method	
… padding	The same in the message extensions
Signature	The signature of all the fields in the AODV packet that are before this field

and second 25 (according to an uniform distribution) and ended at the end of the simulation. The simulation time was of 100 seconds, and the connections where constant bit rate (a packet of 512 each 0.25 seconds).

The nodes in the simulations have used as routing protocols: plain AODV, SAODV with RSA, SAODV with ECC (Elliptic Curve Cryptography), and SAODV with delayed verification (SAODV2 in the figure) with ECC. There is no point in using delayed verification with RSA because its verification time is completely negligible (delayed verification reduces the amount of verifications that have to be done). That means that SAODV with RSA with or without delay verification will give practically identical results. RSA, DSA, and ECC have been used with key lengths that provide equivalent security (1368 bits for RSA and DSA, and 160 bits for ECC).

Table 10.9 shows the times for signing/verifying in a Compaq iPAQ 3670 (206 MHz, 16 M ROM, 64 M RAM) according to [27]. DSA is not used in the simulations as it presents the worst of RSA and ECC (slow signature and verification, and fast increase of computational overhead as the key length needs to be bigger).

In the simulations, end-to-end delay of the packets, packet delivery fraction, and normalized routing load were measured. Figure 10.8 shows

Table 10.9   Times for a Compaq iPAQ 3670

	RSA	DSA	ECC
Key length	1368	1368	160
Sign	210	90	42
Verify	6	110	160

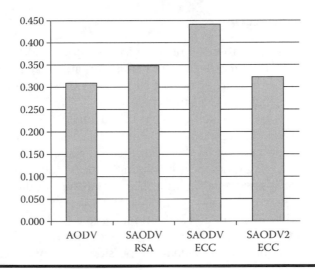

Figure 10.8 Simulation results. Average end-to-end delay, measured in milliseconds.

the averaged result of the end-to-end delay in data packet transmission. There were practically no differences among the routing protocols in packet delivery fraction (that was around 90 percent) and in normalized routing load (that was around 1).

One could expect quite different results with some other simulation scenarios, but almost always having SAODV with delayed verification and ECC as the best of the SAODV options and with a performance very close to plain AODV.

One could argue that, in scenarios in where the routes have more hops, the results of SAODV with delayed verification will be quite worse. But, actually, the results do not depend that much on the number of hops. This is due to the fact that intermediate nodes forward the RREP before verifying the signatures of the RREQ and RREP. Therefore, it is most probable that by the time the node that forwards the RREP to the final destination verifies the signatures of the RREQ and RREP, all the nodes of the route will also have verified them.

In the future, when longer keys are needed, ECC results will look even better than with the key lengths used in these simulations. This is due to the fact that, as the key size increases, the computational overhead of ECC increases in a much slower manner than for RSA.

Therefore, these simulations have shown that SAODV used with delayed verification and ECC performs better than the other combinations with SAODV and that the performance penalty it introduces is almost negligible.

10.11   Open Issues

Although it is true that there is no way to preclude a node of inventing many identities, that cannot be used to create an attack against the secure routing algorithm. An attacker cannot supplant another node, and a node can always prove that it is the same node.

Delayed verification makes possible that a malicious node creates invalid route requests that could flood the network. But, the same malicious node can flood the network with perfectly valid route requests, and there would be no easy way to know if it is trying to flood the network or if it is just trying to see if any of its friend nodes are present in the network (for instance).

As explained before, an attacker cannot forge a public/private key pair from an IP address, so the identity token becomes the IP address itself. Users of nodes might have a mechanism outside the network to bind their public key to their physical identity.

With the current technology, SAODV with delayed verification and ECC provides security features to AODV with an almost negligible performance penalty.

In the future, when longer keys are required, the gain of using delayed verification in conjunction to ECC compared to other SAODV options will be even bigger than it is now. This is due to the fact that as key length gets bigger, the cost of signing/verifying in RSA and other cryptoalgorithms increases exponentially as in ECC (for the equivalent key length): it increases in a logarithmic way.

References

[1]   R. Anderson and M. Kuhn. Tamper resistance—a cautionary note. Proceedings of the Second Usenix Workshop on Electronic Commerce, November 1996.

[2]   R. Anderson and M. Kuhn. Low cost attacks on tamper resistant devices. In *IWSP: International Workshop on Security Protocols, LNCS*, 1997.

[3]   F. Anjum.   Location dependent key management using random key-predistribution in sensor networks. In *WiSe '06: Proceedings of the 5th ACM Workshop on Wireless Security*, pp. 21–30, New York, 2006, ACM Press.

[4]   S. Basagni, K. Herrin, D. Bruschi, and E. Rosti.   Secure pebblenets.   In *Proceedings of the 2001 ACM Iternational Symposium on Mobile Ad Hoc Networking & Computing, MobiHoc 2001*, pp. 156–163, Long Beach, CA, October 4–5, 2001.

[5]   E. Biham and A. Shamir. Differential fault analysis of secret key cryptosystems. In *CRYPTO*, pp. 513–525, 1997.

[6]   S. Cheshire and B. Aboba.   Dynamic configuration of Ipv4 link-local addresses.   IETF INTERNET DRAFT—Work in progress, Zeroconf Working Group, June 2001, draft-ietf-zeroconf-ipv4-linklocal-03.txt.

[7] W. Du, J. Deng, Y. S. Han, S. Chen, and P. K. Varshney. A key management scheme for wireless sensor networks using deployment knowledge. In *INFOCOM*, 2004.

[8] L. Eschenauer and V. Gligor. A key management scheme for distributed sensor networks, 2002.

[9] M. Guerrero Zapata and N. Asokan. Securing ad hoc routing protocols. In *Proceedings of the 2002 ACM Workshop on Wireless Security (WiSe 2002)*, pp. 1–10, September 2002.

[10] R. Hauser, A. Przygienda, and G. Tsudik. Reducing the cost of security in link state routing. In *Symposium on Network and Distributed Systems Security (NDSS '97)*, pp. 93–99, San Diego, California, February 1997, Internet Society.

[11] R. Hinden and S. Deering. IP Version 6 Addressing Architecture. RFC 2373, July 1998.

[12] Y. C. Hu, D. Johnson, and A. Perrig. SEAD: Secure efficient distance vector routing for mobile wireless ad hoc networks. In *4th IEEE Workshop on Mobile Computing Systems and Applications* (WMCSA '02), June 2002, pp. 3–13, June 2002.

[13] Y. C. Hu, A. Perrig, and D. Johnson. Ariadne: A secure on-demand routing protocol for ad hoc networks. Technical report TR01-383, Rice University, December 2001.

[14] IANA. Special-use IPv4 Addresses. RFC 3330, September 2002.

[15] R Laboratories. Elliptic Curve Cryptography Standard. Public-Key Cryptography Standards (PKCS) January 13, 1998.

[16] D. Liu and P. Ning. Location-based pairwise key establishments for static sensor networks. In *SASN '03: Proceedings of the 1st ACM Workshop on Security of Ad hoc and Sensor Networks*, pp. 72–82, 2003. ACM Press, New York.

[17] S. Marti, T. J. Giuli, K. Lai, and M. Baker. Mitigating routing misbehavior in mobile ad hoc networks. In *Proceedings of the 6th Annual International Conference on Mobile Computing and Networking*, pp. 255–265, 2000.

[18] A. J. Menezes, P. C. van Oorschot, and S. A. Vanstone. *The Handbook of Applied Cryptography*, CRC Press, Boca Raton, FL, 1996.

[19] G. Montenegro and C. Castelluccia. Statistically unique and cryptographically verifiable (SUCV) identifiers and addresses. Network and Distributed System Security Symposium (NDSS '02), February 2002.

[20] G. O'Shea and M. Roe. Child-proof authentication for MIPv6 (CAM). *ACM Computer Communication Review*, April 2001.

[21] P. Papadimitratos and Z. J. Haas. Secure routing for mobile ad hoc networks. SCS Communication Networks and Distributed Systems Modeling and Simulation Conference (CNDS 2002), January 2002.

[22] A. Perrig, R. Szewczyk, J. D. Tygar, V. Wen, and D. E. Culler. SPINS: Security protocols for sensor networks. *Wireless Networks*, 8(5): 521–534, 2002.

[23] Y. Rekhter, B. Moskowitz, D. Karrenberg, G. J. de Groot, and E. Lear. Address allocation for private internets. RFC 1918, February 1996.

[24] R. Rivest, A. Shamir, and L. Adleman. A method for obtaining digital signatures and public-key cryptosystems. *Communications of the ACM*, 21(2), February 1978.

[25] S. Thomson and T. Narten. IPv6 stateless address autoconfiguration. IETF, RFC 2462, December 1998.

[26] U.S. National Institute of Standards and Technology, Computer Systems Laboratory. Digital Signature Standard (DSS). Federal Information Processing Standards Publication (FIPS PUB) 186, May 1994.

[27] J. Walter, J. Oleksy, and J. Kong. The role of ECDSA in wireless communications. Masters thesis. Computer Science Department. University of California, 2002.

[28] L. Zhou and Z. J. Haas. Securing ad hoc networks. *IEEE Network Magazine*, 13(6): 24–30, November/December 1999.

SECURITY STANDARDS, APPLICATIONS, AND ENABLING TECHNOLOGIES

Chapter 11

Security in Wireless PAN Mesh Networks

Stefaan Seys, Dave Singelée, and Bart Preneel

Contents

In this chapter we analyze the security issues related to wireless personal area mesh networks. We start with a general introduction on wireless PAN networks and discuss the two most common technologies: Bluetooth and ZigBee. Subsequently, we discuss the security architecture and the security weaknesses of Bluetooth and ZigBee, offering advice on how these weaknesses could be mitigated. Finally, we conclude with challenging open research issues.

11.1 Introduction

As more and more mobile devices (i.e., digital cameras, cell phones, GPS receivers) became available on the market, it became apparent that enabling these devices to communicate over wireless links would allow these devices to work together and augment their functionality. In response to this demand of a low-power wireless transmission medium, the Bluetooth Special Interest Group (SIG) was founded in 1998. Bluetooth is essentially a cable-replacement technology that allows for a limited number of devices to communicate with each other via a wireless link.

With further miniaturization of electronic devices, it now becomes possible to manufacture tiny sensor and actuator nodes programmed to provide specific information (e.g., room temperature, light intensity, etc.) or perform specific tasks (e.g., toggle lights, turn on sprinkler systems, etc.). If these sensors can communicate using wireless links and automatically set up large ad hoc networks, this would drastically reduce the costs of deployment. Engineers and researchers soon discovered that Bluetooth or WiFi would not be suitable for this task due to many reasons, the most important being power consumption and the lack of autonomous self-organized operation. This resulted in the IEEE 802.15.4 standard (among others) that was completed in 2003. The ZigBee standard that specifies a set of higher layer protocols to operate on top of IEEE 802.15.4 was released to the public in 2005.

The main differences between Bluetooth and ZigBee are (1) ZigBee is more efficient and allows longer battery lifetime at the cost of lower transmission speeds (100 to 1000 days for ZigBee compared to a couple of days for Bluetooth); (2) Bluetooth only supports networks up to 8 nodes, while ZigBee supports up to 65,536 nodes; and (3) the range of ZigBee (30 m) is larger than the range of Bluetooth (10 m). These differences show that ZigBee is targeted at large control and monitoring networks that should be able to operate for years without maintenance, while Bluetooth is a cable replacement technology that is used between devices that can be regularly recharged.

It is clear that providing security for both types of networks is essential as wireless links are easy to eavesdrop undetected. The fact that these networks run on battery-operated devices with limited processing power means that the security solutions should be as efficient as possible and avoid intensive use of expensive cryptographic operations such as public key encryption or digital signatures. Moreover, these networks normally operate autonomously without access to online key servers or certification authorities. This means that conventional means of key establishment are not always applicable to these networks. To make things even more difficult, ZigBee networks allow multi-hop routing and node mobility. This means that nodes do not have a clear idea of the continuously changing network topology. These specific properties present interesting challenges when designing security and privacy solutions in these environments. In this chapter, we investigate how Bluetooth and ZigBee have implemented their security architecture.

11.1.1   Basic Principles of Bluetooth

In February 1998, the Bluetooth SIG [1] was founded by major players in the telecommunications and network industries: Ericsson, IBM, Intel, Nokia, and Toshiba. In the next six years, several other companies joined the SIG and now there are already more than 3000 members. The major task of this organization was the creation of the Bluetooth specification which describes how mobile phones, computers, PDAs, headsets, and other mobile devices can communicate with each other over a wireless link. In 2000, the Bluetooth standard was included in IEEE 802.15 [2], the Wireless Personal Area Network (WPAN) Working Group. The specifications have been updated several times: the latest version is v2.0, which was published in 2004.

The Bluetooth wireless technology [3,4] realizes a low-cost, short-range wireless voice- and data-connection through radio propagation. The primary use of Bluetooth is cable replacement, most suited for small networks with relatively high load of communication over short distances. With a normal antenna, the maximal range is about 10 m. The Bluetooth wireless technology uses the 2.4 GHz band, which is unlicensed, and can be used by many other types of devices such as cordless phones, microwave ovens, WiFi [5], and baby monitors. Any device designed for use in an unlicensed band should provide robustness in the presence of interference, and the Bluetooth wireless technology has many features to achieve this, including spread spectrum and frequency hopping. Every time a Bluetooth wireless link is formed, it is within the context of a piconet. A piconet consists of maximally eight devices that occupy the same physical channel. In each piconet, there is exactly one *master*, the other devices are called *slaves*. The theoretical maximum bandwidth is 1 Mbps. The real bandwidth is lower because of error correction. One of the main differences between Bluetooth and some other wireless technologies is the ability to connect different types of devices (e.g., a mobile phone with a PDA).

It is possible to configure the "visibility" of a Bluetooth device. When a device is in *non-discoverable mode*, it does not respond to inquiries of other devices. When the device is in *limited discoverable mode*, it is discoverable only for a limited period of time, during temporary conditions or for a specific event. And finally, when it is in *general discoverable mode*, it is discoverable (visible) continuously. Each device is characterized by a factory-established 48-bit identifier, unique for every device: the *Bluetooth hardware address*.

11.1.2 Basic Principles of ZigBee

ZigBee [6] is a specification set of high-level communication protocols that operate on top of the low-power Media Access Control (MAC) and Physical (PHY) layers described in the IEEE 802.15.4 standard for WPANs [2]. In 2003, the IEEE 802.15.4-2003 standard [7] was approved by the TG4 Task Group of the IEEE 802.15 Working Group. The ZigBee v1.0 specifications were ratified in 2004, based on the IEEE 802.15.4-2003 standard. The TG4 Task Group put itself into hibernation in 2004, after forming the TG4b Task Group. The task of TG4b is to write a revision for specific enhancements and clarifications of the IEEE 802.15.4-2003 standard. The ZigBee alliance is now working on the v1.1 specifications that will benefit from these improvements proposed by the 802.15.4b Task Group.

ZigBee is aimed at extending battery lifetimes of low-power devices. The primary use of ZigBee is control and monitoring in wireless sensor networks, most suited for large networks with small load of communication over short distances. The maximum range is about 30 m and the theoretical maximum bandwidth is 250 kbps. ZigBee operates in the same unlicensed 2.4 GHz radio band as Bluetooth. The radios use direct-sequence spread spectrum coding to avoid interference. The technology is intended to be simpler and cheaper than other WPANs such as Bluetooth. The most capable ZigBee node type is said to require only about 10 percent of the software of a typical Bluetooth or Wireless Internet node, while the simplest nodes are about 2 percent. However, actual code sizes are much higher, more like 50 percent of Bluetooth code size. ZigBee chip vendors have announced 128-kilobyte devices. ZigBee uses two kinds of addressing: a 64-bit IEEE address that can be compared to the IP address on the Internet and a 16-bit short address. The short addresses are used once a network is set up. A network can consist of maximally $2^{16} = 65,536$ devices.

There are three different types of ZigBee devices:

1. **ZigBee coordinator**: The most capable device, the coordinator, forms the root of the network tree and might bridge to other networks. There is exactly one ZigBee coordinator in each network. It is able to store information about the network, including acting as the repository for keys. It configures the security level of the network and the address of the trust center. Each network has exactly one ZigBee trust center. This device is trusted by all other devices within the ZigBee network and is responsible for distributing and establishing keys in the network. By default, the ZigBee coordinator is the ZigBee trust center. The coordinator can always designate an alternate trust center. Section 11.3.2 will focus more on the role of the ZigBee trust center.

2. **A ZigBee router** can act as an intermediate router, passing data from other devices.

3. **A ZigBee end** device contains just enough functionality to talk to its parent node (either the coordinator or a router). It cannot relay data from other devices. It requires the least amount of memory, and therefore can be less expensive to manufacture than a ZigBee router or coordinator.

11.1.3 Designing a WPAN Security Architecture

A security architecture is a collection of building blocks and security policies that make up a complete security solution. When designing a security architecture for a specific technology (e.g., Bluetooth or ZigBee), one usually starts by performing a threat analysis. The resulting security requirements of this threat analysis are a set of inputs that are used to obtain the complete set of *functional requirements* for the security architecture. Other inputs include specific user requests, requirements to fulfill an existing API of existing applications, etc. Next to the functional requirements, the design of the security architecture has to take into account the specific *properties and limitations* of the platform it will run on.

Usually a security architecture consists of a layered structure of different building blocks, where a higher layer builds on the services offered by the lower layer. The bottom layer consists of specific implementations (in hardware or software) of cryptographic *algorithms* such as the Bluetooth SAFER+ block cipher. One level up, we can make an abstraction of the specifics of the cipher and supply abstract cryptographic *primitives* such as "block cipher" or "digital signature algorithm" to the higher layer. Using these primitives, it is possible to implement cryptographic *services* such as authentication, encryption, non-repudiation, etc. Going one layer higher, we can use these cryptographic services to build more advanced *security mechanisms* such as end-to-end security (e.g., SSH or IPSec), electronic payment schemes, digital credentials, PKI, key management, etc. Note that the latter two, PKI and key management, are built on top of cryptographic services such as authentication and encryption, but are also required to exchange the keys that are used by the cryptographic services. Finally, one can build complete *applications* on top of the provided security mechanisms. Obviously, the services we can offer at one layer are limited by the services offered by the lower layer. Thus the choice of cryptographic algorithms implemented on a certain platform reflects on the final security mechanisms that can be offered. For example, one cannot offer digital credentials when no digital signature algorithm is available at the bottom layer.

The major constraint when designing a security architecture for mobile devices with limited resources (the target devices of both Bluetooth and ZigBee) is available energy (i.e., battery power) and speed of the CPU. Today this prohibits the use of public key cryptography in the core of the security architecture. Even elliptic curve-based algorithms are still orders of magnitude slower than, for example, the Advanced Encryption Standard (AES) [8,9]. A second important design factor for Bluetooth and 802.15.4 is the fact that they are situated at the MAC layer of the OSI model. The MAC layer has limited functionality concerning communications, and the security architecture should not out grow this functionality.

For the IEEE 802.15.4 standard the design is very clear. It provides the four basic security services: message authentication, message integrity, message confidentiality, and replay protection. These services are all based on the AES block cipher. A higher layer can request four different security settings: no security, encryption only, authentication only, and encryption and authentication (using AES-CCM). Obviously these services require cryptographic keys to operate, but establishing these keys is not part of the IEEE 802.15.4 security architecture and must be provided by the higher layers.

Bluetooth, also a MAC-layer system, does not provide the four basic security services, but does include a mechanism to bootstrap the system based on a shared PIN-code (see Section 11.2.2). Bluetooth does not provide message authentication, meaning that an adversary could alter messages without detection or replay previous messages. However, it can protect the confidentiality of messages. Next to this, there are also differences in the implementation of the Bluetooth and IEEE 802.15.4 security algorithms. Bluetooth uses the E_0 stream cipher (instead of the AES block cipher) for data encryption. E_0 was designed to achieve a high energy efficiency with a small hardware footprint, rather than for speed. Next to this stream cipher, Bluetooth also uses the SAFER+ block cipher for key derivation (it is common practice to use block ciphers for key derivation). Normally E_0 is implemented in hardware, while SAFER+ is implemented in software as it is only used when a new key needs to be negotiated.

ZigBee operates at higher layers (up to the application layer) on top of the IEEE 802.15.4 standard. The ZigBee security architecture provides nodes with a mechanism to establish keys with other nodes in the network. Essentially, two different keys are known in ZigBee: a networkwide broadcast key and link keys that allow two devices to set up end-to-end security (note that in practice there are more keys; see Section 11.3.2). These keys are always established using a third party: the trust center of the network (note that Bluetooth slaves establish keys with each other without the use of the master in the piconet). Another important aspect of ZigBee security is that every layer originating a frame is responsible for securing it. This simplifies the system, because multiple layers are not responsible for securing the same frame. Next to this, all layers are allowed to use the same key that is shared between source and destination (*open trust model*). Finally, ZigBee limits the encryption mode of IEEE 802.15.4 to *CCM** (see Section 11.3.1).

11.2   Bluetooth Security

11.2.1   *Bluetooth Cryptographic Primitives*

Bluetooth uses the synchronous stream cipher [10] E_0 to encrypt data packets. This encryption engine of Bluetooth is schematically depicted in Figure 11.1 [11,12]. E_0 is an autonomous Finite State Machine (FSM).

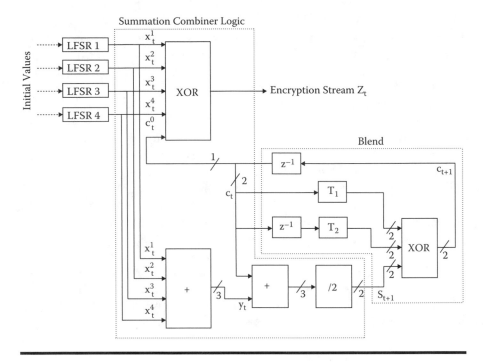

Figure 11.1   Schematics of the E_0 encryption engine.

On every clock cycle, it moves to a new state c_t and produces a single output bit of the key stream Z_t. E_0 makes use of four Linear Feedback Shift Registers (*LFSR1*, ..., *LFSR4*) of lengths $L_1 = 25$, $L_2 = 31$, $L_3 = 33$, and $L_4 = 39$ bits with the following feedback polynomials:

$$LFSR_1 : f_1(t) = t^{25} + t^{20} + t^{12} + t^8 + 1,$$

$$LFSR_2 : f_2(t) = t^{31} + t^{24} + t^{16} + t^{12} + 1,$$

$$LFSR_3 : f_3(t) = t^{33} + t^{28} + t^{24} + t^4 + 1,$$

$$LFSR_4 : f_4(t) = t^{39} + t^{36} + t^{28} + t^4 + 1.$$

The total length of the registers is 128 bits. These primitive polynomials have been chosen as they exhibit the best trade-off between hardware implementation constraints and excellent statistical properties of the output sequences (the polynomials are maximum length *windmill polynomials* [13,14]). Let x_t^i denote the tth symbol of *LFSR$_i$*. The value y_t is the sum over the integers of the four-tuple $x_t^1, x_t^2, x_t^3, x_t^4$. Thus y_t can take the values 0, 1, 2, 3, or 4. The output of the summation generator is obtained by the

Table 11.1 E_0 Linear Bijections

x	$T_1[x]$	$T_2[x]$
00	00	00
01	01	11
10	10	01
11	11	10

following equations:

$$Z_t = x_t^1 \oplus x_t^2 \oplus x_t^3 \oplus x_t^4 \oplus c_t^0,$$

$$S_{t+1} = \left(S_{t+1}^1, S_{t+1}^0\right) = \left\lceil \frac{y_t + c_t}{2} \right\rceil,$$

$$c_{t+1} = \left(c_{t+1}^1, c_{t+1}^0\right) = S_{t+1} \oplus T_1[c_t] \oplus T_2[c_{t-1}],$$

where $T_1[.]$ and $T_2[.]$ are two different linear bijections over $GF(4)$, summarized in Table 11.1, and c_t^0 is the least significant bit of c_t. The stream cipher E_0 needs to be initialized with the initial values for the four LFSRs (altogether 128 bits) and the four bits that specify the values of c_0 and c_{-1}. The 132-bit initial value is derived from three inputs: the encryption key K_C, the Bluetooth hardware address, and the clock of the master (see also Section 11.2.2). With the key stream generator, 200 stream cipher bits are generated, of which the last 128 are fed back into the key stream generator as the initial values of the four LFSRs. The values of c_0 and c_{-1} are kept.

Bluetooth makes use of the key derivation algorithms E_1, E_{21}, E_{22}, and E_3 to map a 128-bit input to a 128-bit output. All of them are based on the SAFER+ block cipher. This is an improved version of the SAFER block cipher, which only works on 64-bit data blocks. An important improvement in SAFER+ is the introduction of the *Armenian Shuffle* permutation, which boosts the diffusion of single bit modifications in the input data. It is a permutation of 16 bytes. SAFER+ consists of:

- A key scheduling algorithm that produces 17 different 128-bit subkeys
- 8 identical rounds
- An output transformation, which is implemented as a bitwise XOR between the output of the last round and the last subkey

Each SAFER+ round calculates a 128-bit word out of two subkeys (the last subkey is used in the SAFER+ output transformation) and a 128-bit input word from the previous round. The central components of the SAFER+ round are the 2-2 Pseudo Hadamard Transform (PHT) [15], the Armenian

Shuffles, and the substitution boxes denoted E and L [11,16]. The PHT takes two input bytes and produces two output bytes, as follows:

$$PHT\,[a,\,b] = [(2a + b) \bmod 256,\, (a + b) \bmod 256]\,.$$

The two mappings E and L introduce nonlinearity and are defined as follows:

$$E\,[x] = (45^x \bmod 257) \bmod 256,$$

$$L\,[x] = y \quad such\ that \quad x = E[y].$$

The structure of one SAFER+ round can be found in [17,18]. For a summary of recent cryptanalytic results, see [19].

11.2.2 Key Agreement Protocol in Bluetooth

The Key Agreement Protocol [20] is a crucial part of the security architecture of Bluetooth [21]. Suppose that two Bluetooth devices, called A and B, want to communicate securely (in the rest of this chapter, we will assume that A initiates the communication). Initially, these devices do not share a secret. They perform a Key Agreement Protocol to generate a *link key* and an *encryption key*. The latter is fed to the stream cipher E_0. The process of generating a shared secret is called *pairing* (two Bluetooth devices are paired when they share a key which can be used to communicate securely).

11.2.2.1 Generation of the Unit Key

When a Bluetooth device is turned on for the first time, it calculates a *unit key*. This is a key that is unique for every device and that is almost never changed. It is stored in non-volatile memory. The unit key is only used if one of the devices does not have enough memory to store session keys (see also Section 11.2.2 for more details). The unit key is based on a random number and the Bluetooth hardware address of the device.

11.2.2.2 Generation of the Initialization Key

At the start of a communication session, the Bluetooth devices do not yet share a session key, and will have to establish one. This is achieved in different steps. First, an *initialization key* is generated. This temporary key is a function of a random number *IN_RAND* (generated by A and sent to B in clear), a shared PIN, and the length L of this PIN. The PIN should be entered in both devices by a user or it can be fed from a higher layer into the pairing procedure. The length of the PIN can be chosen between 8 and 128 bits. Typically, it consists of four decimal digits. If one of the devices does not have an input interface, a fixed PIN can be used (often,

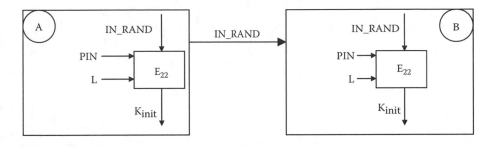

Figure 11.2   Generation of the initialization key.

the default value is 0000). This procedure is shown in Figure 11.2. The result is a temporary shared key: the initialization key. Note that a low-entropy shared secret (the PIN) is used to generate the initialization key. As a consequence, an eavesdropper, which is present during initialization, will know the random number *IN_RAND*.

11.2.2.3   Mutual Entity Authentication

Each time a new shared key is generated (an initialization key or a link key), both devices perform a mutual authentication protocol. The authentication scheme is based on a challenge-response protocol. This protocol is performed twice. First, *B* authenticates itself to *A*, as shown in Figure 11.3. If this authentication is successful, the roles are switched (*B* becomes the verifier and *A* the prover). The authentication goes as follows. *A* generates a random number *AU_RAND* and sends this to *B*. This random number is called the *challenge*. Both devices now compute a response $SRES = E_1(ADDR_B, K_{link}, AU\_RAND)$. $ADDR_B$ is the Bluetooth hardware address of *B* and K_{link} is the shared key (initialization key or link key). *B* sends its response to *A*. If this response corresponds to the value that *A* has calculated, then the authentication is successful. The value *ACO* (*Authenticated Ciphering Offset*) is used for the generation of the encryption key.

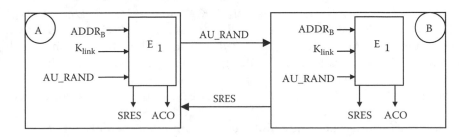

Figure 11.3   Mutual entity authentication protocol.

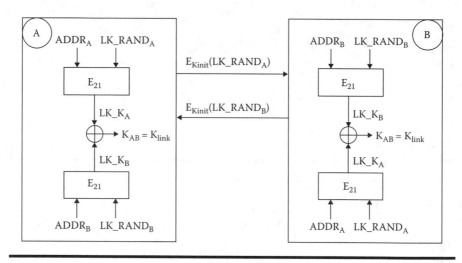

Figure 11.4   The link key is a combination key.

Algorithm E_1 is based on the SAFER+ block cipher, with some small modifications [11].

11.2.2.4   Generation of the Link Key

Both devices now share an initialization key. This key will be used to agree on a new, semi-permanent key (called the link key). The *link key* will be stored on both devices for future communication. Depending on the memory constraints of both devices, the link key can be the unit key of the memory-constrained device or a combination key derived from the input of both devices (Figure 11.4).

If the unit key of device A is the link key, it is transmitted encrypted from A to B. This encryption is done by XORing the unit key of A with the initialization key.

If the link key is a combination key, then both devices first generate a random number *LK_RAND*. These random numbers are encrypted with the initialization key and sent to the other device. Now they both compute $LK\_K_A = E_{21}(LK\_RAND_A, ADDR_A)$ and $LK\_K_B = E_{21}(LK\_RAND_B, ADDR_B)$. The combination key K_{AB} is the XOR of LK_K_A and LK_K_B. This is shown in Figure 11.4. Algorithm E_{21} is based on the SAFER+ block cipher, with some small modifications. After the generation of the link key, the (old) initialization key is definitively discarded and a mutual authentication is started, using the exchanged link key that is shared between both devices (this has already been discussed). The procedure shown in Figure 11.4 is also carried out when a new link key is computed. The only difference is that the random numbers *LK_RAND* are encrypted with the old link key. After the generation of the new link key, the old one will be discarded.

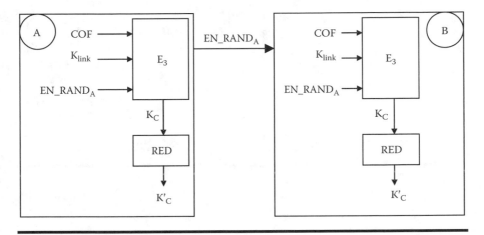

Figure 11.5   Generation of the encryption key.

11.2.2.5   Generation of the Encryption Key and the Key Stream

After a successful generation of the link key and execution of the mutual authentication protocol, the encryption key can be generated. Device A generates a random number EN_RAND_A and sends this to B. Both devices generate the encryption key $K_c = E_3(EN\_RAND_A, K_{link}, COF)$. The COF value (*Ciphering Offset Number*) is the ACO value which was generated during the mutual authentication protocol. However, if the encryption key is used for broadcast, then the COF is the concatenation (denoted by $||$) of the Bluetooth hardware address $ADDR$ of the sender and itself (so $COF = (ADDR || ADDR)$). The encryption key K_C has a length of 128 bits, but its length can be reduced to a truncated encryption key K'_C if necessary. This procedure is shown in Figure 11.5.

Finally, the encryption key K_C (or the truncated key K'_C) is fed to the encryption scheme E_0 together with the Bluetooth hardware address and the clock of the master. These values are used to initialize the four LFSRs of the stream cipher E_0. The output of the cipher is the key stream K_{cipher} (see Figure 11.6). The master clock is used to make the key stream harder to guess.

11.2.3   Security Weaknesses in the Bluetooth Security Architecture

There are several security weaknesses in the Bluetooth standard [21,22]. We now give an overview of the most important security problems.

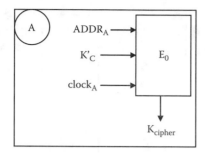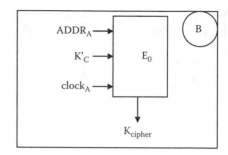

Figure 11.6   Generation of the key stream.

11.2.3.1   Unit Key

The unit key is employed if one of the Bluetooth devices does not have enough memory to store session keys. This key is stored in non-volatile memory and almost never changed. As already described in Section 11.2.2, the unit key is sent encrypted (with the initialization key) to the other device. The result is the following weakness: if A has sent its unit key to device B, then B knows the key of A and can impersonate itself as A to a device C. This impersonation attack is impossible to detect. It is strongly recommended to avoid the use of unit keys!

11.2.3.2   Location Privacy

When two or more Bluetooth devices are communicating, the transmitted packets always contain the Bluetooth hardware address of the sender and the destination (or an identifier which is directly related to these addresses). When an attacker eavesdrops on the transmitted data, he knows the Bluetooth addresses of these devices. The attacker does not have to be physically close to the communicating devices, he can use a device with a stronger antenna (e.g., it is very easy to construct an antenna which can intercept Bluetooth communication from more than one mile away [23,24]) or just place a small tracking device near the two Bluetooth devices.

This way, the attacker can keep track of the place and time these devices were communicating. This is a violation of the privacy of the user. The location information can be sold to other persons or used for location dependent commercial advertisements (e.g., a shop can send advertisements to everybody that is near the shop). It should be possible for the user to decide when his location is revealed and when not.

11.2.3.3   Security Depends on Security of PIN

The initialization key is a function of a random number *IN_RAND*, a shared PIN, and the length L of the PIN. The random number is sent in clear and hence known by an attacker who is eavesdropping during the initialization phase. This means that only the PIN is unknown to the attacker. If an attacker obtains the PIN, he knows the initialization key. Worse yet, because all the other keys are derived from the initialization key, they will also be known by the attacker. Hence the security of the keys used in Bluetooth depends on the security of the PIN. If this value is too short or weak (e.g., 0000), it is very easy for an attacker to guess the PIN (and hence the initialization key). Unfortunately, it is very cumbersome for a user to remember long (and random) numbers.

Note that it is possible to verify a guess of the PIN. The reason is that a mutual authentication protocol is executed after the generation of the initialization key. If an attacker observes this protocol, he obtains a challenge and the corresponding response. The attacker calculates for every guess of the PIN the corresponding response and when this is equal to the observed response, the guess of the PIN was correct. The shorter the PIN, the faster this brute-force attack can be carried out. Shaked and Wool showed that this attack can be optimized by employing an algebraic representation of SAFER+, the cryptographic primitive used in the mutual authentication protocol [16] (see Section 11.2.1). The authors state that a PIN of four digits can be cracked in less than 0.06 seconds on a standard PC. This is a very critical security problem.

11.2.3.4   Denial-of-Service Attacks

Mobile networks are always vulnerable to denial-of-service (DoS) attacks. They consist of mobile devices, and these devices are often battery powered. Bluetooth is no exception. An attacker can send dummy messages to a mobile device. When this device receives a message, it performs some computations, which consumes battery power [25]. After some time, all battery power will be consumed. This exhaustion of the battery power is called the *sleep deprivation attack* [26]. This attack is almost impossible to prevent.

There are also some more advanced DoS attacks, caused by implementation decisions. A nice example is the *black list*, which is used during the mutual authentication protocol. To avoid that a device would start the authentication protocol over and over again (and eventually guess the correct PIN), each device has a black list of the Bluetooth addresses of the devices which failed to authenticate themselves correctly. These devices cannot start an authentication procedure during some period. Each consecutive time the authentication procedure fails, this period is increased exponentially (until a pre-determined upper limit is reached). Candolin discovered that this mechanism can be exploited in several DoS attacks [26]. An attacker can try to authenticate itself to device *A*, but change its Bluetooth hardware address every time. All these authentication attempts will fail and the black list of A will become quite large. If there is no upper limit on this black list, the entire memory of A will be filled with the entries of the black list and device A will crash.

This is not the only DoS attack. Suppose device B wants to authenticate itself to A. After A has sent a challenge to B, the attacker sends a wrong response to A using the Bluetooth hardware address of B. The authentication will fail, B will be put on the black list of A, and the (correct) response of B will be ignored by A. The attacker keeps repeating this attack and B will never be able to authenticate itself successfully to A. Note that the same result could be obtained by jamming the radio signal, but the DoS attacks described above are much easier to perform.

11.2.3.5 Encryption Algorithm E_0

Bluetooth uses the stream cipher E_0 for data encryption. This stream cipher has some security flaws [27–32]; note though that most of the published attacks do not work on the implementation of E_0 in Bluetooth.

The attacks with the lowest complexity are the algebraic attacks [28]. E_0 is vulnerable to algebraic attacks because of the possibility to recover the initial value by solving a system of non-linear equations of degree 4 over the finite field $GF(2)$. This system can be transformed by linearization into a system of linear independent equations with at most 2^{23} unknowns. Fortunately, this attack does not work in Bluetooth because it needs a long key stream during the initialization and E_0 in Bluetooth only uses small packets (the payload ranges from zero to a maximum of 2745 bits [4]).

There are, however, some attacks which can be implemented on the E_0 algorithm in Bluetooth. Most of them are not very efficient, but recently Vaudenay found a practical known-plaintext attack [33]. This is the fastest attack on the Bluetooth encryption scheme. The attack is based on a recently detected flaw in the resynchronization of E_0, as well as the investigation of conditional correlations in the FSM governing the keystream output of E_0. This attack finds the original encryption key for two-level E_0 using the first 24 bits of $2^{23.8}$ frames, requiring 2^{38} computations.

11.2.3.6 Bluejacking

When two Bluetooth devices are paired, these devices will send their "name" to each other. The default name of a device is typically the brand name (e.g., "NOKIA 6110"). The user can, however, change this name in an arbitrary string (up to 248 characters) and this user-defined name will be displayed on the output interface of the other device. The goal of this name is to facilitate the pairing process. First, the device displays a list of all the names of the discoverable devices in the neighborhood. The user then selects the name of the device that it wants to pair its device with. The Bluejacking attack [34] exploits this name to send advertisements to other

Bluetooth devices. The name of the malicious sender is the advertisement itself (e.g., "buy product X now"). A malicious user can try to start a pairing process with all the discoverable devices in the neighborhood and this forces its name to be displayed on the other devices. This is not really a critical security problem, but it can become annoying (e.g., think of the amount of SPAM e-mails a user receives daily). By choosing a misleading name, a malicious device could try to force a pairing process with another device.

11.2.3.7 Implementation Errors

Implementation errors can result in critical security problems. A good example is the *Bluesnarf attack* [35]. It is possible, on some mobile phones, to connect to the device without alerting the owner of the target device of the request, and gain access to restricted portions of the stored data in the phone, including the entire phone book (and any image or other data associated with the entries), calendar, real-time clock, business card, properties, change log, IMEI (*International Mobile Equipment Identity*, which uniquely identifies the phone to the mobile network, and is used in illegal "phone cloning"), etc. This is normally only possible if the device is in *discoverable mode*, but there are tools available that allow even this safety net to be bypassed.

The Bluesnarf attack can also be extended by combining it with a *backdoor attack* [35]. The result of this combined attack is that not only the private data of the mobile phone can data be retrieved, but other services such as access to the Internet, WAP [36], and GPRS gateways or even sending an SMS are available for the attacker without the owner's knowledge. These attacks are caused by implementation errors and hence can be fixed by the vendors.

11.2.3.8 Other Security Problems

There are also some security problems in the challenge-response protocol, which uses the algorithm E_1 and is based on the SAFER+ block cipher. Kelsey et al. [37] discovered a weakness in the key schedule of SAFER+ that allows a key search to be performed slightly faster than by exhaustive search. This attack is only a theoretical issue and does not really endanger the security of Bluetooth. But it indicates that it would be better to replace the SAFER+ block cipher by, for example, AES.

Another security flaw is the lack of integrity checks on the Bluetooth packets. An attacker can always modify a transmitted Bluetooth packet without being detected. Note that encryption in itself does not offer any integrity protection.

Man-in-the-middle attacks are also not prevented in Bluetooth. The reason is that the data is never authenticated by the sender. And there are almost no time stamps or nonces in the protocols, so the freshness of the messages is not guaranteed. Suppose that an attacker has obtained a link key used by two devices. The attacker can now establish a new link with each of the devices, pretending to be the other device. The two devices still believe that they are talking to each other, but in fact they are communicating with the attacker.

To make things even worse, a user can switch off security. Often, the default configuration is no security at all. This certainly has to be avoided.

11.2.4   Bluetooth Security in Practice

Although there are several security problems in the Bluetooth standard, it is certainly possible to use Bluetooth in security-critical applications. Here are some recommendations for designers of Bluetooth applications:

- Avoid the use of unit keys, as this will jeopardize the security.
- Provide data integrity protection in one of the layers on top of Bluetooth. This means that the integrity of the payload cannot be checked in the MAC layer, and that the received data has to be passed to the higher layer. This is, however, still a lot better than no data integrity protection at all.
- If one uses IP over Bluetooth, and the mobile devices are not energy constrained (e.g., a laptop), one can employ standardized solutions like IPSec to protect the security of the Bluetooth link.
- In all the other scenarios, one can implement an advanced pairing protocol [38–41] to securely establish a session key between the mobile devices that want to communicate.
- The use of pseudonyms can make the system robust against tracking. This requires, however, a modification of the Bluetooth standard or specialized hardware.
- Finally, make sure that security is always turned on, certainly in the default configuration (as users tend to use this configuration the most).

11.3   ZigBee Security

ZigBee is a set of communication protocols that operate on the application (APL) and network (NWK) layer. It works on top of the low-power MAC and PHY layer, which are standardized in the IEEE 802.15.4 standard for WPANs. One of the design principles of ZigBee is that the layer that originates a frame is responsible for securing it. So, if an NWK command frame needs protection, NWK layer security shall be employed. Figure 11.7 shows an example of the security fields that may be included in an NWK frame.

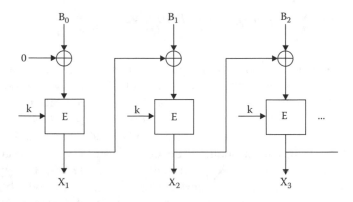

Figure 11.7 (Part of) ZigBee frame with security at the NWK level.

The auxiliary header contains security information (security control, frame counter, etc.), the payload can be encrypted or not, and the Message Integrity Code (MIC) is used to protect the integrity of both header fields and the payload (the *security control field* in the auxiliary header specifies the level of security that is applied to the frame). Both encryption and message integrity are provided by one building block: the CCM^* algorithm. Security information is stored in Access Control Lists (ACLs). Each ACL entry contains the following security information: destination address, security control field, key, nonce, and the key and frame counter. The frame counter is incremented by one for every outgoing frame. The maximum value is $2^{32} - 1$. When a new key is used, the frame counter is reset to 0. There is always a default ACL entry which is used if there is no specific ACL entry for the destination. There can be maximally 255 ACL entries. The exact amount of ACL entries is vendor specific.

ZigBee uses the *open trust model* [6]. This implies that all different layers of the communication stack, and all applications running on a single device, trust each other. Keys can be reused in each layer. To simplify interoperability, the security level used by all devices in a given network and by all layers of a device shall be the same. If protection from theft of service is required, NWK layer security shall be used for all frames. The network key (NWK key) is a broadcast key that is used by all devices in the same network. As a consequence, using an NWK key does not prevent insider attacks. The NWK key is updated regularly and is stored in the default ACL entry. To distinguish between the different NWK keys and to make sure that every device in the network is using the most recent NWK key, a sequence number (called the *key counter*) is assigned to every NWK key. The NWK key is only used in the NWK layer. If application layer security is applied, a link key is used to protect outgoing frames. Link keys are employed to enable end-to-end security (between source and destination device).

11.3.1 ZigBee Cryptographic Primitives

11.3.1.1 CCM* Algorithm

*CCM** is a generic combined encryption and authentication block cipher mode. *CCM** is only defined for use with block ciphers with a 128-bit block size. The block cipher that is used in the ZigBee specification is the AES-128. The *CCM** mode is a minor modification of the *CCM* mode specified in the IEEE 802.15.4 MAC layer specification [7]. *CCM** includes all of the features of *CCM* and additionally offers encryption-only and integrity-only capabilities. In total, there are eight possible security levels: the payload of a frame can be encrypted or not, and the length of the MIC, which protects the integrity of the header fields and the payload of a frame, can be 0, 32, 64, or 128 bits. The security control field in the header specifies which security level is used to secure the frame. As the *CCM* mode, the *CCM** mode requires only one 128-bit key. Together with this key, a unique 104-bit nonce N is used. This nonce is a function of the security control field, the frame counter, and the address of the sender. Within the scope of a key, the nonce value should be unique. The frame counter prevents reusing a nonce under the same key.

An authentication tag T is computed as follows (see also Figure 11.8):

$$T = X_{t+1},$$

$$X_{i+1} = E(key, X_i \oplus B_i) \quad for \quad i = 0, \dots, t.$$

E is the block cipher AES-128, $B_1 \| \dots \| B_t$ are the t data blocks that have to be integrity protected (each block has a length 128 bits), B_0 is a data block that contains the nonce N and some constants, and X_0 is a 128-bit block containing only 0s. The authentication tag T holds the M left-most bits of the output X_{t+1}. The value M specifies the length (in bytes) of the MIC. Note that the block cipher is used in Cipher-Block Chaining (CBC) mode [10].

NWK Header	Auxiliary Header	(Encrypted) NWK Payload	MIC

Figure 11.8 *CCM** **authentication block cipher mode.**

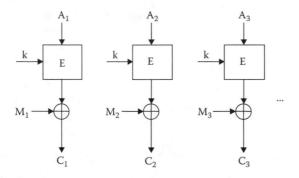

Figure 11.9 *CCM** encryption block cipher mode.

Encryption is performed as follows:

$$A_i = Flags\|N\|i \quad for \quad i = 1, \ldots, t, \quad\quad (11.1)$$

$$C_i = E(key, A_i) \oplus M_i \quad for \quad i = 1, \ldots, t, \quad\quad (11.2)$$

$$S_0 = E(key, A_0).$$

First, the 128-bit blocks A_i are computed. They contain the constant value *Flags* (8-bit representation of the value 1), the nonce N, and a 16-bit counter i. These blocks are fed to the block cipher AES-128. The output is XORed with the t data blocks M_i that have to be encrypted (each block has a length of 128 bits), and the result is the t cipher text blocks C_i (see also Figure 11.9). The M left-most bits of block S_0 are XORed with the authentication tag T. The result is the encrypted authentication tag U. The MIC is equal to T or U (depending on if encryption is applied or not), and the encrypted payload to $C_1\| \ldots \|C_t$.

11.3.1.2 The AES Algorithm

AES is a symmetric block cipher with a block-length of 128 bits and three different key sizes: 128, 192, and 256 bits. The three resulting algorithms are referred to as AES-128, AES-192, and AES-256. The cipher is based on a round operation that is repeated a number of times. Each round has two inputs: a round-key of 128 bits and the result of the previous round. The round-keys can be pre-computed or generated on-the-fly out of the input key. Every round consists of four steps: Byte Substitution, Shift Rows, Shift Columns, and Add Round Key (this simply XORs the round-key with the current block). The number of rounds depends on the size of the key: 9, 11, and 13 rounds for 128-, 192-, and 256-bit keys, respectively. Due to its regular structure, AES can be implemented very efficiently in hardware and software. Computational performance of software implementations often differs between encryption and decryption because the inverse operations in the round function are more complex than the according operation for encryption. For further information, we refer to [8].

11.3.2 Security Architecture of ZigBee

11.3.2.1 Key Hierarchy

Several types of keys are used in ZigBee, forming a key hierarchy. Typically, the security manager of a device (situated in the application layer) will perform the following steps:

1. Obtain the trust center master key: Initially, each device shares a trust center master key with the trust center. The device can obtain this trust center master key (together with the address of the trust center) in two ways: the device acquires the trust center master key via insecure key-transport (e.g., it is sent in clear from the trust center to the device at low power) or it acquires this key via pre-installation (e.g., factory installation or based upon data entered by a user). It is very important that no other device can obtain this trust center master key, as the security of all other keys used in ZigBee depends on the confidentiality of the trust center master key.

2. Establish link key with trust center: The trust center and the device share a trust center master key and will execute the Symmetric-Key Authenticated Key Agreement (SKKE) protocol to establish a link key with each other. First, both devices generate a random 128-bit challenge (QEU and QEV, respectively) and send it to the other device. These challenges are fed, together with the trust center master key, to a key derivation function. The result is two 128-bit keys: the *MacKey* and the *KeyData*. The former is the key of an MIC, used to mutually authenticate the challenges QEU and QEV. After a successful authentication, both devices will use the *KeyData* key as shared link key. This link key will be employed to secure the communication between the trust center and the device.

3. Compute key-load key: The key-load key is derived from the link key as follows:

$$\text{key-load key} = HMAC_{\text{link key}}(0 \times 02).$$

Here, *HMAC* is a keyed message authentication code [10]. This type of MAC function uses a cryptographic hash function in combination with a secret key. The trust center uses the key-load key to transport an application master key securely to a device.

4. Compute key-transport key: The key-transport key is derived from the link key as follows:

$$\text{key-transport key} = HMAC_{\text{link key}}(0 \times 00).$$

The trust center uses the key-transport key to transport an application link key or an NWK key securely to a device.

5. Obtain the NWK key: The trust center puts the NWK key (that is currently being used in the network) in a specially constructed command frame, secures it with the key-transport key, and transmits it to the device. The NWK key is used to encrypt broadcast communication in the network. Note that command frames are always encrypted and integrity protected (with a 128-bit MIC).

6. Obtain the application link key: When two devices in a network want to communicate securely (end-to-end), they need an application link key. One way to obtain such an application key is as follows: the trust center generates the application link key and puts it in a specially constructed command frame. This frame is sent securely to each device. The security of the frame is protected by employing the key-transport key. The advantage of the trust center sending out the application link keys directly is that key-escrow can be implemented.

    a. Obtain the application master key: Instead of directly transmitting the application link key to both devices, the trust center can also generate an application master key. It puts this key in a specially constructed command frame, and sends this securely to both devices. The security of this frame is protected by employing the key-load key.

    b. Establish application link key with other devices: After the devices obtained the application master key, they execute the SKKE protocol. This is done exactly as described above. The only difference is that the application master key is used to derive the link key, instead of the trust center master key. The output of the SKKE protocol is the application link key, which is used for end-to-end security between both devices.

The above is only valid if the trust center is working in commercial mode. When the trust center works in residential mode, the device will not establish a link key with other devices. A more detailed discussion on the modes of operation of the ZigBee trust center is now presented.

11.3.2.2 ZigBee Trust Center

There is always exactly one trust center in each secure ZigBee network. This device is often the ZigBee coordinator and is trusted by all devices in the network. It is responsible for the distribution of keys (link keys and NWK keys) among the ZigBee devices. The ZigBee trust center also enforces the policies in the network. These policies state how a device can join or leave the network (securely or insecurely), if and when keys have to be updated, etc. The trust center can be configured to operate in either *commercial* or *residential* mode:

■ The *commercial mode* of the trust center is designed for high-security commercial applications. In this mode, the trust center maintains a list of devices, master keys, application link keys, and NWK keys that it needs to control. It also enforces the policies of NWK key updates and network admittance. In this mode, the memory required for the trust center grows with the number of devices in the network. When the trust center works in commercial mode, it shall follow the steps of the key hierarchy described above.

■ The *residential mode* of the trust center is designed for low-security residential applications. In this mode, the trust center maintains a list of the NWK keys and controls the policies of network admittance. It does not have to maintain a list of devices, master keys, or application link keys. When operating in residential mode, the NWK key is never updated, and therefore the memory required for the trust center does not grow with the number of devices in the network. This limits the implementation complexity, but also reduces the security. When the trust center works in residential mode, it shall not follow the steps of the key hierarchy described above. Instead, it will just send the NWK key to a device joining the network via insecure key transport. This key is used to secure communication. Master keys and link keys are not employed.

11.3.3 Security Weaknesses in the ZigBee Security Architecture

Improper use of the security mechanisms in ZigBee can cause several security problems [42,43]. ZigBee has, however, solved some security issues that were present in the IEEE 802.15.4 standard [6], e.g., limiting the encryption mode to CCM^* in ZigBee avoids the employment of dangerous security modes, like AES-CTR. We now give an overview of the most important security problems that still remain in ZigBee. Designers of ZigBee applications should take this into account during implementation.

11.3.3.1 IV (Nonce) Management Problems

As already discussed in the previous section, security information is stored in ACLs. Each ACL entry contains the following security information: destination address, security control field, key, nonce, and the key and frame counters. The nonce is a function of the security control field, the frame counter, and the address of the sender. Only the frame counter is really variable, and as a consequence, the nonce is derived directly from the frame counter. Suppose one would encrypt two messages (M_1 and M_2) with the same key and the same nonce. According to Equation 11.1, reusing a nonce results in reusing the block A_i. If we apply Equation 11.2, one obtains the following result:

$$C_1 \oplus C_2 = E(key, A_i) \oplus M_1 \oplus E(key, A_i) \oplus M_2 = M_1 \oplus M_2 .$$

This should certainly be avoided! Fortunately, the frame counter prevents reusing a nonce under the same key. There is, however, a problem if a key is used in two different ACLs (because in this case, the frame counter in each ACL is updated independently and this could result in the reuse of a nonce) or if a nonce is reused in the same ACL (without the key being updated). The latter can occur when a power failure arises. If the frame counter is stored in volatile memory, and the key in non-volatile memory, then the frame counter would be reset to zero after the power failure. The key, however, would remain the same, and one would reuse the nonce under the same key. To avoid this problem, the frame counter and the key should be stored together in non-volatile memory. The same problem would occur if one would use a key that has been employed before, but the probability of such an event to occur is very low.

11.3.3.2 Improper Support of Group Keying

ZigBee does not support group keying. The reason is that each ACL can only contain the address of one destination. Let us assume that one would use multiple ACLs, one for each destination in the group. Then the probability of reusing a nonce would become very large. As explained above, a nonce should never be reused under the same key. If one would use one ACL for the entire group, then one always has to update the address of the destination beforehand (otherwise, the device cannot find the correct ACL entry in its memory). This is not possible, because one would have to know in advance which device is going to send the next message, and normally a device does not have this knowledge. Another problem would be that each device in the group has to update the frame counter every time a message is sent to one of the group members, also when it was not intended for the device itself. So ZigBee only supports secure unicast and broadcast communication, and no secure multicast communication.

11.3.3.3 Key Management

The ZigBee standard states that there can be maximally 255 ACL entries. The exact amount of ACL entries is vendor specific and often much lower than 255. As an example, the Chipcon CC2420 has support for only two ACL entries [43]. The number of application link keys a device can maximally share with other devices is equal to the number of ACL entries. So in the best case, it can only share a key with 255 other ZigBee devices, which is considerably less than the maximum amount of 65,536 devices in a ZigBee network. A better support for secure end-to-end communication is needed.

11.3.3.4   Replay Attacks

Every time a message is transmitted to another device, the frame counter is incremented by one. This prevents replay attacks, as frames with a lower frame counter than stored in the ACL will be discarded. This can, however, cause a security problem in broadcast communication. In a ZigBee network, broadcast communication is secured with the NWK key, which is stored in the default ACL. Every time a message is broadcasted, each device in the network should increment the frame counter in its default ACL. If a device goes to *sleep mode* and does not receive broadcast messages for a certain time, it cannot send any broadcast message anymore. The frame counter in its default ACL will have a lower value than the one in the default ACL of the other devices, and a message with a lower frame counter will be discarded by the other devices, as they wrongfully detect this event as a replay attack. As a consequence, a device can never go to sleep mode, and this can have an important influence on the battery lifetime of a ZigBee device. Requiring each device in the network to update its frame counter regularly causes some key management problems and is not very practical. It would be better not to increment the frame counter in case of broadcast communication, but this would enable replay attacks.

11.3.3.5   Initialization Procedure

The secure initialization and installation of the master key determines the security of the other keys. When an attacker obtains the trust center master key, this would compromise the security of the other keys used in ZigBee, as they are all derived from the trust center master key.

A device can obtain the trust center master key (and the address of the trust center) in two ways: via insecure key-transport or via pre-installation. The former is the easiest method, but also the most insecure one. Transmitting a key at low power, as suggested in the ZigBee standard, does not provide sufficient protection. The attacker can build a ZigBee device with a strong directional antenna and intercept communication from a long distance. Assuming that there is no attacker present during the insecure key-transport is a very dangerous assumption. Theoretically, insecure key-transport is only secure when it is conducted in a Faraday cage. This is, however, not very practical. That is why it is recommended to obtain the trust center master key via pre-installation. This is more awkward, but provides more security. For example, one could install the trust center address and master key during the fabrication of the ZigBee device. There are, however, some practical problems. One does not always know in advance in which network the ZigBee device will be employed. Deriving the trust center master key from data entered by a user (a password) can be dangerous. Users tend to use low-entropy passwords, and an attacker can try all passwords or perform a dictionary attack. Because the SKKE protocol, used to establish a link key, contains a key confirmation step, an attacker can easily verify every guess of the password.

That is why ZigBee needs a secure initialization procedure (e.g., install the keying information via out-of-band mechanisms [38–41,44]). This is a critical security problem that has yet to be solved.

11.3.3.6   Location Privacy

The header of a ZigBee frame, which is never encrypted, contains the address of the source and destination device. This address is either the 64-bit IEEE address, or a 16-bit short address (used once the network is set up). When an attacker eavesdrops on the transmitted data, he knows the addresses of the devices that were communicating. It is possible for an attacker to construct a stronger antenna to intercept ZigBee communication from a further distance. As a consequence, an eavesdropper does not have to be physically close to the communicating devices.

This way, the attacker can keep track of the place and time that ZigBee devices are communicating. This is a violation of privacy. The problem, however, is less critical than in Bluetooth. In contrast to Bluetooth devices, ZigBee devices do not always belong to a specific user, but are usually used in small sensor networks. In that case, information about the place and time a ZigBee device is communicating might not be very interesting for an attacker.

11.3.3.7   Insufficient Integrity Protection

In total, there are eight security levels that can be employed to secure a frame. The payload can be encrypted or not, and the frame can contain an MIC of 0, 32, 64, or 128 bits. As a consequence, it is possible to apply encryption and no integrity protection on a frame. This is a dangerous mode of security and should never be used. Encryption in itself does not provide integrity protection. As shown in Equation 11.2, the cipher text C_i is the XOR of the plaintext message M_i and the encryption of a block A_i. This means that if the attacker changes the jth bit of C_i, the same bit will change in the message M_i. This can have important consequences. Fortunately, the ZigBee standard states that all ZigBee command frames should be encrypted and integrity protected with a 128-bit MIC.

11.4 Conclusion and Open Issues

We have evaluated the security architectures of both the Bluetooth and ZigBee standards. We can conclude that both Bluetooth and ZigBee have some (minor) security weaknesses. However, it is still possible to use these systems in a secure way, if the necessary precautions are taken. The security weaknesses in Bluetooth range from design problems (e.g., the use of unit keys) to problems with the cryptographic algorithms that are used (e.g., weaknesses in the E_0 and SAFER+ ciphers). Many of the problems can be mitigated using some practical guidelines (see Section 11.2.4). The problems with the cryptographic ciphers can only be solved by replacing these ciphers or by "patching" them, for example, by switching keys before an adversary has enough data to determine the key. ZigBee already solves a number of the security problems of IEEE 802.15.4 by only allowing the CCM^* mode, but still has a number of security problems that should be solved in the next version of the standard.

The main difference between the Bluetooth and ZigBee security architectures is that Bluetooth is limited to the MAC layer, but the ZigBee standard also includes the application layer. This results in the fact that Bluetooth only allows the establishment of link keys between two nodes that are within range, but ZigBee allows *any two nodes* to establish a shared key. Therefore, ZigBee is more tailored toward wireless mesh networks than Bluetooth.

One important issue that has not been solved by either Bluetooth or ZigBee is location privacy. Both standards allow an adversary to track the location of devices using the unique identity of the source that is included in every frame. To solve this, advanced solutions are required that hide the identity of the devices by employing one-time pseudonyms instead of the fixed identifiers.

A second important open issue is how to securely initialize the security mechanisms that are available in a WPAN. Bluetooth only offers the use of a PIN that has to be manually entered by the user. One potential solution here could be the use of more advanced pairing protocols. For large scale ad hoc networks such as ZigBee, initializing the security mechanisms is even harder. An ideal initialization procedure should be very efficient (meaning that extensive use of public key cryptography should be avoided), user friendly (no or very limited user interaction required), and flexible to many different scenarios in which these networks will be deployed.

References

[1] Bluetooth Special Interest Group (http://www.bluetooth.com/).
[2] The Wireless Personal Area Network Working Group, IEEE 802.15 (http://www.ieee802.org/15/).

[3] J. Haartsen, M. Naghshineh, J. Inouye, O. Joeressen, and W. Allen, Bluetooth: Visions, goals and architecture, *ACM SIGMOBILE Mobile Computing and Communications Review*, Volume 2, Issue 4, 1998, pp. 38–45.

[4] Bluetooth Specification (https://www.bluetooth.org/spec/).

[5] The Wi-Fi Alliance (http://www.wi-fi.org/).

[6] The ZigBee Alliance (http://www.zigbee.org/).

[7] IEEE 802.15.4-2003 Standard, Wireless Medium Access Control and Physical Layer Specifications for Low-Rate Wireless Personal Area Networks, 2003.

[8] J. Daemen and V. Rijmen, *The design of Rijndael—AES: The Advanced Encryption Standard*, Springer-Verlag, 2002.

[9] S. Seys, Cryptographic Algorithms and Protocols for Security and Privacy in Wireless Ad Hoc Networks, Ph.D. thesis, Katholieke Universiteit Leuven, 2006.

[10] A. Menezes, P. Van Oorschot, and S. Vanstone, *Handbook of applied cryptography*, CRC Press, 1996.

[11] C. Gehrmann, J. Persson, and B. Smeets, *Bluetooth security*, Artech House, 2004.

[12] E. Filiol, Zero-knowledge-like Proof of Cryptanalysis of Bluetooth Encryption, 2006.

[13] B. Smeets and W. Chambers, Windmill generators—A generalization and an observation of how many there are, Advances in Cryptology EUROCRYPT 1988, Lecture Notes in Computer Science, Vol. 330, Springer-Verlag, 1988, pp. 325–330.

[14] B. Smeets and W. Chambers, Windmill PN-sequence generators, *Computers and Digital Techniques*, Volume 136, Issue 5, 1989, pp. 401–404.

[15] H. Lipmaa, On differential properties of Pseudo-Hadamard Transform and related mappings, Progress in Cryptology, INDOCRYPT 2002, Lecture Notes in Computer Science 2551, Springer-Verlag, 2002, pp. 15–18.

[16] Y. Shaked and A. Wool, Cracking the Bluetooth PIN, 3rd International Conference on Mobile Systems, Applications, and Services (MobiSys '05), 2005, pp. 39–50.

[17] J.L. Massey, G.H. Khachatrian, and M.K. Kuregian, SAFER+, Cylink Corporation's Submission for the Advanced Encryption Standard, 1998.

[18] J.L. Massey, On the Optimality of SAFER+ Diffusion, Proceedings of the 2nd Advanced Encryption Standard Candidate Conf (AES2), 1999.

[19] NESSIE Project, New European Schemes for Signatures, Integrity, and Encryption (http://www.cryptonessie.org/).

[20] G. Lamm, G. Falauto, J. Estrada, and J. Gadiyaram, Security Attacks against Bluetooth Wireless Networks, Second Annual IEEE Workshop on Information Assurance and Security, 2001, pp. 265–272.

[21] D. Singelée and B. Preneel, Review of the Bluetooth security architecture, *Information Security Bulletin*, Volume 11, Issue 2, 2006, pp. 45–53.

[22] M. Jakobsson and S. Wetzel, Security Weaknesses in Bluetooth, Cryptographer's Track at the RSA Conference (CT–RSA '01), Lecture Notes in Computer Science 2020, Springer-Verlag, 2001, pp. 176–191.

[23] DEF CON, Computer Underground Hackers Convention (http://www.defcon.org).

[24] H. Cheung, The Bluesniper Rifle, 2004.

[25] A. Hodjat and I. Verbauwhede, The Energy Cost of Secrets in Ad-Hoc Networks, IEEE Workshop on Wireless Communications and Networking (CAS '02), 2002.

[26] C. Candolin, Security Issues for Wearable Computing and Bluetooth Technology, 2000.

[27] C. De Cannière, T. Johansson, and B. Preneel, Cryptanalysis of the Bluetooth Stream Cipher, COSIC internal report, Department of Electrical Engineering, Katholieke Universiteit Leuven, 2001.

[28] N. Courtois and W. Meier, Algebraic Attacks on Stream Ciphers with Linear Feedback, Advances in Cryptology—EUROCRYPT 2003, Lecture Notes in Computer Science 2656, Springer-Verlag, 2003, pp. 345–359.

[29] S. Fluhrer and S. Lucks, Analysis of the E0 Encryption System, 8th Annual International Workshop of Selected Areas in Cryptography (SAC 2001), Lecture Notes in Computer Science 2259, Springer-Verlag, 2001, pp. 38–48.

[30] J. Golic, V. Bagini, and G. Morgari, Linear Cryptanalysis of Bluetooth Stream Cipher, Advances in Cryptology—EUROCRYPT 2002, Lecture Notes in Computer Science 2332, Springer-Verlag, 2002, pp. 238–255.

[31] M. Hermelin, and K. Nyberg, Correlation Properties of the Bluetooth Combiner Generator, 2nd International Conference on Information Security and Cryptology (ICISC '99), Lecture Notes in Computer Science 1787, Springer-Verlag, 1999, pp. 17–29.

[32] F. Armknecht, J. Lano, and B. Preneel, Extending the Resynchronization Attack, 11th Annual International Workshop of Selected Areas in Cryptography (SAC 2004), Lecture Notes in Computer Science 3357, Springer-Verlag, 2004, pp. 19–38.

[33] Y. Lu, W. Meier, and S. Vaudenay, The Conditional Correlation Attack: A Practical Attack on Bluetooth Encryption, Advances in Cryptology — CRYPTO 2005, Lecture Notes in Computer Science 3621, Springer-Verlag, 2005, pp. 97–117.

[34] Bluejacking (http://www.bluejackq.com/).

[35] A. Laurie and B. Laurie, Serious Flaws in Bluetooth Security Lead to Disclosure of Personal Data, 2003.

[36] D. Singelée and B. Preneel, The Wireless Application Protocol (WAP), *International Journal of Network Security*, Volume 1, Issue 3, 2005, pp. 161–165.

[37] J. Kelsey, B. Schneier, and D. Wagner, Key Schedule Weaknesses in SAFER+, 2nd Advanced Encryption Standard Candidate Conference, 1999, pp. 155–167.

[38] D. Balfanz, D. Smetters, P. Stewart, and H. Wong, Talking to Strangers: Authentication in Ad hoc Wireless Networks, Network and Distributed System Security Symposium (NDSS 2002), The Internet Society, 2002.

[39] J. H. Hoepman, The Ephemeral Pairing Problem, Financial Cryptography, Lecture Notes in Computer Science 3110, Springer-Verlag, 2004, pp. 212–226.

[40] J. H. Hoepman, Ephemeral Pairing on Anonymous Networks, 2nd International Conference on Security in Pervasive Computing (SPC 05), Lecture Notes in Computer Science 3450, Springer-Verlag, 2005, pp. 101–116.

[41] D. Singelée and B. Preneel, Improved Pairing Protocol for Bluetooth, in Proceedings of the 5th International Conference on Ad-Hoc Networks and Wireless (ADHOC-NOW 2006), Lecture Notes in Computer Science 4104, T. Kunz, and S. S. Ravi (Eds.), Springer-Verlag, 2006, pp. 252–265.

[42] F. Perez, Security in Current Commercial Wireless Networks: A Survey, 2006, http://www.hig.no/imt/file.php?id=1098/.

[43] N. Sastry and D. Wagner, Security Considerations for IEEE 802.15.4 Networks, ACM Workshop on Wireless Security (WISE 04), 2004, pp. 32–42.

[44] F. Stajano and R. Anderson, The Resurrecting Duckling: Security Issues in Ad Hoc Wireless Networks, 7th International Workshop on Security Protocols, Lecture Notes in Computer Science 1796, Springer-Verlag, 1999, pp. 172–182.

Chapter 12

Security in Wireless LAN Mesh Networks

Nancy-Cam Winget and Shah Rahman

Contents

A technology that is sure to affect our lives significantly over the next few years is wireless mesh networking. Wireless mesh as a technology has been around almost as long as wireless LANs, but has only recently become more popular. As the popularity of wireless mesh networks grows, end users are demanding higher bandwidth, greater coverage, improved reliability, and robust security. The industry has come together at various IEEE 802 work groups to standardize wireless mesh networks with the right ingredients and the right framework. Security is one of the cornerstones of making the disruption which is believed to be a reality with WLAN mesh networks. The WLAN mesh networking task group at IEEE codenamed TGs has reached the first-draft specification stage, where security specification is now essential. Security aspects of WLAN mesh networks entail a vast array of features and requirements to ensure that robust security is achieved at every link of the mesh network. The roadmap for TGs is to develop a full, official Extended Service Set or ESS mesh standard including mesh transport security (versus end-to-end security) specifications targeted to complete around 2009 [1].

12.1   Introduction

Wireless mesh networks have drawn a lot of attention in various market segments, including home and small business networks, medium and large enterprise networks, public safety, emergency and first-responder networks, service providers and wireless broadband networks, municipal and public access networks, and military and tactical networks. One of the core components in making WLAN mesh networks successful and an enabler into all these different markets is security. A core challenge in securing the WLAN mesh network is the large number of communication links over the air; as each mesh device is mobile and deployed outdoors, each mesh link presents an exposure and vulnerability into the mesh network.

Original mesh architectures emerged from mobile ad hoc networks (MANETs) for military networks. The IETF MANET Work Group has been developing various MANET protocols for almost a decade [2–5]. MANETs were envisioned to be military and tactical networks where peer nodes could either come with or gain mutual trust between them. Mesh networks are different from MANETs in that there is more infrastructure communication rather than direct, peer-to-peer communication with mesh networks becoming a popular deployment in public spaces. Especially in the metropolitan space, existing IEEE networks' security standards 802.1X [6] and 802.11i-2007 [29] based security mechanisms lack the specificity for securing the WLAN mesh network. Even though many vendors are using strong 128-bit encryption to relay client and infrastructure traffic over the air, as previous wireless LAN attacks have shown, a cunning hacker may not necessarily need to crack the key to get user information or damage the network. Security researcher Shawn Merdinger says that municipal metro deployments are going to be "a very serious security challenge to many people" [8].

The rest of the chapter walks through the links and definitions in WLAN mesh networks from the security perspective; challenges and possible attacks in WLAN mesh networks; mesh client security; mesh infrastructure security; authentication, authorization, and access control; confidentiality and privacy in mesh networks; and key management in WLAN mesh networks.

12.2   WLAN Mesh Primer

It is important to carefully define WLAN mesh components and segments for examining the security implications on the overall mesh network. From a security perspective, there are two major components of a mesh network:

1. A wired or bridged segment: The network attached to a mesh network and that operates over the wire, e.g., Ethernet or fiber. One or more of these segments may be attached to a mesh network.
2. A wireless or mesh segment: The all-wireless network that may or may not be attached to a wired or bridged segment. The transport media of this segment is IEEE 802.11 for WLAN mesh networks. This segment is commonly referred to as a mesh network.

Wired and bridged segments of the network are generally considered outside the scope of a WLAN mesh network. However, they may impact security in a mesh network by launching attacks or injecting carefully crafted frames into it. Hence, it is important to secure the entry points from these segments into a mesh network.

The mesh segment of the network requires careful security considerations as it is exposed to attackers as frames are transmitted over the air. There are two major sub-components of this segment:

1. Mesh backhaul: A mesh backhaul consists only of mesh nodes and mesh links. This is an all-wireless, multi-hop network helping WLAN client traffic to traverse over 802.11 links to and from a wired entry point or other WLAN clients in mesh.

2. Mesh access: A mesh access consists of mesh nodes co-located with WLAN access points and WLAN clients. This single-hop network allows end users to connect to a mesh network.

A mesh node is a physical or logical entity in a mesh network participating in formation of a mesh. TGs define mesh nodes as either a mesh point (MP, capable of forming links between mesh nodes only) or a mesh access point (MAP, capable of forming links between mesh nodes as well as links between mesh nodes and WLAN clients). There is a special mesh node, which interfaces a mesh network to wired or non-WLAN bridged networks, called mesh portal or MPP. Common mesh node architectures include:

■ Single-radio node: A mesh node consisting a single IEEE 802.11b/g or 802.11a radio. This node commonly is an MAP allowing user access on the same radio where mesh backhaul links are formed. An MP with a single-radio allows only mesh backhaul links over its radio.

■ Dual-radio node: A mesh node consisting of two IEEE 802.11b/g or 802.11a radios (in any combination), one dedicated for forming mesh backhaul links, the other dedicated for allowing user access. This architecture is common today where lower-capacity and lower-cost radio (such as 802.11b) is used for client access and higher-capacity radio (such as 802.11a) is used for mesh backhaul.

■ Multi-radio node: A mesh node consisting of multiple IEEE 802.11b/g or 802.11a radios (in any combination). Multiple radios may be used for allowing user access and multiple radios may be used for mesh backhaul. Typically, mesh backhaul forming on different radios dedicates one radio for frame transmission and another for frame reception. Another common division of labor occurs for separating upstream and downstream traffic of mesh backhaul to and from an MPP.

A mesh link is a logical 802.11 WLAN link between two MPs or MAPs. An access link is a logical 802.11 WLAN link between an MAP and a WLAN client. A mesh network consists of both types of links, whereas a mesh backhaul consists only of mesh links. Typically, access links are simple radio-links set up and operated according to IEEE 802.11 standards. The mesh links are more complicated and two mesh nodes can have connections over multiple radios. Such links are common in mesh networks where

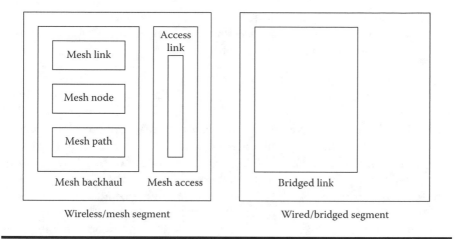

Figure 12.1   Links, nodes, and segments for WLAN mesh security.

multi-radio mesh nodes are deployed. Figure 12.1 shows how all different components and segments come together in a WLAN mesh network.

12.3   Security in WLAN Mesh Networks

Because WLAN mesh networks are based on original WLAN networks, we first look at WLAN security protocol standards and how they are deployed. We then examine how and where these protocols are not sufficient for WLAN mesh security.

12.3.1   WLAN Security Background

The first IEEE 802.11 standard included a weak security protocol called WEP (Wired Equivalent Privacy), which failed to provide the goal of wired equivalence [26,28]. These flaws and the adoption of NIST-approved ciphers were addressed by the ratification of the IEEE 802.11i [7] amendment in 2004 and its inclusion in the base IEEE 802.11-2007 specification [29]. Prior to the ratification, the industry also embraced an early version of 802.11i to provide a migration path to 802.11i. The wireless alliance WiFi embraced this migration path and referred to it as WPA (Wireless Protected Access).

The major weaknesses of WEP include:

1. Lack of mutual authentication
2. No access control
3. No replay prevention

4. No message modification detection
5. Compromised message privacy due to IV reuse, RC4 weak keys, and possibility of direct key attacks

For details on these weaknesses, look at Chapter 6 of [9], which provides an in-depth analysis of them; alternately [28] provides a comprehensive summary. IEEE 802.11i defines a new type of wireless network called an RSN (Robust Secure Network). To allay industry concerns for already-deployed systems, the WiFi alliance took a subset of 802.11i and created WPA while allowing the IEEE 802.11 standards body to focus on a sound, longer-term solution. As WPA is a subset of 802.11i, they both provide a framework referred by 802.11i as RSN. The framework allows for the negotiation of authentication, key management, and cipher suites used to ultimately protect the 802.11 link. While the RSN framework enables proprietary mechanisms to coexist, it defines the following components:

1. Authentication and key management: The mandatory-to-implement mechanism is based on IEEE 802.1X to enable Extensible Authentication Protocol (EAP) methods to be used for authentication. Similarly, IEEE 802.1X is used to employ a key management mechanism to allow the client and access point to mutually derive the keying material needed to protect the 802.11 link and subsequent 802.1X key management functions. Optionally, an RSN also enables the use of pre-shared keys as a replacement to EAP for those systems that do not have the back-end infrastructure for identity management.
2. Cipher suite: The mandatory-to-implement cipher suite is based on AES-CCM and Temporal Key Integrity Protocol (TKIP) is provided to allow already-deployed systems to allay the vulnerabilities of WEP.

12.3.2   WLAN Mesh Security Primer

In the past, security architectures were often developed based on the assumption that the core parts of the network were not physically accessible to an enemy. Attacks were only expected to be launched in well-defined places such as connections to the public Internet. Firewalls and intrusion detection systems were deemed sufficient to keep valuable electronic assets in a corporation or personal data from being stolen, exposed, or compromised. WLAN networks break this conventional assumption in network security. Because data now passes over radio waves, ready and easy access to data becomes trivial. Original WLAN technology was targeted for indoor LAN networks, keeping the sphere of exposure somewhat limited, although unpredictable radio waves do propagate outside the buildings. War-driving and sniffing near buildings may allow an attacker to see much of the data traveling inside the buildings, too. Sniffing is defined as simply using a software and WLAN radio card to read and store all frames flowing over a WLAN channel.

Outdoor WLAN networks exacerbate security exposure by deliberately transporting data over radio waves through open air in metropolitan and rural areas; that is, exposing the physical access points in the open public. In other words, now an attacker does not need to drive closer to the buildings anymore. Anyone can see those radio waves and its data at will from anywhere in a city or rural area wherever those radio waves traverse or access the exposed access points from the street. Whether indoor or outdoor, mesh networks may take the strategy of re-using 802.11i for mesh access. But this leaves mesh backhaul not secured and there is no standard mechanism for securing mesh backhaul today. There is also the need to secure peripheral devices attached to the wired interfaces of mesh nodes. Finally, mobility of WLAN clients and mesh nodes makes mesh security a great challenge in defining an interoperable standard. Vendors are currently offering proprietary mechanisms for backhaul and bridge security restricting single vendor mesh deployments presenting a hurdle toward widespread adoption of secure WLAN mesh networks.

12.4 Possible Attacks on WLAN Mesh Networks

This section examines possible attacks and threat models in WLAN mesh networks. Many of these attacks are similar to that of attacks in WLAN networks. Attacks on wireless networks can be classified into five broad categories: eavesdropping, forgery, masquerading, man-in-the-middle (MIM), and denial of service (DoS). The first category of attack is also known as passive, the other three are known as active attacks. Some in-depth attack scenarios and analysis of those scenarios would be useful in understanding and deriving the mechanisms needed to prevent these attacks and protect the network against them.

12.4.1   Types of Attacks

Eavesdropping is accessing information without detection of either the data originator or the intended receiver. More importantly, it is information to which the attacker does not have legal access. Such information may include confidential company data, personal financial and medical information, etc. An attacker may sniff data over 802.11 channels in either a mesh access or backhaul network. Especially in a wireless medium, this form of vulnerability enables an attacker to gain information without detection from any of the communicating parties and is typically referred to as a passive attack.

Forgery is the ability to change any content of a frame without detection. Such modification can cause a frame to be redirected to a different source or, more damaging, change the original information to the intended receiver. Although protection from eavesdropping can help, equally damaging is the ability for an attacker, for example, to forge a stock transaction from a buy to a sell order.

Masquerading (sometimes referred to as spoofing) occurs when an attacking network device impersonates a valid device. Depending on whether a device is accessing a mesh node using its MAC or IP address, an attacker may either use IP address spoofing or MAC address spoofing. Notorious attacks, such as evil twin attacks, can potentially allow hackers to steal personal information such as credit cards or any personal identity information.

Man-in-the-middle can be another form of a forgerer, a masquerader, and even an eavesdropper. An MIM attacker interjects communication by pretending to be the network to the client and the client to the network. By interjecting the communication, neither the client nor the network may be aware that the MIM can now gain identity information from the client and potentially launch other attacks against the network.

DoS attacks work with the principle of causing damage to the target device or the overall network itself. In wireless, DoS attackers can simply jam the radio frequency. In general though, DoS attackers often target some nodes in a network and overwhelm them with traffic, eventually causing them to reboot or melt down. ICMP flood or Ping of Death are examples of classic DoS attacks, which the Internet experienced in the 1990s. A variation of DoS, distributed DoS (DDoS) attacks are more effective where attackers launch DoS traffic from several zombie computers from different locations. While DoS and DDoS attacks are easy to mount in WLAN networks and in mesh networks, they are almost impossible to prevent. Because most WLAN mesh networks run in the unlicensed 2.4 and 5 GHz bands, hackers may not even need to use WiFi to conduct DoS attacks against these networks. Especially in municipal networks where free WLAN infrastructures are now in place outdoors, more and more esoteric attacks will come into play. For example, widespread Bluetooth attacks and Bluetooth spamming are real possibilities with WLAN mesh networks combined with small PCs like GumStix with Bluetooth.

Although it may be more challenging to ward off all DoS attacks, WLAN security must address protection from eavesdropping, forgery, masquerading, MIM, and, where feasible, DoS attacks.

In further providing security mechanisms, attacks on such protective means must also be addressed. As most systems employ the use of a known secret referred to as a key, considerations for the threats against the very cryptographic tools used to provide security also merit description. These attacks are categorized as:

1. Attacks to recover the secret key
2. Attacks with limited or no knowledge of the secret keys

12.4.2 Attacks on the Keys

The challenge in any cryptographic tool employing shared keys is to ensure that these keys are strong enough and not susceptible to its recovery. Because the shared key is used to gain access to the network or to protect the communication with the network, it is critical that it be very difficult to recover these keys; otherwise, knowledge of the key often represents a full breach in security [9]. In real-world use, these keys may oftentimes be required to be manually entered, especially when used as a means to identify a user. In this scenario, these keys are often referred to as passwords as people usually choose something that can be easily remembered.

As passwords tend to be derived from a language source of finite vocabulary, tools based on dictionary attacks can be readily employed to break such keys. Other, more complicated attacks can analyze the actual functions used to derive the keys, or how the keys are actually employed to recover the actual key. The original (flawed) IEEE 802.11 security protocol WEP constructed its protocol in such a way that it was easy to recover the key [26]. Though such attacks require some data sampling, this requirement is trivialized in WLAN mesh networks as the data is easily obtained by capturing the signals over the air.

Attacks on keys are beneficial and worth pursuing especially if the strength (e.g., entropy) of a key is known to be weak. Some techniques of attacking on the keys include:

1. Brute-force method: An attacker tries every possible key until he finds a match. Guessing passwords is an example of such attacks. The time taken for a brute-force attack depends on key entropy. Hence, making the key-size longer does not always solve the problem (it only takes longer to break the key).
2. Dictionary method: An attacker uses a dictionary, or database, containing all the likely passwords/keys. Sometimes known as an off-line attack, an adversary can take known matching ciphertext and plaintext and run a computer and a dictionary loaded to find the keys, which produces the ciphertext from the given plaintext. IEEE 802.11i key derivation makes keys dynamic and usable only for a single session to reduce the chance of such attacks. WLAN mesh networks should not be susceptible to dictionary attacks if similar session key derivation mechanisms are used.
3. Algorithmic method: Adversaries also have the actual cryptographic algorithms and frame constructions from which they can analyze, as was shown by Fluhrer et al. [26] to demonstrate weaknesses in

the algorithm and aid in key recovery. There are also optimizations on dictionary attacks that enable smaller or more exhaustive dictionaries and variations to be used by trading memory and space [27].

As many tools for cracking WEP are now readily available and with the wider adoption of IEEE 802.11i, WLAN mesh networks must not consider WEP for either infrastructure, access, or ad hoc security. With the level of exposure in metro and outdoor areas, cracking WEP would be trivial for attackers of WLAN mesh networks. Note that some WLAN client devices like cameras (e.g., D-Link IP Camera and Linksys Wireless-G Internet Video Camera) and video game consoles (e.g., Linksys Wireless-B Game Adapter and Xbox 360 Wireless Networking Adapter) continue to implement WEP-based encryption only. These devices should not be allowed to connect to WLAN mesh networks. Fortunately, many client devices like Cannon SD430 Powershot Camera now support advanced 802.11 encryption, e.g., AES-CCMP which is part of the IEEE 802.11i standard. Over time, all WLAN client devices should migrate to these more-robust encryption methods.

12.4.3 Attacks without Requiring Knowledge of the Secret Keys

Ironically, all five types of attacks described earlier in the section can be conducted without or with limited knowledge of these keys. Even encrypted traffic can reveal information such as how, when, and by which devices the network is being used. Another example is that of management frames, especially beacons and probe responses as they are never encrypted and where an attacker can readily learn the SSID being broadcast by mesh node or manufacturer, model, and other device information of the node encoded in 802.11 information elements. The attacker may exploit any known vulnerabilities in that particular model hardware or software. For example, there may be open-source security software libraries (e.g., openSSH [10] and openSSL [11]) in cheaper mesh nodes and the attacker may have the knowledge of public-domain vulnerabilities which can be easily exploited. The attacker can also perform sophisticated traffic analysis by studying message externals, e.g., frequency of communication, size of payload, traffic load on a device, etc. Finding a correlation of TCP acknowledgment frames or DHCP discover messages, which are of fixed length and might occur at regular intervals, provides a wealth of information to the attacker. Typically, such information is useful in conjunction with other techniques, such as modification.

In a secure WLAN where packets are encrypted, forgery and MIM attacks are difficult to mount against networks because the attacker must intercept transmission from either end (AP or client) and relay it without giving any clue to the receiver about the compromise. This is done in turn for both ends creating a relay or repeater node in between the AP and client. In a WLAN mesh network, an MIM attack can be launched between MP links as well. MIM between mesh nodes would be more damaging compared to a compromised AP and client link because now all backhaul traffic over the compromised mesh link is affected. A carefully crafted attack may get an MP in thinking of a rogue device to be valid and relay traffic to and from it. In this attack, the adversary can either direct traffic to its intended destination or mess with the data. Both strategies impact the services of a mesh network, more so if the MIM is in between mesh nodes.

Another threat emerges from the ability to replay messages either to the network or to the endpoint device. The attack could be maliciously or fraudulently repeated by either the originator or an MIM.

DoS and DDoS attacks do not require knowledge of the shared secret, especially in WLANs. An attacker or a group of attackers launch these attacks simply to bring down a network or its services. WLAN mesh networks are particularly susceptible to these attacks and present a great challenge. A special type of DoS attack known as RF jamming against WLAN networks is very difficult to detect and prevent. More damaging is the current lack of protection for 802.11 management frames. Two such frames, Disassociation and Deauthentication, permit using the broadcast MAC address as the target and are easy means to disrupt WLAN service to all connected clients of the victim access point. These frames may also be directed to a specific station, denying service to targeted victims. Similarly, an attacker may observe the victim station's MAC address and send an Association Request to a different AP on the same wired LAN. This association request is accepted as if the station is roaming and the wired network now forwards all traffic to the attacker. In some networks, the victim station may be disconnected from the AP it was attached to and, depending on the security method negotiated, the adversary may not be required to re-authenticate with the new AP. Yet another example is where an adversary uses a station simulator tool, such as the Veriwave WLAN Simulator, and congests an AP with bogus stations exhausting its available resources over the air, eventually causing the victim AP to stop accepting new clients or, in some implementations, to reboot. Clever attackers may continually keep loading bogus stations on the AP, completely taking it out of service.

All these classic DoS/DDoS attacks are more easily applicable to WLAN mesh networks because adversaries now have visibility into client traffic streams from anywhere in a mesh deployed area. Many other possible attacks on WLAN and ad hoc networks without keys are described in [21–23].

12.5 Attacks on WLAN Mesh Protocols

WLAN mesh networks face another array of security challenges that emerge from its multi-hop nature. The default routing protocol in TGs is Hybrid Wireless Mesh Protocol or HWMP, which provides the ability for a mesh node to learn routes to another mesh node using a broadcast route discovery mechanism. Broadcast-based route discovery mechanisms are traditionally susceptible to DoS attacks as they use exhaustive re-broadcasting methods. An attacker may snoop frames over a WLAN mesh backhaul and learn about MAC addresses or various mesh nodes in the network. Because HWMP is based on the IETF's AODV [12], an open-source AODV software stack can be used to continually generate route request (RREQ) frames keeping all mesh nodes in the network busy re-broadcasting those. This may cause one or more mesh nodes to melt down, reboot, or stop servicing the network.

Other attacks possible on an unprotected RREQ include:

■ Route disruption by changing message type, destination address, source address, or originator address
■ Route invasion by increasing RREQ-ID, originator sequence number, or destination sequence number by at least one

Attacks on route replies (RREP) are possible when the attacking node drops all routing frames, causing the routes to take longer and sub-optimal paths. Often an attacking device positioned in between valid devices may cut off some routes all together. MIM attacks are possible if particular route destinations can be lured to an adversary's device followed by a detour somewhere over the Internet. The attacker may do so by sending fake RREPs with a large enough destination sequence number or short hop count.

Attacks on route errors (RERR) are not as severe because the result is route disruption. Yet, generating bogus RERRs can cause many nodes to attempt to repair processing and re-discover valid routes. Another point to note is that most fields of RRER, RREP, and RERR, e.g., ID, Hop Count, Metric, Sequence Number, etc., are vulnerable to modification and forgery. Most damaging is the vulnerability of an MAC address, as an adversary can impersonate an MP by simply using its MAC address; an adversary can simply form part of mesh forwarding paths and launch any attack from there. Note that similar attacks are also possible against RA-OLSR, which is the optional path selection protocol in IEEE 802.11s draft standard.

12.5.1 Approaches against Attacks on WLAN Mesh Protocols

Even when mesh nodes are authenticated before joining a WLAN mesh network, many aspects of a mesh are controlled via broadcast frames.

In a broadcast environment, all parties can discern the information and often can affect other members of the group. An insider attack is a form by which an adversary may be able to join the mesh by exploiting weaknesses in the mesh authentication mechanism and exploit the broadcast environment to launch attacks. Broadcast protocols in the IEEE 802.11s draft standard do not have any mechanism for protecting themselves from insider attacks. There are techniques for protecting HWMP by using methods such as authenticated broadcast of RREQ, authenticated unicast of RREP, and authenticated broadcast of RERR. On top of node-based authentication of routing nodes, individual message integrity and authenticity are also needed to limit and prevent the attacks described earlier. SAODV [13] is a secure version of the original AODV protocol, which combines these techniques and more (e.g., digital signature for static fields in headers and hash chains to protect Hop Count). While SAODV is appropriate for ad hoc networks, it comes with some costs for WLAN mesh networks. Even though hash chains are efficient for Hop Count authentication, a malicious node can still choose not to increase it. Other drawbacks of SAODV include PKI infrastructure usage and key distribution, too frequent signature computations, and extra overhead for exchanging signatures, which can be up to two signatures per message, becomes computationally prohibitive. At the time of this publication, IEEE 802.11 TGs is evaluating these techniques and may incorporate some subset of SAODV for securing the default path selection protocol, HWMP.

ARAN [14] and Ariadne [15] are two other published techniques for securing AODV, which can be adapted for securing HWMP.

12.5.2 Advanced Attacks on WLAN Mesh Protocols

In addition to the attacks previously discussed in this chapter, attacks targeted to peer-to-peer or mesh networks may also be applied to WLAN mesh networks. These attacks can be summarized as follows:

- Sybil attacks: An adversary presents itself as being multiple illegitimate identities to the mesh network. Thus, given a single faulty entity, it can masquerade as many other entities and control a part of the network. This attack requires that each MP be provisioned with strong authentication identification and authentication of the traffic being routed within the mesh.
- Sinkhole attacks: An attacking node lures all traffic around it by installing an attractive node. Powerful transmitters and high-gain antennas may allow the device to emerge as high-quality routes. Sinkhole attacks open doors for further ugly attacks and tampering with application data. Detection of sinkholes is difficult without higher-layer protections such as asking for acknowledgments from

the final destinations for all messages (TCP and HTTP implement acknowledgments as part of the base protocols). Sinkhole devices are often referred to as honeypots.

■ Black hole/gray hole attacks: An attacking node drops all frames it receives (black) or drops selective frames it receives (gray). In black hole attacks, mesh nodes can protect themselves by requesting explicit acknowledgment for routing protocol and application frames. Gray hole attacks are more challenging to detect because the attacking node appears as a valid forwarder. Higher-layer protocols end up suffering from the dropped frames, which may degrade application quality (e.g., for UDP streams) or cause excessive retransmissions and shrinkage of data burst windows used by transport layer protocols, e.g., sliding window in TCP.

■ Wormhole attacks: An attacker may leverage multiple attacking nodes and create low-latency and high-speed route tunnels between them. This strategy will make attacker's tunnel appear attractive over a multi-hop path and cause a wide area of nodes to attempt to use the tunnel. Black hole/gray hole/sinkhole attacks might follow. Unfortunately, wormhole attacks are effective even if the protocol/system provides authenticity and confidentiality.

Given the use of strong identification credentials, e.g., strong entropy keys and unique identities, IEEE 802.11 TGs may be able to address some of the above attacks, but may still be susceptible to insider attacks.

12.6 Other Security Issues in WLAN Mesh Networks

In addition to the various WLAN and WLAN mesh attacks described in previous sections and approaches in solving those, there are further security-related issues that exist in practical WLAN mesh networks:

■ Mesh node hijacking
■ Threats from bridged networks
■ Unfairness from greedy nodes
■ No real mutual authorization
■ Supplicant-authenticator dilemma
■ Authentication server location
■ Management frame security

12.6.1 Mesh Node Hijacking

In a WLAN mesh network, route paths and topologies can be arbitrarily established independent of the path selection protocols: HWMP or OLSR. Because there is no administrative boundary or domain enforced by these protocols, different ISP networks that can see each others' mesh nodes may end up proliferating into each others' network. A greedy network owner may attempt to leverage other owners' mesh nodes for forwarding its own traffic. A hostile network owner may attempt to leverage neighbor owners' mesh nodes for forwarding its own traffic and take one step further that protects its own mesh nodes by proprietary means. HWMP should consider defining administrative boundaries like routing protocols used in the Internet, e.g., Border Gateway Protocol (BGP) or Open Shortest Path First (OSPF).

12.6.2 Threats from Bridged Networks

In a WLAN mesh network, many nodes are equipped with Ethernet or fiber-wired interfaces. A greedy network owner may install a large wired LAN to its mesh node and connect to the network. Because there is no standard method of authenticating the devices connected to these interfaces of a mesh node, this poses a security challenge on these open ports. Unless there is an authentication server (AS) in the mesh node, it will have to reach out to some remote AS inside or outside the WLAN mesh to authenticate the devices connected to these interfaces. If there is a reachable AS, the node may employ IEEE 802.1x [6] port control mechanisms on it. There are still open issues as to which devices should be authenticated and how many, as there may be an entire switched or bridged LAN behind those wired interfaces.

Another threat from bridged networks occurs when there are two wired LANs connected to the same WLAN mesh network and they start using the mesh as a wireless bridged network. Because there is an inherent mismatch between wire speed of wired LANs and shared media in WLAN mesh, this may seriously starve traffic in a WLAN mesh network or even simulate a DoS-attacked WLAN mesh.

12.6.3 Unfairness from Greedy Nodes

As mesh nodes may relay traffic for their own clients as well as for other mesh nodes, throughput obtained by them may significantly vary depending on their position in the network. This is particularly true for a hierarchical mesh where most communication occurs to and from a limited number of MPPs. Usually, nodes further away from the portals suffer highly unfair and degraded throughput. This implies degrading quality of service for the clients farther away from MPPs. Currently, there is no solution to this problem in the IEEE 802.11s draft standard.

An attacker with knowledge of a mesh hierarchy may exploit the fact and start installing greedy nodes anywhere in the hierarchy with a mission of further starving or completely blocking out access to nodes farther from

MPPs. They may appear as hidden nodes to the suffering nodes, winning (or jamming) the channel and causing excessive collisions.

12.6.4 No Real Mutual Authorization

In a WLAN mesh network, it is difficult to ascertain what service data forwarding, service clients, etc., the nodes are authorized to. Even though server-based policies can be used to provide proper authorization for a new mesh node, there is no mechanism for the mesh node to authorize other members in the network or to learn of their peer authorizations. This may result in a mesh node to join an alien network and become a slave.

12.6.5 Supplicant–Authenticator Dilemma

The EAP security mechanisms [16] are widespread not only in the IP-based data communication world, but also in cellular and other parts of the wireless communications world. EAP works based on a three-party model attempting to authenticate a node in a network (supplicant) via an already authenticated node (authenticator) by an AS. If there is an AS present in the network, whichever node has an active connection to the AS takes up the role of authenticator and the other becomes a supplicant. In a mutual authentication scenario, the roles would have to be swapped for the nodes to be fully and mutually authenticated using an EAP method. This scheme requires implementing both supplicant and authenticator stacks in every node, causing code and other resource bloats, such as system memory.

One alternative to avoid this problem is to use a fixed authenticator in the network, e.g., a portal device, and let authenticated nodes pass through for nodes which join the network. This method requires implementing only the supplicant stack on mesh nodes while implementing authenticator stack at selective mesh nodes, such as a portal. Another alternative is to avoid the use of EAP for authentication and use a peer-based mutual authentication method.

12.6.6 Authentication Server Location

AS location and setup is another open issue in WLAN mesh networks. An AS can be located inside or outside a WLAN mesh network. The location of the AS affects re-authentication unless there is optimization to avoid involving the AS in the re-authentication process. If the AS is located inside the mesh, all mesh nodes must be aware of where it is. If the AS is outside the mesh, only portals need to know where it is. The number of ASs and orientation also affects WLAN mesh security. For example, a centralized

AS can be used for authentication, authorization, and access control (AAA) of all mesh nodes. Similarly, a distributed AS model can be used where multiple ASs provide AAA services in mesh. TGs is not specifying any particular AS deployment model for WLAN mesh networks.

12.6.7 *Management Frame Security*

The final topic we examine in WLAN mesh security issues is the securing of management frames as these frames are the foundation for many DoS attacks against early 802.11 WLAN networks. IEEE 802.11 has already formed a Task Group W to address this need for the general 802.11 management frames. The objective for management frame security in a WLAN mesh is to assure authenticity, integrity, and privacy (where appropriate) of the management frames sent and received among MPs on a link-by-link basis. The IEEE 802.11i-based link level authentication model can be leveraged to support authentication, key distribution, and encryption for management frames. There is unlikely to be any separate management frame specific authentication and encryption architecture. Management frames should have the same level of security and use the same mechanisms as data frames. Wherever possible, the security mechanisms defined by the Task Group 802.11w [19] will be utilized. WLAN mesh management frame protection is used for the following purposes in a WLAN mesh network:

1. Forgery protection
2. Confidentiality protection
3. Compatibility with 802.11i key hierarchy
4. Incremental inclusion of new management frames
5. Protection only after key establishment
6. Fragmentation support for management frames

When considering security, the mesh management frames as well as 802.11 standard [20] management frames can be classified in two broad categories:

1. Those sent prior to authentication
2. Those sent once 802.11 link layer is secured

The management frames sent prior to authentication are Mesh Beacon, Probe Request/Response, 802.11 and 802.1X Authentication Request/Response, Association Request/Response, and the 802.11i four-way handshake. When 802.1X EAP is used, the management frames used are not protected at the link layer. The management frames sent and received after authentication are Mesh Beacon, Reassociation Request/Response, ATIM, Disassociation, Deauthentication, action management frames and mesh-specific management frames. All these frames should be secured using 802.11w and derivative techniques.

12.7 WLAN Mesh Security Requirements

Now that typical security threats and attacks in WLAN mesh networks have been discussed and analyzed, WLAN mesh security requirements can be derived in a methodical manner. From a high-level perspective, they can be first categorized into the following four broad categories:

1. Infrastructure security: Data, control, and management traffic security that flows over the infrastructure mesh nodes and mesh links. This is often termed "backhaul security."
2. Network access security: Data, control, and management traffic security that flows between a WLAN client and MAP.
3. Ad hoc security: Data, control, and management traffic security that flows between two WLAN clients over a multi-hop path in a mesh network. In many cases, MAPs and clients may be mobile and susceptible to dynamic topology changes in mesh backhaul or network.
4. Application security: Security of the applications run by WLAN clients in a mesh network, such as VoIP, database, etc.

Among these security categories, ad hoc security is by far the most challenging of all. Application security is typically not addressed within the network stack and is implemented by the applications at network endpoints. With respect to the other three categories, the WLAN security requirements can be stated as follows:

1. Mesh node and client authentication: A mesh node should authenticate a requesting WLAN client before servicing it. The WLAN client should also authenticate the mesh node to avoid joining rogue mesh nodes. This mutual authentication requirement is needed to prevent unauthorized network access from both mesh node and client perspectives.
2. Mesh node and client key agreement: A mesh node and client should undergo handshakes to establish a fresh shared key to encrypt, authenticate, and integrity protect all traffic flowing between the mesh node and the client. This key must be a short-lived key that is freshly derived when the session is initiated and deleted once the communication between the mesh node and client is terminated.
3. Mesh node and mesh node authentication: A mesh node should authenticate another mesh node before forwarding traffic to and from it. The joining mesh node should also authenticate any other mesh

node it is forming a peer relationship with to avoid joining rogue mesh nodes. This mutual authentication requirement is needed to prevent unauthorized mesh nodes from joining mesh networks.

4. Mesh node and mesh node authorization: A mesh node should authorize the authenticating mesh node before forwarding traffic to and from it. The joining mesh node should also authorize its peer. By also obtaining authorization, both peers are assured that the mesh node they are joining is authorized to perform the services of a mesh node.

5. Mesh node and mesh node key agreement: Two mutually authenticated mesh nodes should undergo handshakes to establish a fresh shared key to encrypt, authenticate, and integrity protect all traffic flowing between them. This key must be a short-lived key that is freshly derived when the session is initiated and deleted once the communication between the mesh nodes is terminated.

6. Location privacy: Security between mesh node and client as well as two mesh nodes should be agnostic about location of the devices in question. Identities of mesh devices and clients should have no correlation with physical locations of those devices.

7. Signaling authentication: Management and control frame protection is important in mesh backhaul as well as mesh access. Such broadcast frames must be distinguishable from those announced by an attacker.

8. Service availability: A mesh node must be protected from DoS attacks and continue to offer services under such attacks. Even better is if such attackers can be located and mitigated in case of service disruption. A mesh client cannot be excluded by a DoS attacker.

9. Secure routing: Because multi-hop and multi-path routings are used inside, upstream, and downstream traffic forwarding from a wired portal, any routing protocol in operation must be secure against malicious attacks.

10. Secure MAC: The MAC protocol employed in mesh backhaul as well as access must be sufficiently resilient against RF and media-access attacks.

11. Secure bridging: Because a mesh network can be interworked with other 802 LAN networks, any bridging protocol in use must be secure against any malicious attacks launched from those LAN networks.

Some of the requirements above are discussed in detail in [17] and attempt to derive theoretical models of the attacks which may be launched when these requirements are not met by a WLAN mesh network.

We look at the proposals which were presented at TGs next.

12.8   Security in IEEE 802.11s WLAN Mesh

12.8.1   The Original IEEE 802.11s Proposal

12.8.1.1   Overview

The original proposal uses the IEEE 802.11i concepts and mechanisms for mesh discovery and mesh association. It supports distributed and centralized models for AS functions. It utilizes optional additional security mechanisms to support scalable security for data and management traffic.

Scalable security for data and management traffic allows pre-shared multicast keys so that information may be broadcast to all neighbors of an MP. IEEE 802.11i mechanisms are used to distribute the required 802.11i keys and optional keys. These multicast keys are either unique to each MP (Neighbor Master Keys [NMK] and Neighbor Temporary Keys [NTK]) or pre-shared among all MPs (Group Master Key [GMK] and Group Temporary Keys [GTK]).

IEEE 802.11i required keys are pair-wise keys for securing the link between a client and AP (PMK, PTK) and group keys for all nodes (GTK). The optional keys for mesh networks are local multicast keys and global multicast group keys (MMK/MTK). The local multicast keys support one key per neighbor transmitting the data. The global multicast group keys support one multicast encryption key per multicast group.

Basic 802.11i functions are extended to provide multi-hop encryptions for unicast and multicast data or control frames. The extensions occur at the neighbor security associations in mesh beacon or neighbor discovery Hello functions.

12.8.1.2   Security Framework

The original IEEE 802.11 TGs security proposal is based on 802.11i RSNA security and supports both centralized and distributed IEEE 802.1x-based authentication and key management. In a WLAN mesh, an MP performs both the supplicant and the authenticator roles, and may optionally perform the role of an AS. The AS may be co-located with an MP or be located in a remote entity to which the MP has a secure connection (this is assumed and not specified by the 802.11s proposal). Figure 12.2 shows the security framework in a WLAN mesh network. A node establishes RSNA in one of three ways:

1. Centralized 802.1x authentication model
2. Distributed 802.1x authentication model
3. Pre-shared key authentication model

The first two use 802.1x EAP-based authentication followed by an 802.11i-based four-way handshake. A central AS is used in the first model whereas it is presumed that each MP in the MP–MP perform mutual

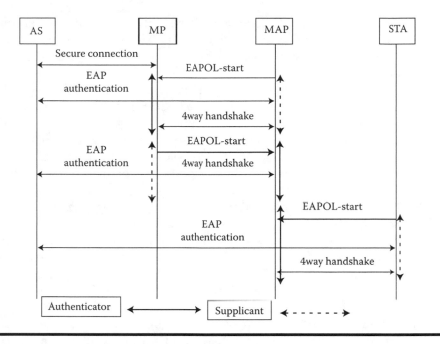

Figure 12.2 Example security exchanges in WLAN mesh.

authentication in the second model. The pre-shared model, where a single key is shared among the mesh does not quite scale to mesh networks where multi-hop routing is required. In particular, it is infeasible to secure routing functionality when a pre-shared key is used in a mesh with more than two nodes, because it is no longer possible to reliably determine the source of any message. Alternatively, each MP may be provisioned with its unique pre-shared key, but then this also presents an unscalable model as every MP must be provisioned with all of the MPs in the mesh.

IEEE 802.11 TGs is effectively taking a different approach to solving the WLAN mesh security. At the time of writing this chapter, there were two proposals which were presented at the IEEE 802 Plenary meeting at San Diego, California, in July 2006.

12.8.2 Current IEEE 802.11s Security Proposals

At the time of publication, two security proposals were evaluated by TGs; since, the core of Intel's proposal has been adopted into the TGs base specification though many security issues still remain to be stabilized. Both proposals are preceded by an almost common security framework. We first discuss that framework.

- Discovery: Each MP advertises its security policy in the beacons and probe responses it generates. Other MPs within range interpret received beacons and probe responses to learn the security policy of the message source.
- IEEE 802.11 authentication: When used, this performs peer authentication and implicit authorization to perform mesh forwarding.
- Role determination: The security policy is determined by an algorithm that also determines which party plays the role of IEEE 802.1X authenticator and which plays the role of supplicant for each link instance. This algorithm executes prior to beginning the link establishment procedure.
- Link security policy selection: This involves the supplicant selecting among the pairwise cipher suites and authenticated key management protocols advertised by the authenticator in its beacons and probe responses. The supplicant asserts its selection through the WLAN mesh link establishment procedure. The IEEE 802.1x entity closes its controlled port when the secure link establishment procedure begins.
- Authentication and key management: After link establishment is asserted, the authenticator initiates IEEE 802.1X authentication followed by a variant of the authenticated key management process defined in Clause 8.5 of [6] to enable the authenticator and supplicant to mutually authenticate and establish fresh keys to secure the 802.11 link. IEEE 802.1X authentication may be null if a pre-shared key is optionally employed.
- Secure link operation: Once authenticated key management completes successfully, the IEEE 802.1X entity opens its controlled port to allow data to flow, which is now protected.

When security is enabled, mutual authentication between the two parties must be achieved and thus at least one of IEEE 802.11 authentication or authentication and key management is required.

12.8.2.1 Proposal from Intel Corporation

One of the two proposals originates from a group of security researchers from Intel Corporation. The proposal leverages IEEE 802.11i to secure the mesh transport and is summarized in this section.

When a mesh node wants to utilize IEEE 802.1X to authenticate and authorize with other MPs, it shall advertise its security policy by including the RSN information element into its beacons and probe responses. An MP shall also set bits 7 and 8 of the RSN Capabilities field in the RSN information element as follows:

- Bit 7: The mesh node shall set this bit to 1 if it uses the mesh default role determination scheme. Otherwise, the node shall set this bit to 0 if it uses some other role determination scheme, such as a proprietary scheme. The specification of other schemes is outside the scope of this proposal and the TGs standard.
- Bit 8: This bit is meaningful when bit 7 is set to 1. The mesh node shall also set bit 8 to 1 if the mesh node can execute the role of the IEEE 802.1X authenticator; otherwise, it sets this bit 0. Because a mesh node must relay on an authentication database, it must either provision it locally or be able to reach an 802.1X authentication server. Thus, if either case is true, then bit 8 may be set to 1; otherwise setting this bit to 0 indicates that this mesh node has no access or means to 802.1X authenticate its peers.

When an MP wishes to use 802.1X for authentication and authorization of different mesh roles, it inspects beacons and probe responses from the other MPs. When it receives a beacon or probe response from another MP, the receiving MP shall examine whether bit 7 of the Capabilities field of the RSN information element from the message is set to 1. If both MPs have advertised the ability to employ the proposed role determination by both setting bit 7 to 1, then the proposed standard is employed. Otherwise, if one of the mesh peers has not set bit 7 to 1, then based on the MPs policy, a non-standard role determination may be negotiated or otherwise the MPs fail to establish a secure link.

If an IEEE 802.1X-based authentication and key management method is used, the MP playing the role of the IEEE 802.1X supplicant shall include an RSN information element in the association request specified by this mechanism. In the RSN information element, the supplicant MP shall specify exactly one pairwise cipher suite and one authenticated key management suite.

In a wireless mesh network, all mesh nodes must utilize the same group cipher suite. Therefore, a supplicant MP must include the same group cipher suite as advertised by the other MPs, especially the authenticator MP; similarly, the supplicant MP shall reject association requests from the authenticator MP (with status code 41), if the group cipher suite advertised by the authenticator MP does not match its own.

The authenticator MP shall also reject the association request from the supplicant MP if either the pairwise cipher suite (with status code 42) or authenticated key management suite (with status code 43) selected by the supplicant is not included in the corresponding lists of pairwise cipher suites and authenticated key management suites specified in its own beacons and probe responses. The authenticator MP may also reject the supplicant MP's association request for other reasons unrelated to security. The authenticator MPs may accept the association request if the supplicant selected pairwise and authenticated key management suites from among those specified by the authenticator in its beacons and probe responses.

Once the role of the supplicant and authenticator is established between two MPs, the logic followed for the security negotiation, 802.1X authentication, and key establishment is the same as that defined in IEEE 802.11i. The proposal provided by Intel allows for as much of the re-use of IEEE 802.11i with the modifications and enhancements to include the role and authorizations of the peers to behave as mesh nodes.

12.8.2.2   Proposal from Tropos Networks and Earthlink

The second proposal is called Comminus, jointly proposed by Tropos® Networks and Earthlink. This proposal attempts to provide peer authentication prior to full authorization and key management. Comminus attempts to partition the steps of authentication, authorization, and secure link establishment as a means to allow flexibility in requiring access to an authentication server or provisioning of a full authentication database. By using the standard 802.11 authentication mechanism versus 802.1X, Comminus obviates the need to negotiate the supplicant and authenticator roles. Comminus begins with the requirements of dynamically generating ephemeral session keys, not being susceptible to active or passive attacks, ability to provide some level of DoS resistance, providing implicit or no authorization, and providing authorization as an overlay.

Comminus protocol is based on SKEME [18], a well-known key agreement protocol that is known to be secure. It is based on Diffie–Hellman and to achieve mutual authentication can employ pre-shared keys or certificate-based authentication. The Diffie–Hellman authenticated key agreement uses the 802.11 authentication frames and can provide mutual authentication between two nodes (no notion of supplicant or authenticator or need of an AS). Comminus provides perfect forward secrecy. However, to achieve such mutual authentication, each MP must now be provisioned with all of its peer MPs' pre-shared keys or a means to validate their certificates, if provided. Without the use or means to authenticate such credentials, e.g., pre-shared keys or public keys (e.g., certificates), the result is only in a secure key agreement with two unauthenticated parties. That is, there are assurances that there is no MIM but no gains on authentication. Lastly, there is no means to complete the authorization between the two MPs. However, once a key has been secured among the two MPs, though maybe lacking in authentication and authorization, it can provide the following additional properties:

1. Resistant to passive and DoS attacks, limited active attacks possible
2. May allow using ephemeral keys for management frame protection after authentication is complete
3. No authorization, it is to be used for mesh formation only

4. No master key exposure issues as shared secret is known by only two nodes

The Diffie–Hellman computation is generally expensive to perform in hardware even though there are optimized versions of the algorithm now available. There is also no real re-authentication or key refresh mechanisms built into Comminus, nor is there a means to address mobility. Further, it is not clear about the lifetime of the session keys in case of link or node outage between the two nodes sharing the same secret. Comminus does not provide a full WLAN mesh security solution. Hence, it proposes to use EAP methods along with AAA/RADIUS for a mesh node to servicing additional mesh functionalities, such as routing and bridging on top of mesh link formation.

12.9 Discussion and Conclusion

Security is often an afterthought in new technology evolutions. But to make these technologies a commercial success, security problems need to be solved up front with careful considerations into topics like authentication, authorization, and access control of all members of the network; data and management frame confidentiality, privacy, authenticity, and integrity; intrusion detection and prevention; rogue member detection and prevention; malicious attack detection and prevention; and damage containment and mitigation plans. Especially for multi-hop wireless networks, e.g., a WLAN mesh, it is necessary to address end user concerns over these requirements. This chapter discussed many security issues, threats, and solution approaches for WLAN mesh networks with some highlights of the current security proposals discussed within the IEEE 802.11 TGs. Further, there are open issues that remain:

- Centralized AAA and AS schemes are not scalable in WLAN mesh networks.
- There is no single efficient and reliable security solution suitable for WLAN mesh as many of those solutions may be compromised due to vulnerabilities of channels and nodes in shared media, absence of reliable links to infrastructure, and dynamic topology changes.
- Attackers may launch MIM and modification attacks against routing protocols, such as AODV and OLSR.
- Without strong authorization, attackers may enter into the network and impersonate legitimate nodes and not follow protocol rules.
- Attackers may create sinkholes, black holes, gray holes, and wormholes to disrupt network traffic and take shortcuts.
- Greedy nodes may utilize MAC back-off procedures and NAV for virtual carrier sense mechanisms of 802.11 MAC and cause congestions in the network.

- Availability of an AS and mechanisms to authenticate in its exchanges using peer-based mutual authentication schemes need security analysis for WLAN mesh.
- Group key management remains a challenge in the absence of a central authority, trusted third party, or server to manage the keys. Some distributed and self-organizing key management schemes may be needed for WLAN mesh.

Most WLAN mesh security technologies (inclusive of the ones proposed at IEEE TGs) are attempting to leverage existing EAP and IEEE 802 security mechanisms and embed mesh-specific extensions as needed. However, techniques for security monitoring, response systems to detect attacks, monitoring service disruption, responding quickly to attacks, and mitigating/containing damage in WLAN mesh networks are still limited [24,25]. TGs focuses only in addressing the link and network layer security problems as it presumes use of other security mechanisms such as IPSec, VPN, and other technologies for securing the higher layers. Unfortunately, there is very little focus on cross- and multi-layer coordinated security protocols to combat simultaneous attacks on different protocol layers. Much work remains to develop a framework for building systems that can actually battle multi-protocol attacks as well as detect and prevent intrusion in WLAN mesh networks.

References

[1] IEEE P802.11s/D0.01, Amendment X: ESS Mesh Networking, IEEE, Draft Standard, March 2006, work in progress.
[2] http://www.ietf.org/html.charters/manet-charter.html
[3] http://www.ietf.org/html.charters/nemo-charter.html
[4] R. Ogier, F. Templin, and M. Lewis, Topology Dissemination Based on Reverse-Path Forwarding, RFC 3684, IETF, February 2004.
[5] I. Chakeres, E. Belding-Royer, and C. Perkins, Dynamic MANET On-Demand (DYMO) Routing, draft-ietf-manet-dymo-03, IETF, Internet Draft, October 2005, work in progress.
[6] IEEE Std 802.1X-2004, 802.1X: Port-Based Network Access Control, IEEE, LAN/MAN Standard, 2004.
[7] IEEE Std 802.11i-2004, 802.11i: Amendment 6: Medium Access Control (MAC) Security Enhancements, IEEE, LAN/MAN Standard, 2004.
[8] D. Jones, Metro-Mesh: A Hacker's Paradise, May 2006, available at http://www.darkreading.com/document.asp?doc_id=95609
[9] J. Edney and W.A. Arbaugh, *Real 802.11 security: Wi-Fi protected access and 802.11i*, Addison-Wesley Reading, MA, 2004.
[10] http://www.openssh.org
[11] http://www.openssl.org

[12] C. Perkins, E. Belding-Royer, and S. Das, Ad hoc On-Demand Distance Vector (AODV) Routing, RFC 3561, IETF, July 2003.

[13] M. Zapata and N. Asokan, Securing Ad hoc Routing Protocols, ACM Workshop on Wireless Secuirty (WiSe), September 2002.

[14] K. Sanzgiri, B. Dahill, B.N. Levine, C. Shields, and E.M. Belding-Royer, A Secure Protocol for Ad hoc Networks, IEEE International Conference on Network Protocols (ICNP), 2002.

[15] Y. Hu, A. Perrig, and D. Johnson, Ariadne: A Secure On-demand Routing Protocol for Ad hoc Networks, ACM Annual International Conference on Mobile Computing and Networking (MOBICOM), September 2002.

[16] B. Aboba, L. Blunk, J. Vollbrecht, J. Carlson, and H. Levkowetz, Eds., Extensible Authentication Protocol (EAP), RFC 3748, IETF, June 2004.

[17] Y. Zhang and Y. Fang, ARSA: An attack-resilient security architecture for multi-hop wireless mesh networks, *IEEE Journal on Selected Areas in Communications*, 4th Quarter, 2006.

[18] H. Krawczyk, SKEME: A Versatile Secure Key Exchange Mechanism for the Internet, August 1995.

[19] IEEE P802.11w/D0.0, Amendment 11: Protected Management Frames, IEEE, Draft Standard, March 2006, work in progress. TGs has since progressed and, as of this publication is working towards a new draft version 2.0.

[20] IEEE Std 802.3-2002, 802.3: Carrier Sense Multiple Access with Collision Detection (CSMA/CD) Access Method and Physical Layer Specifications, IEEE, LAN/MAN Standard, 2002.

[21] N. Borisov, I. Goldberg, and D. Wagner, Intercepting Mobile Communications: The Insecurity of 802.11, ACM Annual International Conference on Mobile Computing and Networking (MOBICOM), September 2002.

[22] L. Buttyan and J.-P. Hubaux, Report on a working session on security in wireless ad hoc networks, *ACM Mobile Computing and Communications Review*, 7, 1, 2002.

[23] V. Gupta, S. Krishnamurthy, and M. Faloutsos, Denial of Service Attacks at the MAC Layer in Wireless Ad hoc Networks, IEEE Military Communication Conference (MILCOM), 2002.

[24] H. Yang, H. Luo, F. Ye, S. Lu, and L. Zhang, Security in Mobile Ad hoc Networks: Challenges and Solutions, *IEEE Wireless Communications*, 11, 1, 3847, 2004.

[25] J.-P. Hubaux, L. Butttan, and S. Capkun, The Quest for Security in Mobile Ad hoc Networks, ACM International Symposium on Mobile Ad Hoc Networking and Computing (MOBIHOC), 2001.

[26] S. Fluhrer, I. Mantin, and A. Shamir, Weaknesses in the Key Scheduling Algorithm of RC4, Eighth Annual Workshop on Selected Areas in Cryptography, 2001.

[27] P. Oechslin, Making a Faster Cryptanalytic Time-Memory Trade-Off, Proceedings of Crypto, 2003.

[28] S. Cam-Winget, R. Housley, D. Wagner, and J. Walker, Security flaws in 802.11 data link protocols, *Communications of the ACM*, Volume 46, Issue 5, May 2003.

[29] IEEE STD 802.11-2007. Wireless Local Area Networks, IEEE, WLAN Standard, 2002.

Chapter 13

Security in IEEE 802.15.4 Cluster-Based Networks

Moazzam Khan and Jelena Misic

Contents

The recently adopted IEEE 802.15.4 standard is poised to become the key enabler for low complexity, ultra-low power consumption, low data rate wireless connectivity among inexpensive devices such as sensors. This standard will play an important role in sensitive applications including habitat monitoring, burglar alarms, inventory control, medical monitoring, emergency response, and battlefield management which needs reliable and secure data transfer.

Two network topologies are allowed by the standard, but both of them rely on the presence of a central controller device known as the PAN coordinator. In the peer-to-peer topology, devices can communicate with one another directly, as long as they are within the physical range. In star-based topology, the devices must communicate through the PAN coordinator. The network uses two types of channel access mechanism: one based on a slotted CSMA-CA algorithm in which the slots are aligned with the beacon frames sent periodically by the PAN coordinator, and another based on unslotted CSMA-CA in which there are no beacon frames. The beacon-enabled mode and the star-based[1] hierarchical topology appear to be better suited to sensor network implementation than their peer-to-peer counterparts because the PAN coordinator can act as both the network controller and the sink to collect the data from the sensor nodes. Within one cluster, time is organized in superframes which are delineated by beacons sent by the PAN

---

[1] In the text that follows we will refer to star-based topology as cluster-based topology.

coordinator. A superframe is further organized in active part, where nodes can transmit using CSMA-CA or TDMA (called guaranteed time slots), and inactive part, where all nodes sleep. Larger areas under surveillance can be efficiently covered by interconnecting clusters in mesh topology through their coordinators. This feature is enabled through the existence of the inactive superframe part because the coordinator can then switch to another cluster and communicate as an ordinary node. When communication in a foreign cluster is finished, the coordinator returns to its own cluster.

Wireless devices used for sensing the environment are low in computational power and memory resources. The bandwidth offered by IEEE 802.15.4 standard is low, because the standard allows the PAN to use either one of three frequency bands: 868 to 868.6, 902 to 928, and 2400 to 2483.5 MHz with raw data rates of 20, 40, and 250 kbps, respectively. However the bandwidth available to the application is further decreased due to CSMA-CA access with small back-off windows (default back-off window sizes without power saving mode are 8, 16, 32, 32, 32, respectively, for five allowed back-off attempts). Also, in downlink communications, the PAN coordinator first has to advertise the packet in the beacon, then the node has to send the request packet asking for downlink transmission, and finally, downlink transmission can commence. Therefore, in the presence of many nodes in the cluster, effective bandwidth left to the application is less than 20 percent of the raw bandwidth [15].

Providing security services in such wireless sensor networks is a technical challenge. Algorithms for key exchange which naturally include authentication elements and addition of packet signature will further decrease the bandwidth available to the sensing application. Besides, complex computations often involved in public key cryptography might consume too much energy and memory resources. Therefore, the goal of designing low-power sensor devices forces security mechanism to fit under processing, memory, and bandwidth constraints.

This chapter is organized as follows. In Section 13.1, we explain the relationship between the sensor network architecture and its availability for both data collection and event sensing applications. We believe that network availability for sensing applications has the same importance as data integrity and to some extent data confidentiality. Section 13.2 explains the need of security in wireless sensor networks and which types of security techniques are considered in such networks. A detailed description of security features of IEEE 802.15.4-based [3] sensor networks is presented in Section 13.3. Section 13.4 discusses keying models currently used in WPANs. Security issues addressed by the ZigBee alliance specifications [4] are discussed in Section 13.5. Finally, Section 13.6 concludes this chapter.

13.1   Cluster-Based Networks and Network Lifetime

One of the most significant benefits of sensor networks is that they extend the computation capability to physical environments where human beings cannot reach. However, energy possessed by sensor nodes is limited, which becomes the most challenging issue in designing sensor networks. The main power consumptions in sensor networks are computation and communication between sensor nodes. In particular, the ratio of energy consumption for communication and computation is typically in the scale of 1000 [12]. Therefore it is critical to enable collaborative information processing and data aggregation to prolong the lifetime of sensor networks. The choice of network topology in wireless sensor networks is still an open question. However, it seems that the choice of topology is an issue of trade-off between node simplicity and homogeneity versus the duration of network lifetime. For sensor networks covering large geographic areas, it is difficult to replace sensor batteries when they are exhausted, and therefore when nodes close to the sink die the whole network is unavailable. Therefore, from the aspect of availability, long network lifetimes become an important security aspect.

Wireless sensor networks can carry two different types of sensing. The first kind of sensing is data collection where nodes in the network frequently communicate to report measurements that lead to continuous flow of data from nodes. Depending on the application requirements, some collective sleep technique for all the nodes in the cluster can be used to extend the network lifetime. Data collection applications exploit spatial correlation of sensed data and, to save bandwidth, perform some kind of data aggregation. In peer-to-peer IEEE 802.15.4 architectures, aggregation is performed in nodes which are conveying sensed data toward the sink. In cluster-based architectures, aggregation occurs at the PAN coordinator and aggregated packets are conveyed to the next coordinator along the path, possibly over a more powerful link (GTS) compared to the link type which is available to ordinary nodes (CSMA-CA). From the aspect of available bandwidth, the presence of GTS links between the cluster coordinators gives the cluster-based networks an advantage over the peer-to-peer networks. Also, the aggregation done by the coordinator can be made much more secure than the aggregation in peer-to-peer networks because the coordinator is always aware of the identities of the nodes which participate in the aggregation (because this is done in the attachment process), while the set of neighbors in the peer-to-peer network might depend on the type of query. From the aspect of lifetime, it is reasonable to assume that PAN coordinators will have higher power resources than the ordinary nodes, which, combined with the GTS access, will extend the lifetime of the network (because they will relay packets).

In the second kind of sensing, communication occurs only when some important event occurs and data is communicated in bursty fashion from nodes toward the sink. For applications where event detection is the target (e.g., enemy troops movement, detection of noise level), sensors are required to be vigilant most of the time, which means that collective sleep of the nodes is prohibited. Event detection requires reporting only when an event occurs in contrast to data collection where communication of measurements is more frequent. In this case aggregation is avoided and it is important to deliver the sensed data to the sink within some time bound (time bounds are not important for data collection due to time correlation of sensed data). In event-detection applications, network availability and data integrity are much more critical than in data collection applications. Again, we argue that a cluster-based architecture where PAN coordinators have higher power resources, GTS links for communication, and reliable information about cluster members offers better availability and data integrity than a peer-to-peer architecture.

Nodes in wireless sensor networks can directly communicate with nearby nodes. Nodes that are not within direct communication range use other nodes to relay messages between them. Routing in such a multi-hop network is challenging due to the lack of central control and the high dynamics of the network. Recent work has focused on discovering and maintaining routes that keep the connectivity between the nodes or that minimize the number of hops on a path. One important restriction of a wireless sensor network is that nodes are energy-constrained as they are normally powered by batteries. However, the algorithms that aim to minimize the path length may ignore fairness in routing, for example, the shortest-path routing is likely to use the same set of hops to relay packets for the same source and destination pair. This will heavily load those nodes on the path even when other feasible paths exist. Such an uneven use of the nodes may cause some nodes to die earlier, thus creating holes in the network, or worse, leaving the network disconnected.

Low available bandwidth to nodes, CSMA-CA access, data aggregation, and routing in wireless sensor networks based on IEEE 802.15.4 make the implementation of security a technical challenge. Even at the MAC layer it is possible to launch a denial-of-service attack which will drastically increase the number of collisions and prevent data communication (due to CSMA-CA access and small back-off windows). The processing, communication, and aggregation cost of secure packets first increases both computational and communication overhead. To decrease this overhead all the security parameters and keying models under which the network will work are selected with great care so that the objectives of both secure communication and longer network life are achieved. These two objectives are competing and trade-off between them is necessary. For implementation of secure sensor network we have to compromise on network life to some extent and vice versa.

13.2 Security in Wireless Sensor Networks

Radio is a shared medium; everything that is transmitted or received over a wireless network can be intercepted in such an environment. An adversary can gain access to information by monitoring the communication among nodes. For example, few wireless receivers placed outside a house might be able to monitor the light and temperature sensor readings of a sensor network inside the house, thus revealing detailed information about the occupant's daily personal activity. Similarly, an attacker can obtain a commodity sensor node and present it as a legitimate node inside the network; once an attacker has a few nodes like that in a network, he can launch a different types of attack, for example, denial of service, falsification of sensed data, dropping of sensed data, etc.

13.2.1 Security Techniques

Different security techniques are employed to safeguard threats of such eavesdropping, and we will discuss such techniques next.

13.2.1.1 Data Confidentiality

All nodes in a sensor network communicate through one wireless medium, and listening to this medium is easy. Hence a network should not leak sensor data to any neighboring network or any node that is not part of the network. The standard approach for keeping sensitive data secret is to encrypt the data with a secret key that is carried by the intended receivers only.

13.2.1.2 Data Authentication

Data authentication allows the receiver to verify that the data was really sent by the claimed sender. Authentication also prevents an attacker from modifying a hacked device to impersonate another device. Because an adversary can easily inject messages, the receiver needs to ensure that the data used in any decision-making process originates form a trusted source. Data authentication is usually achieved through a symmetric mechanism where sender and receiver share a key to compute the Message Authentication Code (MAC). The data is appended along with its MAC, and once the receiver gets the data, it recalculates the MAC. If the same MAC is calculated that it received from same sender, it shares the key. Authentication can be achieved both at the cluster level and the device level. Cluster-level authentication is achieved using a common network key, whereas device-level authentication is achieved by using unique pairwise keys for each link in the network.

13.2.2 Data Integrity

Data integrity allows the receiver to verify that the data received is the same as the data sent by the sender and is not changed during its transmission to the receiver. If the MAC calculated by the receiver is the same as received, it means that the data was not altered during transmission to receiver. Message authentication codes must be hard to forge without the secret key. Consequently, if an adversary alters a valid message or injects a bogus message, he will not be able to compute the corresponding MAC, and authorized receivers will reject these forged messages. In sensor networks data integrity is usually achieved in symmetric fashion and is again relied on the appended MAC, hence integrity and authentication options allow trade-off between message protection and message overhead.

13.2.3 Replay Protection

An adversary that eavesdrops on a legitimate message sent between two authorized nodes and replays it at some later time engages in replay attack. Because the message originated from an authorized sender, it will have a valid MAC, so the receiver will accept it again. Replay protection prevents these types of attacks. The sender typically assigns a monotonically increasing sequence number to each packet and the receiver rejects packets with a smaller sequence number than it has already seen.

In symmetric mechanisms sender and receiver share one common key and rely on different security techniques for the secrecy of these keys. Hence the whole security model revolves around the secrecy of symmetric keys that can be either at the network level or a link level.

13.3 Overview of IEEE 802.15.4 Security Operations

IEEE 802.15.4, a link layer security protocol, provides four basic security services: access control, message integrity, message confidentiality, and replay protection. The security requirements can be tuned by setting the appropriate control parameters of the protocol stack. If an application does not set any parameters, then security is not enabled by default. An application must explicitly enable security features.

13.3.1 Addressing

For unique identification in a network or cluster, addressing in IEEE 802.15.4 is accomplished via a 64-bit node identifier and a 16-bit network identifier.

IEEE 802.15.4 supports a few different addressing modes. For example, a 16-bit truncated address may be used in place of the full 64-bit node identifier in certain cases. This allows the size of the source and destination addresses to vary between 0 and 10 bytes, depending on whether truncated or full addresses are used, and whether or not the node sends to broadcast address.

The specification defines four packet types for the media access control layer:

1. Beacon packets
2. Data packets
3. Acknowledgment packets
4. Control packets

The specification does not support security for acknowledgment packets although security is optional for other packet types, depending on the need of application. Depending on the threat environment, the application has a choice of security suites that control the type of security protection provided for the transmitted data. Each security suite offers a different set of security properties and results in different packet formats. The IEEE 802.15.4 specification defines eight different security suites outlined in Table 13.1.

We can classify the suites by the properties they offer:

- No security
- Encryption only (AES-CTR)
- Authentication only (AES-CBC-MAC)
- Encryption and authentication (AES-CCM)

Table 13.1   Security Suites Supported by 802.15.4

Identifier	Security Suite Name	Access Control	Data Encryption	Frame Integrity	Description
0 × 00	None	—	—	—	No security
0 × 01	AES-CTR	X	X	—	Encryption only
0 × 02	AES-CCM-128	X	X	X	Encryption and 128-bit MAC
0 × 03	AES-CCM-64	X	X	X	Encryption and 64-bit MAC
0 × 04	AES-CCM-32	X	X	X	Encryption and 32-bit MAC
0 × 05	AES-CBC-MAC-128	X	—	X	128-bit MAC
0 × 06	AES-CBC-MAC-64	X	—	X	64-bit MAC
0 × 07	AES-CBC-MAC-32	X	—	X	32-bit MAC

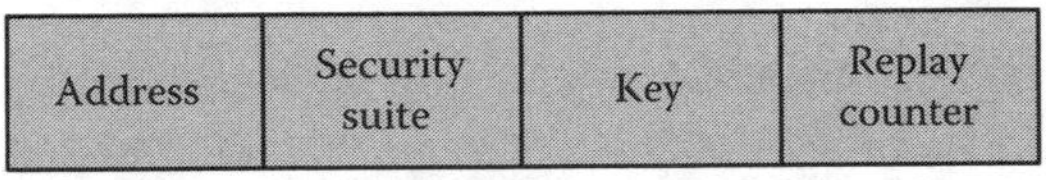

Address	Security suite	Key	Replay counter

Figure 13.1   Access control list entry. (From M. Khan, F. Amini, and J. Mišić, in *Mobile Ad-hoc and Sensor Networks,* **Springer, 2006. With permission.)**

The specification supports MAC of sizes that can be either of 4, 8, or 16 bytes long. The security feature of authentication is directly proportional to the length of MAC and it is very difficult for an adversary to break or guess a MAC of longer size. For example, with a 16-byte MAC, an adversary has a 2^{-128} chance of forging the MAC. The trade-off is a larger packet size for increased protection against authenticity attacks. The choice of secure authentication is tied with the addressing of devices in IEEE 802.15.4 devices. Hence security suites are based on source and destination authentication addresses. Every device supporting IEEE 802.15.4 has an access control list (ACL) that controls what security suite and keying information is used by each device. Each device can support up to 255 ACL entries. Each entry contains an 802.15.4 device address, a security suite identifier, and security material as shown in (Figure 13.1).

The security material is the persistent state necessary to execute the security suite. It consists of

■ Cryptographic key
■ Security suite identifier
■ Nonce state must be preserved across different packet encryption invocations

13.3.1.1   Outgoing Frame Packet and Use of ACL

If security is enabled, the media access control layer looks up the destination address in its ACL table. If there is a match ACL entry, the security suite and nonce specified in that ACL entry are used to encrypt or authenticate the outgoing packets. On the other hand, in case of broadcast type of data packet where no specific destination address is mentioned, a default ACL entry is used, and this default entry matches all destination addresses.

13.3.1.2   Incoming Frame Packet and Use of ACL

On packet reception the media access control defined by IEEE 802.15.4 examines flag fields in the packet to determine if any security suite has been applied to that packet. If no security was applied, the packet is passed to an upper layer. Otherwise, the media access control layer finds an appropriate ACL entry corresponding to the sender's address. It then applies the

| Frame counter | Key counter | Encrypted payload | Encrypted MAC |

Figure 13.2 Frame format after adding security features. (From M. Khan, F. Amini, and J. Mišić, in *Mobile Ad-hoc and Sensor Networks*, Springer, 2006. With permission.)

appropriate security suite and replay counter to the incoming packet. The general structure of secured frame is shown in (Figure 13.2).

We will now provide more detail about the categories of security suites.

13.3.2 No Security

This is the simplest security suite. Its inclusion is mandatory in all radio chips. It does not have any security material and operates as the identity function. It does not provide any security guarantees.

13.3.3 AES-CTR

This suite provides confidentiality protection using the AES (Advanced Encryption Standard) block cipher with counter mode. To encrypt data under counter mode, AES block cipher breaks the plaintext packet into 16-byte blocks $p1....., pn$ and computes $c_i = p_i \oplus E_k(x_i)$. Each 16-byte block uses its own varying counter, which we call x_i. The recipient recovers the original plaintext by computing $p_i = c_i \oplus E_k(x_i)$. Clearly the recipient needs the counter value x_i to reconstruct p_i.

The x_i counter, known as nonce or IV, is composed of

- a static flag field,
- the sender's address, and
- three separate counters: a four-byte frame counter that identifies the packet, a one-byte key counter field (the key counter is under application control and can be incremented if the frame counter ever reaches its maximum value), and a two-byte block counter that numbers the 16-byte blocks within the packet.

The requirement for employing infallible security is that the nonce must never repeat within the lifetime of any single key, hence frame and key counters are introduced to prevent nonce re-use. The two-byte block counter ensures that each block will use a different nonce value.

In summary, the sender includes the frame counter, key counter, and encrypted payload into the data payload field of the packet as shown in (Figure 13.2).

13.3.4   AES-CBC-MAC

This suite provides integrity protection using CBC-MAC. The sender can compute either a 4-, 8-, or 16-byte MAC using the CBC-MAC algorithm, leading to three different AES-CBC-MAC variants. The MAC can only be computed by parties with the symmetric key. The MAC protects packet headers as well as the data payload. The sender appends the plaintext data with the MAC. The recipient verifies the MAC by computing the MAC and comparing it with the value included in the packet.

13.3.5   AES-CCM

This security suite uses CCM mode for encryption and authentication. Broadly, it first applies integrity protection over the header and data payload using CBC-MAC, and then encrypts the data payload and MAC using AES-CTR mode. As such, AES-CCM includes the fields from both the authentication and encryption operations: a MAC and the frame and key counters. These fields serve the same function as above. Just as AES-CBC-MAC has three variants depending on the MAC size, AES-CCM also has three variants.

13.3.6   Replay Protection

A receiver can optionally enable replay protection when using a security suite that provides confidentiality protection. This includes AES-CTR and all of the AES-CCM variants. The recipients use the frame and key counter as a five-byte value, the replay counter, with the key counter occupying the most significant byte of this value. The recipient compares the replay counter from the incoming packet to the highest seen, as stored in the ACL entry. If the incoming packet has a larger replay counter than the stored one, then the packet is accepted and the new replay counter is saved. If, however, the incoming packet has a smaller value, the packet is rejected and application is notified of the rejection. We refer to this counter as the replay counter, even though it is the same counter as the nonce, which is used for confidentiality. The replay counter is not exposed to the application to use.

13.4   Key Management Models

Key management is the process by which keys are generated, stored, protected, transferred, updated, and destroyed. Keying refers to the process of deriving common secret keys among communicating parties. Pre-deployed keying refers to the distribution of key(s) to the nodes before their deployment. Pairwise keying involves two parties agreeing on and communicating with a session key after deployment, and group keying involves more than two parties using a common group key. Group keying is important for multicasting.

The keying model that is most appropriate for an application depends on the threat model that an application faces and what type of resources it is willing to expend for key management. Depending on application types, key management models can be discussed under the following parameters: (1) network architectures such as distributed or hierarchical, (2) communication styles such as pairwise (unicast), groupwise (multicast), or networkwise (broadcast), (3) security requirements such as authentication, confidentiality, or integrity, and (4) keying requirements such as pre-distributed or dynamically generated pairwise, groupwise, or networkwise keys. The constrained energy budgets and the limited computational and communication capacities of sensor nodes make use of public cryptography impractical in large-scale sensor networks. At present, the most practical approach for bootstrapping secret keys in sensor networks is to use pre-deployed keying in which keys are loaded into sensor nodes before they are deployed. Several solutions based on pre-deployed keying have been proposed in the literature, including approaches based on the use of a global key shared by all nodes, approaches in which every node shares a unique key with the base station, and approaches based on random key sharing. In wireless sensor networks, nodes use pre-distributed keys directly, or use keying materials to dynamically generate pairwise and groupwise keys. The challenge is to find an efficient way of distributing keys and keying materials to sensor nodes prior to deployment. Solutions to key distribution problems in WSN can use one of the following popular approaches.

13.4.1 Probabilistic Keying Models

In probabilistic solutions, keychains are randomly selected from a keypool and distributed to sensor nodes. For example, random pairwise key scheme [8] addresses unnecessary storage problems. In this scheme, each sensor node stores a random set of N_p pairwise keys to achieve probability p that two nodes are connected. At key setup phase, each node identity is matched with N_p other randomly selected nodes with probability p. A pairwise key is generated for each node pair, and is stored in every node's keychain along with the identity of its corresponding node. Similarly, [10] also proposed probabilistic key pre-distribution scheme that relies on probabilistic key sharing among the nodes of a random graph and uses a simple shared-key discovery protocol for key distribution, revocation, and node re-keying. This scheme showed that a pair of nodes may not share a key, but if a path of nodes sharing keys pairwise exists between the two nodes at network initialization, the pair of nodes can use that path to exchange a key that establishes a direct link. Therefore, full shared-key connectivity offered by pairwise private key sharing between every two nodes becomes unnecessary.

13.4.2 Deterministic Keying Models

In deterministic solutions, deterministic processes are used to design the keypool and the keychains to provide better key connectivity. For example, [5] suggested that all possible link keys in a network of size N can be represented as an $N \times N$ key matrix. It is possible to store a small amount of information to each sensor node, so that every pair of nodes can calculate a corresponding field of the matrix, and use it as the link key. Multiple space key pre-distribution scheme [9] improves the resilience of Blom's scheme. It uses a public matrix G and a set of ω private matrices D. Polynomial-based key pre-distribution scheme [6] distributes a polynomial share (a partially evaluated polynomial) to each sensor node by using whichever pair of nodes can generate a link key.

13.4.3 Hybrid Keying Models

Finally, hybrid solutions use probabilistic approaches on deterministic solutions to improve scalability and resilience. Polynomial pool-based key pre-distribution scheme [13] considers the fact that not all pairs of sensor nodes have to establish a key. It combines polynomial-based key pre-distribution scheme [6] with the keypool idea in [8,10] to improve resilience and scalability.

13.4.4 Groupwise Keying Models

In hierarchical WSNs, sensor nodes require groupwise keys to secure multicast messages. One approach is to use secure but costly asymmetric cryptography [7], and IKA2 [17] use a Diffie–Hellman-based group key transport protocol. Recently, some works on the public key cryptography protocols (e.g., elliptic curve cryptography) evaluation and efficiency measurements on sensor node platforms showed optimistic results [11,18]. In a hierarchical network, where a base station shares pairwise keys with all the sensor nodes, the base station can intermediate establishment of groupwise keys. Localized Encryption and Authentication Protocol (LEAP) [19] provides a mechanism to generate groupwise keys which follow the LEAP pairwise key establishment phase.

An important design consideration for security protocols based on symmetric keys is the degree of key sharing between the nodes in the system. At one extreme, we can have networkwide keys that are used for encrypting data and for authentication. This key sharing approach has the lowest storage costs and is very energy-efficient because no communication is required between nodes for establishing additional keys. However, it has the obvious security disadvantage that the compromise of a single node will reveal the global keys. At the other extreme, we can have a key sharing approach in which all secure communication is based on keys that are shared pairwise between two nodes. From the security point of view, this approach is ideal because the compromise of a node does not reveal any keys that are used by the other nodes in the network. However, under this approach, each node will need a unique key for every other node that it communicates with. Moreover, in many sensor networks, the immediate neighbors of a sensor node cannot be predicted in advance; consequently, these pairwise shared keys will need to be established after the network is deployed. A unique issue that arises in sensor networks that needs to be considered while selecting a key sharing approach is its impact on the effectiveness of in-network processing. Particular keying mechanisms may reduce the effectiveness of in-network processing.

IEEE 802.15.4-compliant devices can share a network key such that each cluster shares only one key among all devices to exchange data and for authentication purposes. This will ease the key management and memory overhead issues, but this comes at the cost of lower security. Similarly, IEEE 802.15.4-compliant devices can also support pairwise key exchange that improves the overall security of a network where any two devices exchanging data will share a different key. This improved robustness of network security comes at a cost, particularly in the overhead of key management. A device communicating with many devices in a network has to have different keys for each corresponding communicating device, which will increase the memory overhead on resource-scarce devices used in the network.

13.4.5   Key Updates

Key management schemes are at the heart of securing such networks. Key management schemes for sensor networks can be classified broadly into static and dynamic keying based on administrative key updates after network deployment. While static schemes assume no updates, dynamic ones provide for post-deployment key updates. The general security and performance objective of key management schemes include minimizing number of keys stored per sensor node, providing rich logical pairwise connectivity, and enhancing network resilience to node capture.

13.4.5.1   Static Keying Schemes

Static keying management schemes (a.k.a. key-predistribution) perform key management functions statically prior to or shortly after the deployment of the network. Administrative keys are generated at the sensor manufacturing time or by the base station upon network bootstrapping. Key assignment to nodes may be performed on a random basis or may take place based on some deployment information. Once generated and assigned, keys are pre-distributed to nodes. The main feature of static key management is the fact that the above key management cycle takes place only once at or prior to initialization. Accordingly, lost keys due to node capture or failure are not compensated.

13.4.5.2 Dynamic Keying Schemes

The main feature of dynamic key management schemes is repeating the key management process either periodically or on-demand to respond to node capture. After initial keying, key generation, assignment, and distribution might take place (in a process known as re-keying) to create new keys that replace the keys assumed lost or revealed to an attacker so that the network is refreshed and the attacker loses information earned by node capture. Another advantage of dynamic keying is that upon adding new nodes, unlike static keying, the probability of network capture does not necessarily increase. Various dynamic key management techniques have been proposed with different key management responsibility taken by different network components.

13.4.6   Limitations of IEEE 802.15.4 Standard from the Security Aspect

Higher layers will determine when the security is to be used at the MAC layer by any device and provide all keying material necessary to provide the security services. Key management, device authentication, and freshness protection may be provided by the higher layers, but is not addressed in IEEE 802.15.4 standard. The management and establishment of keys is the responsibility of the implementer of higher layers. There is no simple way to group keys in IEEE 802.15.4-enabled WSNs because, as mentioned earlier, the ACL entries are only associated to a single destination address. A detailed analysis of shortcomings of security features is mentioned by [16].

13.5   Security Services Provided by ZigBee Alliance

As explained above, the IEEE 802.15.4 addresses good security mechanisms, but it still does not address what type of keying mechanism will be used to employ supported security techniques.

ZigBee Alliance [4] is an association of companies working together to enable wireless networked monitoring and control products based on IEEE 802.15.4 standard. After the acceptance of 802.15.4 as IEEE standard, ZigBee Alliance is mainly focused on developing network and application layer issues. ZigBee Alliance is also working on application programming interfaces (API) at the network and link layers of IEEE 802.15.4. The Alliance also introduced secure data transmission in wireless sensor networks based on IEEE 802.15.4 specification, but most of this work is in general theoretical descriptions of security protocol at the network layer. There is no specific study or results published or mentioned by ZigBee Alliance in regard to which security suites perform better in different application overheads. ZigBee Alliance has recommended both symmetric and asymmetric key exchange protocols for different networking layers. Asymmetric key exchange protocols that mainly rely on public key cryptography are computationally intensive and their feasibility in wireless sensor networks is only possible with devices that are resource-rich both in computation and power.

13.5.1 Keyed Hash Function for Message Authentication

A hash function is a way of creating a small digital fingerprint of any data. Cryptographic hash function is a one-way operation and there is no practical way to calculate a particular data input that will result in a desired hash value, thus it is difficult to forge. A practical motivation for constructing hash functions from block ciphers is that if an efficient implementation of block cipher is already available within a system (either in hardware or in software), then using it as the central component for a hash function may provide latter functionality at little additional cost. IEEE 802.15.4 protocol supports a well-known block cipher AES, and hence ZigBee Alliance specification also relied on AES. ZigBee Alliance suggested the use of Matyas–Meyer–Oseas [14] as the cryptographic hash function that will be based on AES with a block size of 128 bits.

Mechanisms that provide integrity checks based on a secret key are usually called MACs. Typically, message authentication codes are used between two parties that share a secret key to authenticate information transmitted between these parties. ZigBee Alliance specification suggests the keyed hash message authentication code (HMAC) as specified in the FIPS Pub 198 [2]. A MAC takes a message and a secret key and generates a $MACtag$, such that it is difficult for an attacker to generate a valid (message, tag) pair and is used to prevent attackers forging messages. The calculation of $MacTag$ (i.e., HMAC) of data $MacData$ under key $MacKey$ will be shown as follows:

$$MacTag = MAC_{MacKey}MacData$$

13.5.2 Symmetric-Key Key Establishment Protocol

Key establishment involves two entities, an initiator device and a responder device, and is prefaced by a trust-provisioning step. Trust information (e.g., a master key) provides a starting point for establishing a link key and can

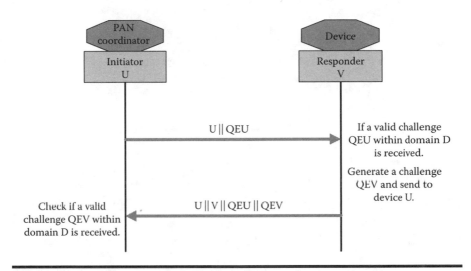

Figure 13.3   Exchange of ephemeral data. (From M. Khan, F. Amini, and J. Mišić, in *Mobile Ad-hoc and Sensor Networks*, Springer, 2006. With permission.)

be provisioned in-band or out-band. In the following explanation of the protocol, we assume unique identifiers for initiator devices as U and for responder device (PAN coordinator) as V. The master key shared among both devices is represented as $Mkey$.

We will divide Symmetric-Key Key Establishment (SKKE) protocol between initiator and responder in the following major steps.

13.5.2.1   Exchange of Ephemeral Data

Figure 13.3 illustrates the exchange of the ephemeral data where the initiator device U will generate the challenge QEU. QEU is a statistically unique and unpredictable bit string of length *challengelen* by either using a random or pseudo-random string for a challenge *Domain D*. The challenge domain D defines the minimum and maximum length of the challenge.

$$D = (minchallengeLen, maxchallengeLen)$$

Initiator device U will send the challenge QEU to a responder device which upon receipt will validate the challenge QEU by computing the bit-length of bit string challenge QEU as *Challengelen* and verify that

$$Challengelen \in [minchallengelen, maxchallengelen]$$

Once the validation is successful, the responder device will also generate a challenge QEV and send it to initiator device U. The initiator will also validate the challenge QEV as described above.

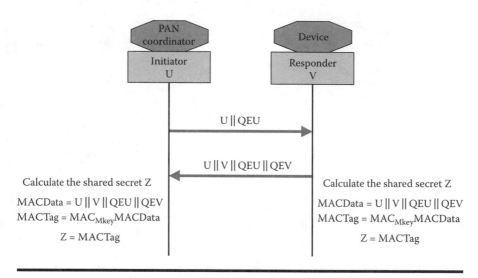

Figure 13.4 Generation of shared secret. (From M. Khan, F. Amini, and J. Mišić, in *Mobile Ad-hoc and Sensor Networks*, Springer, 2006. With permission.)

13.5.3 Generation of Shared Secret

Both parties involved in the protocol will generate a shared secret based on unique identifiers (i.e., distinguished names for parties involved), symmetric master keys, and challenges received and owned by each party, as shown in Figure 13.4.

1. Each party will generate a *MACData* by appending their identifiers and respective valid Challenges together as follows:

$$MACData = U \,||\, V \,||\, QEU \,||\, QEV$$

2. Each party will calculate the *MACTag* (i.e., keyed hash) for $MAC\,Data$ using $Mkey$ (master key for the device) as the key for keyed hash function as follows:

$$MACTag = MAC_{Mkey}MACData$$

3. Now both parties involved have derived the same secret Z. (Note: This is just a shared secret, not the link key. This shared secret will be involved in deriving the link key, but is not the link key itself.)

$$\text{Set } Z = MACTag$$

13.5.4 Derivation of Link Key

Each party involved will generate two cryptographic hashes (this is not the keyed hash) of the shared secret as described in ANSI X9.63-2001 [1].

$$Hash_1 = H(Z||01)$$

$$Hash_2 = H(Z||02)$$

The hash value $Hash_2$ will be the link key among two devices (Figure 13.5). Now for confirming that both parties have reached the same link key ($KeyData = Hash_2$), we will use value $Hash_1$ as the key for generating keyed hash values for confirming the stage of the protocol.

$$MACKey = Hash_1 \tag{13.1}$$

$$KeyData - Hash_2 \tag{13.2}$$

$$KKeyData = Hash_1||Hash_2 \tag{13.3}$$

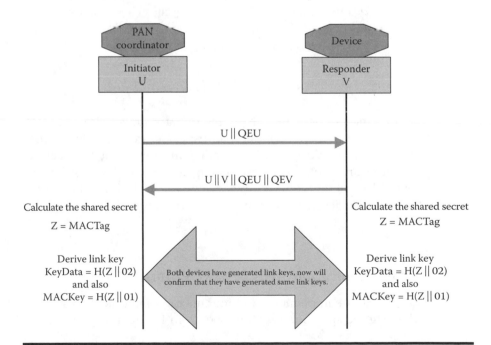

Figure 13.5   Generation of link key.

13.5.5 Confirming Link Key

Up to this stage of protocol, both parties are generating the same values and now they want to make sure that they have reached the same link key values, but they do not want to exchange the actual key at all. For this, they will once again rely on keyed hash functions and now both devices will generate different *MACTags* based on different data values, but will use the same key (i.e., *MACKey*) for generating the keyed hashes (*MACTags*).

1. Generation of *MACTags*: Initiator and responder devices will first generate *MACData* values and based on these values will generate *MACTags*. Initiator device D will receive the $MACTag_1$ from the responder device V and generate $MACTag_2$ and send to device V. We explain the generation of both *MACData* values and MAC $Tags$ as follows. First, both devices will calculate *MACData* values:

$$MACData_1 = 02_{16}||V||U||QEU||QEV$$

$$MACData_2 = 03_{16}||V||U||QEU||QEV$$

From the above *MACData* values both devices will generate the *MACTags* using the key $MACkey$ (Equation 13.1) as follows:

$$MacTag_1 = MAC_{MacKey}MacData_1$$

$$MacTag_2 = MAC_{MacKey}MacData_2$$

2. Confirmation of MACTags: Now the initiator device D will receive $MacTag_1$ from the responder and responder device V will receive $MACTag_2$ from device D and both will verify that the recieved *MACTags* are equal to corresponding calculated *MACTags* by each device. Now if this verification is successful, each device knows that the other device has computed the correct link key, as shown in Figure 13.6.

13.5.6 Communication Steps in SKKE Protocol

SKKE protocol can be implemented in four major communication steps, as described in ZigBee specification [4] as shown in Figure 13.7.

1. SKKE-1: Initiator U will send the challenge QEU and wait for the challenge QEV from responder V.

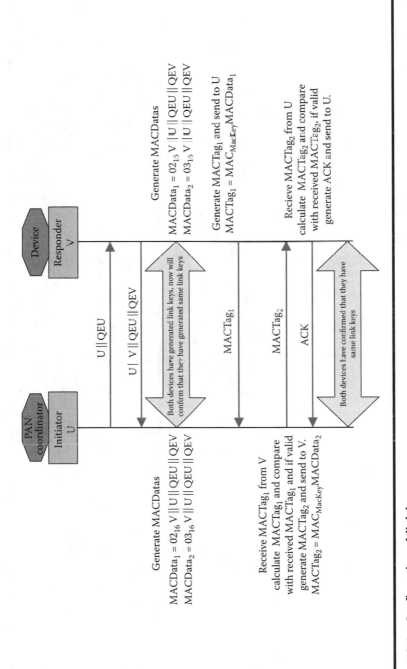

Figure 13.6 Confirmation of link keys.

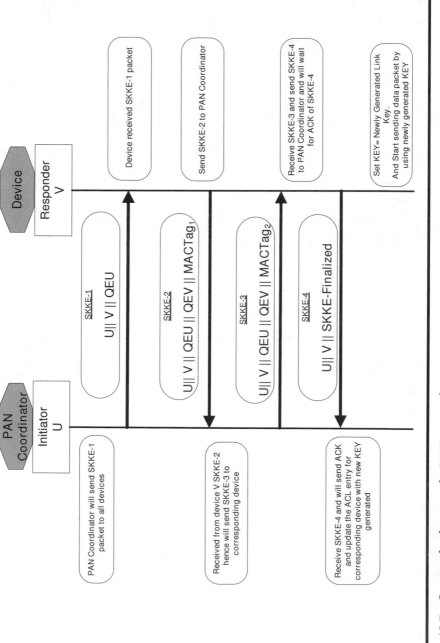

Figure 13.7 Communication steps in SKKE protocol.

2. SKKE-2: Responder V will receive the challenge QEU from initiator U, calculate its QEV, and in the same data packet will send the $MacTag_1$.

3. SKKE-3: Initiator will verify the $MacTag_1$ and if it is verified successfully, will send its $MacTag_2$. Now the initiator has a link key, but will wait for an acknowledgment that its $MacTag_2$ has been validated by the responder V.

4. SKKE-4: Responder will receive and validate the $MacTag_2$ from the initiator. If $MacTag_2$ validates successfully, the responder will send an acknowledgment and now both initiator and responder have link keys. Once initiator receives this SKKE-4 message, keys establishment is complete, and now regular secure communication can proceed using the link key among the initiator and the responder.

Authors have simulated the key exchange process in IEEE 802.15.4 on top of the simulation model of this network and initial results confirm the expected performance decrease of the overall network. They also have provided data encryption by exchanging link keys between each device and clusterhead. The signature payload plays a big role on performance of the cluster. Also we have observed that the total access delay is higher when encryption and decryption are provided.

13.6 Summary

In this chapter we have outlined a number of problems in achieving the target of secure communication in wireless sensor networks. IEEE 802.15.4 cluster-based wireless sensor network provides higher bandwidth links for inter-coordinator communication, and allows higher power resources at the coordinator, but still the implementer of higher layers should make a great deal of effort in choosing the right keying model based on the application requirements. Even though ZigBee Alliance has outlined protocols regarding the key exchange, it is necessary to integrate them with the IEEE 802.15.4 MAC protocol. Because key exchange protocols require downlink communications from the PAN coordinator to the ordinary nodes, it will consume a lot of bandwidth. Therefore, a period of key exchange is a crucial design parameter which has to match both security and bandwidth requirements for the sensing application. Also, addition of the message authentication code at the end of the packet decreases the bandwidth which is left to the application and affects the complexity of the aggregation. We expect that future work in this area (by us and other researchers) will deliver the reasonable trade-off between the level of security and application bandwidth in large sensor networks implemented over interconnected IEEE 802.15.4 clusters.

References

[1] ANSI X9.63-2001, Public Key Cryptography for the Financial Services Industry—Key Agreement and Key Transport Using Elliptic Curve Cryptography. American Bankers Association, 2001.

[2] FIPS Pub. 198, The Keyed-Hash Message Authentication Code (HMAC). Federal Information Processing Standards Publication 198, U.S. Department of Commerce/N.I.S.T., 2002.

[3] Standard for part 15.4: Wireless medium access control (MAC) and physical layer (PHY) specifications for low rate wireless personal area networks (WPAN). IEEE Std 802.15.4, IEEE, 2003.

[4] Z. Alliance. ZigBee specification (ZigBee document 053474r06, version 1.0), Dec. 2004.

[5] R. Blom. An optimal class of symmetric key generation systems. In *Proc. of the EUROCRYPT 84 Workshop on Advances in Cryptology: Theory and Application of Cryptographic Techniques*, pages 335–338, New York, 1985. Springer-Verlag.

[6] C. Blundo, A. D. Santis, A. Herzberg, S. Kutten, U. Vaccaro, and M. Yung. Perfectly-secure key distribution for dynamic conferences. In *CRYPTO '92: Proceedings of the 12th Annual International Cryptology Conference on Advances in Cryptology*, pages 471–486, London, UK, 1993. Springer-Verlag.

[7] M. Burmester and Y. Desmedt. A secure and efficient conference key distribution system. In *In Advances in Cryptology—EUROCRYPT 94, A. D. Santis, Ed., Lecture Notes in Computer Science, vol. 950*, pages 275–286, New York, 1994. Springer-Verlag.

[8] H. Chan, A. Perrig, and D. Song. Random key predistribution schemes for sensor networks. In *SP '03: Proceedings of the 2003 IEEE Symposium on Security and Privacy*, page 197, Washington, DC, 2003. IEEE Computer Society.

[9] W. Du, J. Deng, Y. S. Han, and P. K. Varshney. A pairwise key predistribution scheme for wireless sensor networks. In *CCS '03: Proceedings of the 10th ACM Conference on Computer and Communications Security*, pages 42–51, New York, 2003. ACM Press.

[10] L. Eschenauer and V. D. Gligor. A key-management scheme for distributed sensor networks. In *CCS '02: Proceedings of the 9th ACM Conference on Computer and Communications Security*, pages 41–47, New York, 2002. ACM Press.

[11] V. Gupta, M. Millard, S. Fung, Y. Zhu, N. Gura, H. Eberle, and S. C. Shantz. Sizzle: A standards-based end-to-end security architecture for the embedded Internet (best paper). In *PERCOM '05: Proceedings of the Third IEEE International Conference on Pervasive Computing and Communications*, pages 247–256, Washington, DC, 2005. IEEE Computer Society.

[12] H. Kang and X. Li. Power-aware sensor selection in wireless sensor networks. *Scalable Software Systems Laboratory, Computer Science Department, Oklahoma State University*, 2006.

[13] D. Liu, P. Ning, and R. Li. Establishing pairwise keys in distributed sensor networks. *ACM Trans. Inf. Syst. Secur.*, 8(1):41–77, 2005.

[14] A. Menezes, P. van Oorschot, and S. Vanstone. *Handbook of Applied Cryptography.* Boca Raton, FL, 1997, CRC Press.

[15] J. Mišić, S. Shafi, and V. B. Mišić. Performance of beacon enabled IEEE 802.15.4 cluster with downlink and uplink tarffic. *IEEE Transactions on Parallel and Distributed Systems*, 17(4):361–377, Apr. 2006.

[16] N. Sastry and D. Wagner. Security considerations for IEEE 802.15.4 networks. In *WiSe '04: Proceedings of the 2004 ACM Workshop on Wireless Security*, pages 32–42, New York, 2004. ACM Press.

[17] G. W. M. Steiner, and M. Tsudik. Key agreement in dynamic peer groups. In *IEEE Transactions on Parallel and Distributed Systems*, pages 769–780, Washington, DC, Aug. 2000. IEEE Computer Society.

[18] A. S. Wander, N. Gura, H. Eberle, V. Gupta, and S. C. Shantz. Energy analysis of public-key cryptography for wireless sensor networks. In *PERCOM '05: Proceedings of the Third IEEE International Conference on Pervasive Computing and Communications*, pages 324–328, Washington, DC, 2005. IEEE Computer Society.

[19] S. Zhu, S. Setia, and S. Jajodia. LEAP: efficient security mechanisms for large-scale distributed sensor networks. In *CCS '03: Proceedings of the 10th ACM Conference on Computer and Communications Security*, pages 62–72, New York, 2003. ACM Press.

[20] M. Khan, F. Amini, and J. Mišić. Key exchange in 802.15.4 networks and its implications. In *Mobile Ad-hoc and Sensor Networks*, H. Zhang, S. Olariu, J. Cao, and D. B. Johnson (Eds.), (pp. 497–508). Berlin, 2006. Springer.

Chapter 14

Security in Wireless Sensor Networks

Yong Wang, Garhan Attebury, and Byrav Ramamurthy

Contents

Wireless sensor networks (WSNs) are used in many applications in military, ecological, and health-related areas. These applications often include the monitoring of sensitive information such as enemy movement on the battle-field or the location of personnel in a building. Security is therefore important in WSNs. However, WSNs suffer from many constraints including low computation capability, small memory, limited energy resources, susceptibility to physical capture, and the use of insecure wireless communication channels. These constraints make security challenging in WSNs. In this chapter, we present a survey of security issues in WSNs. First we outline the constraints, security requirements, and attacks with corresponding countermeasures in WSNs. We then present a holistic view of security issues. These issues are classified into five categories: cryptography, key management, secure routing, secure data aggregation, and intrusion detection. Along the way we highlight advantages and disadvantages of various WSN security protocols and further compare and evaluate these protocols based on each of these five categories. We also point out the open research issues in each sub-area and conclude with possible future research directions on security in WSNs.

14.1   Introduction

Advances in wireless communication and electronics have enabled the development of low-cost, low-power, multi-functional sensor nodes. These tiny sensor nodes, consisting of sensing, data processing, and communication components, make it possible to deploy WSNs, which represent a significant improvement over traditional wired sensor networks. WSNs can greatly simplify system design and operation as the environment being monitored does not require the communication or energy infrastructure associated with wired networks [1].

WSNs are expected to be solutions to many applications, such as detecting and tracking the passage of troops and tanks on a battlefield, monitoring environmental pollutants, measuring traffic flows on roads, and tracking the location of personnel in a building. Many sensor networks have mission-critical tasks and thus require security to be considered [2,3]. Improper use of information, or using forged information, may cause unwanted information leakage and provide inaccurate results.

While some aspects of WSNs are similar to traditional wireless ad hoc networks, important distinctions exist which greatly affect how security is achieved. The differences between sensor networks and ad hoc networks [4] are:

- The number of sensor nodes in a sensor network can be several orders of magnitude higher than the nodes in an ad hoc network.
- Sensor nodes are densely deployed.
- Sensor nodes are prone to failures due to harsh environments and energy constraints.
- The topology of a sensor network changes very frequently due to failures or mobility.

- Sensor nodes are limited in computation, memory, and power resources.
- Sensor nodes may not have global identification.

These differences greatly affect how secure data transfer schemes are implemented in WSNs. For example, the use of radio transmission, along with the constraints of small size, low cost, and limited energy, make WSNs more susceptible to denial-of-service attacks [5]. Advanced anti-jamming techniques such as frequency-hopping spread spectrum and physical tamper-proofing of nodes are generally impossible in a sensor network due to the requirements of greater design complexity and higher energy consumption [5]. Furthermore, the limited energy and processing power of nodes makes the use of public key cryptography nearly impossible. While the results from recent studies show that public key cryptography might be feasible in sensor networks [6,7], it remains for the most part infeasible in WSNs. Instead, most security schemes make use of symmetric key cryptography. One thing required in either case is the use of keys for secure communication. Managing key distribution is not unique to WSNs, but again constraints such as small memory capacity make centralized keying techniques impossible. Straight pairwise key sharing between every two nodes in a network does not scale to large networks with tens of thousands of nodes as the storage requirements are too high. A security scheme in WSNs must provide efficient key distribution while maintaining the ability for communication between all relevant nodes.

In addition to key distribution, secure routing protocols must be considered. These protocols are concerned with how a node sends messages to other nodes or a base station. A key challenge is that of authenticated broadcast. Existing authenticated broadcast methods often rely on public key cryptography and include high computational overhead, making them infeasible in WSNs. Secure routing protocols proposed for use in WSNs, such as security protocols for sensor networks (SPINS) [8], must consider these factors. Additionally, the constraint on energy in WSNs leads to the desire for data aggregation. This aggregation of sensor data needs to be secure to ensure information integrity and confidentiality [9,10]. Although this is achievable through cryptography, an aggregation scheme must take into account the constraints in WSNs and the unique characteristics of the cryptography and routing schemes. It is also desirable for secure data aggregation protocols to be flexible, allowing lower levels of security for less important data, saving energy, and higher levels of security for more sensitive data, consuming more energy.

As with any network, the awareness of compromised nodes and attacks is desirable. Many security schemes provide assurance that data remains intact and communication unaffected as long as fewer than t nodes are compromised [11]. The ability of a node or base station to detect when other nodes are compromised enables them to take action, either ignoring the compromised data or reconfiguring the network to eliminate the threat.

The remainder of this chapter discusses the above areas in more detail and considers how they are all required to form a complete WSN security scheme. A few existing surveys on security issues in ad hoc networks can be found in [12–14]. However, only small sections of these surveys focus on WSNs. A recent survey paper on security issues in mobile ad hoc networks also included an overview of security issues in WSNs [15]. However, the paper did not discuss cryptography and intrusion detection issues. Further, it included only a small portion of the available literature on security in WSNs.

The rest of the chapter is organized as follows. Section 14.2 presents background information on WSNs. Section 14.3 discusses attacks in the different network layers of sensor networks, followed by Section 14.4, which focuses on the selection of cryptography in WSNs. Section 14.5 focuses on key management, Section 14.6 on secure routing schemes, Section 14.7 on secure data aggregation, and Section 14.8 on intrusion detection systems. Section 14.9 discusses future research directions on security in WSNs, and Section 14.10 concludes the chapter.

14.2   Background

14.2.1   Communication Architecture

A WSN is usually composed of hundreds or thousands of sensor nodes. These sensor nodes are often densely deployed in a sensor field and have the capability to collect data and route data back to a base station (BS). A sensor consists of four basic parts: a sensing unit, a processing unit, a transceiver unit, and a power unit [4]. It may also have additional application-dependent components such as a location finding system, power generator, and mobilizer (Figure 14.1). Sensing units are usually composed of two sub-units: sensors and analog-to-digital converters (ADCs). The ADCs convert the analog signals produced by the sensors to digital signals based on the observed phenomenon. The processing unit, which is generally associated with a small storage unit, manages the procedures that make the sensor node collaborate with the other nodes. A transceiver unit connects the node to the network. One of the most important units is the power unit. A power unit may be finite, such as a single battery, or may be supported by power scavenging devices, such as solar cells. Most of the sensor network routing techniques and sensing tasks require knowledge of location, which is provided by a location finding system. Finally, a mobilizer may sometimes be needed to move the sensor node depending on the application.

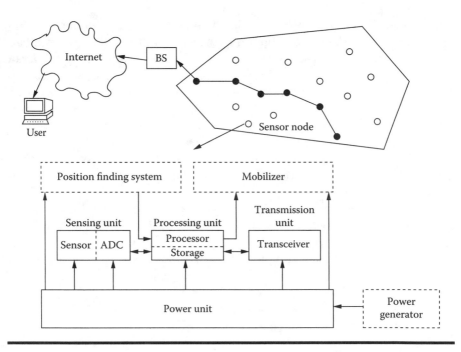

Figure 14.1   The components of a sensor node. (From Y. Wang, G. Attebury, and B. Ramamurthy, *IEEE Communications Surveys and Tutorials*, Vol. 8, no. 2, pp. 2–23, 2006. With permission.)

The protocol stack used in sensor nodes contains physical, data link, network, transport, and application layers defined as follows [4]:

- Physical layer: Responsible for frequency selection, carrier frequency generation, signal deflection, modulation, and data encryption.
- Data link layer: Responsible for the multiplexing of datastreams, data frame detection, medium access, and error control. This layer ensures reliable point-to-point and point-to-multipoint connections.
- Network layer: Responsible for specifying the assignment of addresses and how packets are forwarded.
- Transport layer: Responsible for specifying how the reliable transport of packets will take place.
- Application layer: Responsible for specifying how the data is requested and provided for both the individual sensor nodes and the interactions with the end user.

14.2.2   Constraints in WSNs

Individual sensor nodes in a WSN are inherently resource constrained. They have limited processing capability, storage capacity, and communication bandwidth. Each of these limitations is due in part to the two greatest constraints: limited energy and physical size. Table 14.1 shows several currently available sensor node platforms. The design of security services in WSNs must consider the hardware constraints of the sensor nodes.

14.2.2.1 Energy

Energy consumption in sensor nodes can be categorized into three parts:

1. Energy for the sensor transducer
2. Energy for communication among sensor nodes
3. Energy for microprocessor computation

The study in [20,21] found that each bit transmitted in WSNs consumes about as much power as executing 800 to 1000 instructions. Thus, communication is more costly than computation in WSNs. Any message expansion caused by security mechanisms comes at a significant cost. Further, higher security levels in WSNs usually correspond to more energy consumption for cryptographic functions. Thus, WSNs could be divided into different security levels depending on energy cost [22,23].

14.2.2.2 Computation

The embedded processors in sensor nodes are generally not as powerful as those in nodes of a wired or ad hoc network. As such, complex cryptographic algorithms cannot be used in WSNs.

14.2.2.3 Memory

Memory in a sensor node usually includes flash memory and RAM. Flash memory is used for storing downloaded application code and RAM is used for storing application programs, sensor data, and intermediate computations. There is usually not enough space to run complicated algorithms after loading OS and application code. In the SmartDust project, for example, TinyOS consumes about 3500 bytes of instruction memory, leaving only 4500 bytes for security and applications [20,21]. This makes it impractical to use the majority of current security algorithms [8]. With an Intel Mote, the situation is slightly improved, but still far from meeting the requirements of many algorithms.

14.2.2.4 Transmission Range

The communication range of sensor nodes is limited both technically and by the need to conserve energy. The actual range achieved from a given transmission signal strength is dependent on various environmental factors such as weather and terrain.

Table 14.1 Variety of Real-Life Sensor Nodes

	Berkeley Mote [16]					EYES [17]	Medusa MK-2 [18]	Imote [19]
	WeC	rene2	rene2	dot	mica			
Month/year	09/99	10/00	06/01	08/01	02/02	03/02	09/02	01/03
CPU	AT90LS8535			ATmega163	ATmega103[a]	MSP 430F149	40 MHz ARM THUMB	ARM core 12 MHz
Prog. mem.	8 kb			16 kb	128 kb	60 kb	1 MB	512 kb
RAM	0.5 kb			1 kb	4 kb	2 kb	136 kb	64 kb
Radio	TR1000 916 MHz				TR1000 916 MHz	TR1001 868.35 MHz	TR1000 916 MHz	BT 2.4 GHz
Rate	10 kbps				10/40 kbps	115 kbps	115 kbps	100 kbps

[a]Later versions are an ATmega128 running in 103 mode.

Source: Y. Wang, G. Attebury, and B. Ramamurthy, *IEEE Communications Surveys and Tutorials*, Vol. 8, no. 2, pp. 2–23, 2006. With permission.

14.2.3   Security Requirements

The goal of security services in WSNs is to protect the information and resources from attacks and misbehavior. The security requirements in WSNs include:

- Availability: Ensures that the desired network services are available even in the presence of denial-of-service attacks.
- Authorization: Ensures that only authorized sensors can be involved in providing information to network services.
- Authentication: Ensures that the communicating node is the one that it claims to be.
- Confidentiality: Ensures that a given message cannot be understood by anyone other than the desired recipients.
- Integrity: Ensures that a message sent from one node to another is not altered by unauthorized or unknown means.
- Non-repudiation: Denotes that a node cannot deny sending a message it has previously sent.
- Freshness: Implies that the data is recent and ensures that no adversary can replay old messages.

The security services in WSNs are usually centered around cryptography. However, because of the constraints in WSNs, many already-existing secure algorithms are not practical for use. We discuss this problem in Section 14.4.

14.2.4   Threat Model

In WSNs, it is usually assumed that an attacker may know the security mechanisms that are deployed in a sensor network, and may be able to compromise a node or even physically capture a node. Due to the high cost of deploying tamper-resistant sensor nodes, most WSN nodes are viewed as non-tamper-resistant. Further, once a node is compromised, the attacker is capable of stealing the key materials contained within that node.

Base stations in WSNs are usually regarded as trustworthy. Most research studies focus on secure routing between sensors and the base station. Deng et al. considered strategies against threats which can lead to the failure of the base station [24].

Attacks in sensor networks can be classified into the following categories:

- Outsider versus insider attacks: Outside attacks are defined as attacks from nodes which do not belong to a WSN. Inside attacks occur when legitimate nodes of a WSN behave in unintended or unauthorized ways.

■ Passive versus active attacks: Passive attacks include eavesdropping on packets exchanged within a WSN. Active attacks involve some modifications of the datastream or the creation of a false stream.

■ Mote-class versus laptop-class attacks: In mote-class attacks, an adversary attacks a WSN by using a few nodes with similar capabilities to the network nodes. In laptop-class attacks, an adversary can use more powerful devices such as a laptop to attack a WSN. These devices have greater transmission range, processing power, and energy reserves than the network nodes.

14.2.5 Evaluation

We suggest using the following metrics to evaluate whether a security scheme is appropriate in WSNs.

■ Security: A security scheme has to meet the requirements discussed in Section 14.2.3.

■ Resiliency: In case a few nodes are compromised, a security scheme can still protect against the attacks.

■ Energy efficiency: A security scheme must be energy-efficient to maximize node and network lifetime.

■ Flexibility: The key management needs to be flexible to allow for different network deployment methods such as random node scattering and pre-determined node placement.

■ Scalability: A security scheme should be able to scale without compromising the security requirements.

■ Fault-tolerance: A security scheme should continue to provide security services in the presence of faults such as failed nodes.

■ Self-healing: Sensors may fail or run out of energy. The remaining sensors may need to be re-organized to maintain a set level of security.

■ Assurance: Assurance is the ability to disseminate different information at different levels to end users [25]. A security scheme should offer choices as to desired reliability, latency, and so on.

14.3   Attacks in Sensor Networks

WSNs are vulnerable to various types of attacks. According to the security requirements in WSNs, these attacks can be categorized [3] as:

■ Attacks on secrecy and authentication: Standard cryptographic techniques can protect the secrecy and authenticity of communication channels from outsider attacks such as eavesdropping, packet replay attacks, and modification or spoofing of packets.

- Attacks on network availability: Attacks on availability are often referred to as denial-of-service (DoS) attacks. DoS attacks may target any layer of a sensor network.
- Stealthy attacks against service integrity: In a stealthy attack, the goal of the attacker is to make the network accept a false data value. For example, an attacker compromises a sensor node and injects a false data value through that sensor node.

In these attacks, keeping the sensor network available for its intended use is essential. DoS attacks against WSNs may permit real-world damage to the health and safety of people [5]. In this section, we focus only on DoS attacks and their countermeasures in sensor networks. Section 14.6 discusses attacks on secrecy and authentication and Section 14.8 discusses stealthy attacks and countermeasures.

The DoS attack usually refers to an adversary's attempt to disrupt, subvert, or destroy a network. However, a DoS attack can be any event that diminishes or eliminates a network's capacity to perform its expected function [5]. Sensor networks are usually divided into layers, and this layered architecture makes WSNs vulnerable to DoS attacks, which may occur in any layer of a sensor network.

Previous discussions on DoS attacks in WSNs can be found in [3,5,26,27]. The remainder of this section summarizes possible DoS attacks and countermeasures in each layer of a sensor network.

14.3.1 Physical Layer

The physical layer is responsible for frequency selection, carrier frequency generation, signal detection, modulation, and data encryption [4]. As with any radio-based medium there exists the possibility of jamming in WSNs. In addition, nodes in WSNs may be deployed in hostile or insecure environments where an attacker has easy physical access. These two vulnerabilities are explored in this subsection.

14.3.1.1 Jamming

Jamming is a type of attack which interferes with the radio frequencies that a network's nodes are using [3,5]. A jamming source may either be powerful enough to disrupt the entire network or less powerful and only able to disrupt a smaller portion of the network. Even with lesser-powered jamming sources, such as a small compromised subset of the network's sensor nodes, an adversary has the potential to disrupt the entire network provided the jamming sources are randomly distributed in the network.

Typical defenses against jamming involve variations of spread-spectrum communication such as frequency hopping and code spreading [5]. Frequency-hopping spread spectrum (FHSS) is a method of transmitting signals by rapidly switching a carrier among many frequency channels using a pseudo-random sequence known to both transmitter and receiver. Without being able to follow the frequency selection sequence, an attacker is unable to jam the frequency being used at a given moment in time. However, as the range of possible frequencies is limited, an attacker may instead jam a wide section of the frequency band.

Code spreading is another technique used to defend against jamming attacks and is common in mobile networks. However, this technique requires greater design complexity and energy restricting its use in WSNs. In general, to maintain low cost and low power requirements, sensor devices are limited to single-frequency use and are therefore highly susceptible to jamming attacks.

14.3.1.2 Tampering

Another physical layer attack is tampering [5]. Given physical access to a node, an attacker can extract sensitive information such as cryptographic keys or other data on the node. The node may also be altered or replaced to create a compromised node which the attacker controls. One defense against this attack involves tamper-proofing the node's physical package [5]. However, it is usually assumed that the sensor nodes are not tamper-proofed in WSNs due to the additional cost. This indicates that a security scheme must consider the situation in which sensor nodes are compromised.

14.3.2 Link Layer

The data link layer is responsible for the multiplexing of datastreams, data frame detection, medium access, and error control [4]. It ensures reliable point-to-point and point-to-multipoint connections in a communications network. Attacks at the link layer include purposely introduced collisions, resource exhaustion, and unfairness. This sub-section looks at each of these link layer attack categories [5].

14.3.2.1 Collisions

A collision occurs when two nodes attempt to transmit on the same frequency simultaneously [5]. When packets collide, a change will likely occur in the data portion causing a checksum mismatch at the receiving end. The packet will then be discarded as invalid. An adversary may strategically cause collisions in specific packets such as ACK control messages. A possible result of such collisions is the costly exponential back-off in certain media access control protocols.

A typical defense against collisions is the use of error-correcting codes [5]. Most codes work best with low levels of collisions such as those caused by environmental or probabilistic errors. However, these codes also add additional processing and communication overhead. It is reasonable to assume that an attacker will always be able to corrupt more than what can be corrected. Although it is possible to detect these malicious collisions, no complete defenses against them are known at this time.

14.3.2.2 Exhaustion

Repeated collisions can also be used by an attacker to cause resource exhaustion [5]. For example, a naive link layer implementation may continuously attempt to retransmit the corrupted packets. Unless these hopeless retransmissions are discovered or prevented, the energy reserves of the transmitting node and those surrounding it will be quickly depleted.

A possible solution is to apply rate limits to the admission control in the medium access control protocol such that the network can ignore excessive requests preventing the energy drain caused by repeated transmissions [5]. A second technique is to use time-division multiplexing where each node is allotted a time slot in which it can transmit [5]. This eliminates the need of arbitration for each frame and can solve the indefinite postponement problem in a back-off algorithm. However, it is still susceptible to collisions.

14.3.2.3 Unfairness

Unfairness can be considered a weak form of a DoS attack [5]. An attacker may cause unfairness in a network by intermittently using the above link layer attacks. Instead of outright preventing access to a service, an attacker can degrade it to give them an advantage such as causing other nodes in a real-time medium access control protocol to miss their transmission deadline. The use of small frames lessens the effect of such attacks by reducing the amount of time an attacker can capture the communication channel. However, this technique often reduces efficiency and is susceptible to further unfairness such as an attacker trying to retransmit quickly instead of randomly delaying.

14.3.3 Network and Routing Layer

The network and routing layer of sensor networks is usually designed according to the following principles [4]:

- Power efficiency is an important consideration.
- Sensor networks are mostly data-centric.
- An ideal sensor network has attribute-based addressing and location awareness.

The attacks in network and routing layer are discussed next:

14.3.3.1 Spoofed, Altered, or Replayed Routing Information

The most direct attack against a routing protocol in any network is to target the routing information itself as it is exchanged between nodes. An attacker may spoof, alter, or replay routing information to disrupt traffic in the network [26]. These disruptions include the creation of routing loops, attracting or repelling network traffic from select nodes, extending and shortening source routes, generating fake error messages, partitioning the network, and increasing end-to-end latency.

A countermeasure against spoofing and alteration is to append a MAC after the message. By adding a MAC to the message, the receivers can verify whether the messages have been spoofed or altered. To defend against replayed information, counters or timestamps can be included in the messages [8].

14.3.3.2 Selective Forwarding

A significant assumption made in multi-hop networks is that all nodes in the network will accurately forward received messages. An attacker may create malicious nodes which selectively forward only certain messages and simply drop others [26]. A specific form of this attack is the black hole attack in which a node drops all messages it receives. One defense against selective forwarding attacks is using multiple paths to send data [26]. A second defense is to detect the malicious node or assume it has failed and seek an alternative route.

14.3.3.3 Sinkhole

In a sinkhole attack, an attacker makes a compromised node look more attractive to surrounding nodes by forging routing information [5,26]. The end result is that surrounding nodes will choose the compromised node as the next node to route their data through. This type of attack makes selective forwarding very simple as all traffic from a large area in the network will flow through the adversary's node.

14.3.3.4 Sybil

The Sybil attack is a case where one node presents more than one identity to the network [3,26,27]. Protocols and algorithms which are easily affected include fault-tolerant schemes, distributed storage, and network topology maintenance. For example, a distributed storage scheme may rely on there being three replicas of the same data to achieve a given level of redundancy. If a compromised node pretends to be two of the three nodes, the algorithms used may conclude that redundancy has been achieved although in reality it has not.

14.3.3.5   Wormholes

A wormhole is a low latency link between two portions of the network over which an attacker replays network messages [26]. This link may be established either by a single node forwarding messages between two adjacent but otherwise non-neighboring nodes or by a pair of nodes in different parts of the network communicating with each other. The latter case is closely related to the sinkhole attack as an attacking node near the base station can provide a one-hop link to that base station via the other attacking node in a distant part of the network. Hu et al. presented a novel and general mechanism called packet leashes for detecting and defending against wormhole attacks [28]. Two types of leashes were introduced: geographic leashes and temporal leashes. The proposed mechanisms can also be used in WSNs.

14.3.3.6   Hello Flood Attacks

Many protocols which use Hello packets make the naive assumption that receiving such a packet means the sender is within radio range and is therefore a neighbor. An attacker may use a high-powered transmitter to trick a large area of nodes into believing they are neighbors of that transmitting node [26]. If the attacker falsely broadcasts a superior route to the base station, all of these nodes will attempt transmitting to the attacking node despite many being out of radio range in reality.

14.3.3.7   Acknowledgment Spoofing

Routing algorithms used in sensor networks sometimes require acknowledgments to be used. An attacking node can spoof the acknowledgments of overheard packets destined for neighboring nodes to provide false information to those neighboring nodes [26]. An example of such false information is claiming that a node is alive when in fact it is dead.

14.3.4   Transport Layer

The transport layer is responsible for managing end-to-end connections [4]. Two possible attacks in this layer, flooding and desynchronization, are discussed in this sub-section.

14.3.4.1 Flooding

Whenever a protocol is required to maintain state at either end of a connection it becomes vulnerable to memory exhaustion through flooding [5]. An attacker may repeatedly make new connection requests until the resources required by each connection are exhausted or reach a maximum limit. In either case, further legitimate requests will be ignored. One proposed solution to this problem is to require that each connecting client demonstrates its commitment to the connection by solving a puzzle [5]. The idea is that a connecting client will not needlessly waste its resources creating unnecessary connections. Given an attacker does not likely have infinite resources, it will be impossible for him to create new connections fast enough to cause resource starvation on the serving node. Although these puzzles do include processing overhead, this technique is more desirable than excessive communication.

14.3.4.2 Desynchronization

Desynchronization refers to the disruption of an existing connection [5]. An attacker may, for example, repeatedly spoof messages to an end host causing that host to request the retransmission of missed frames. If timed correctly, an attacker may degrade or even prevent the ability of the end hosts to successfully exchange data causing them instead to waste energy attempting to recover from errors which never really existed.

A possible solution to this type of attack is to require authentication of all packets communicated between hosts [5]. Provided that the authentication method is itself secure, an attacker will be unable to send the spoofed messages to the end hosts.

Table 14.2 shows the possible DoS attacks and countermeasures in WSNs.

In the following sections we discuss cryptography, key management protocols, secure routing protocols, secure data aggregation, and intrusion detection for WSNs. For the remainder of this article we use the following notation:

- A, B are principals such as communicating nodes.
- ID_A denotes the sensor identifier of node A.
- N_A is a nonce generated by A (a nonce is an unpredictable bit string, usually used to achieve freshness).
- K_{AB} denotes the secret pairwise key shared between A and B.
- M_K is the encryption of message M with key K.
- $MAC(K, M)$ denotes the computation of the message authentication code of message M with key K.
- $A \longrightarrow B$ denotes A unicasts a message to B.
- $A \longrightarrow *$ denotes A broadcasts a message to its neighbors.

Table 14.2   Sensor Network Layers and Denial-of-Service Defenses

Network layer	Attacks	Defense
Physical	Jamming	Spread-spectrum, priority messages, lower duty cycle, region mapping, mode change
	Tampering	Tamper-proofing, hiding
Link	Collision	Error-correcting code
	Exhaustion	Rate limitation
	Unfairness	Small frames
Network and routing	Spoofed, altered, or replayed routing information	Egress filtering, authentication, monitoring
	Selective forwarding	Redundancy, probing
	Sinkhole	Authentication, monitoring, redundancy
	Sybil	Authentication, probing
	Wormholes	Authentication, packet leashes by using geographic and temporal information
	Hello flood attacks	Authentication, verify the bidirectional link
	Acknowledgment spoofing	Authentication
Transport	Flooding	Client puzzles
	Desynchronization	Authentication

Source: Y. Wang, G. Attebury, and B. Ramamurthy, *IEEE Communications Surveys and Tutorials,* Vol. 8, no. 2, pp. 2–23, 2006. With permission.

14.4   Cryptography in WSNs

Selecting the most appropriate cryptographic method is vital in WSNs as all security services are ensured by cryptography. Cryptographic methods used in WSNs should meet the constraints of sensor nodes and be evaluated by code size, data size, processing time, and power consumption. In this section, we focus on the selection of cryptography in WSNs. We discuss public key cryptography first, followed by symmetric key cryptography.

14.4.1   Public Key Cryptography in WSNs

Many researchers believe that the code size, data size, processing time, and power consumption make it undesirable for public key algorithm techniques, such as the Diffie–Hellman key agreement protocol [29] or RSA signatures [30], to be employed in WSNs.

Public key algorithms such as RSA are computationally intensive and usually execute thousands or even millions of multiplication instructions to perform a single security operation. Further, a microprocessor's public key algorithm efficiency is primarily determined by the number of clock cycles required to perform a multiply instruction [31]. Brown et al. found that public key algorithms such as RSA usually require on the order of tens of seconds and up to minutes to perform encryption and decryption operations in constrained wireless devices, which exposes a vulnerability to DoS attacks [32]. On the other hand, Carman et al. found that it usually takes a microprocessor thousands of nano-joules to do a simple multiply function with a 128-bit result [31]. In contrast, symmetric key cryptographic algorithms and hash functions consume much less computational energy than public key algorithms. For example, the encryption of a 1024-bit block consumes approximately 42 mJ on the MC68328 DragonBall processor using RSA, and the estimated energy consumption for a 128-bit AES block is a much lower at 0.104 mJ [31].

Recent studies have shown that it is feasible to apply public key cryptography to sensor networks by using the right selection of algorithms and associated parameters, optimization, and low power techniques [6,7,33]. The investigated public key algorithms include Rabin's Scheme [34], Ntru-Encrypt [35], RSA [30], and Elliptic Curve Cryptography (ECC) [36,37]. Most studies in the literature focus on RSA and ECC algorithms. The attraction of ECC is that it offers equal security for a far smaller key size, thereby reducing processing and communication overhead. For example, RSA with 1024-bit keys (RSA-1024) provides a currently accepted level of security for many applications and is equivalent in strength to ECC with 160-bit keys (ECC-160) [38]. To protect data beyond the year 2010, RSA Security recommends RSA-2048 as the new minimum key size, which is equivalent to ECC with 224-bit keys (ECC-224) [39]. Table 14.3 summarizes the execution

Table 14.3   Public Key Cryptography: Average ECC and RSA Execution Times

Algorithm	Operation Time (s)
ECC secp160r1	0.81
ECC secp224r1	2.19
RSA-1024 public key $e = 2^{16} + 1$	0.43
RSA-1024 private key w. CRT[a]	10.99
RSA-2048 public key $e = 2^{16} + 1$	1.94
RSA-2048 private key w. CRT	83.26

[a] Chinese Remainder Theory.
Source: Y. Wang, G. Attebury, and B. Ramamurthy, *IEEE Communications Surveys and Tutorials*, Vol. 8, no. 2, pp. 2–23, 2006. With permission.

time of ECC and RSA implementations on an Atmel ATmega128 processor (used by Mica2 mote) [6]. The execution time is measured on average for a point multiplication in ECC and a modular exponential operation in RSA. ECC secp160r1 and secp224r1 are two standardized elliptic curves defined in [40]. As shown in Table 14.3, by using the small integer $e = 2^{16} + 1$ as the public key, RSA public key operation is slightly faster than ECC point multiplication. However, ECC point multiplication outperforms RSA private key operation by an order of magnitude. The RSA private key operation, which is too slow, limits its use in a sensor node. ECC has no such issues because both the public key operation and private key operation use the same point multiplication operations.

Wander et al. investigated the energy cost of authentication and key exchange based on RSA and ECC cryptography on an Atmel ATmega128 processor [7]. The result is shown in Table 14.4. The ECC-based signature is generated and verified with the Elliptic Curve Digital Signature Algorithm (ECDSA) [41]. The key exchange protocol is a simplified version of the SSL handshake, which involves two parties: a client initiating the communication and a server responding to the initiation [42]. The WSN is assumed to be administered by a central point with each sensor having a certificate signed by the central point's private key using an RSA or ECC signature. In the handshake process, the two parties verify each other's certificate and negotiate the session key to be used in the communication. As Table 14.4 shows, compared with RSA cryptography at the same security level, ECDSA signatures are significantly cheaper than RSA signatures and ECDSA verifications are within reasonable range of RSA verifications. Further, the ECC-based key exchange protocol outperforms the RSA-based key exchange protocol at the server side, and there is almost no difference in the energy cost for these two key exchange protocols at the client side. In addition, the

Table 14.4 Public Key Cryptography: Average Energy Costs of Digital Signature and Key Exchange Computations [mJ]

	Signature		Key Exchange	
Algorithm	Sign	Verify	Client	Server
RSA-1024	304	11.9	15.4	304
ECDSA-160	22.82	45.09	22.3	22.3
RSA-2048	2302.7	53.7	57.2	2302.7
ECDSA-224	61.54	121.98	60.4	60.4

Source: Y. Wang, G. Attebury, and B. Ramamurthy, *IEEE Communications Surveys and Tutorials*, Vol. 8, no. 2, pp. 2–23, 2006. With permission.

relative performance advantage of ECC over RSA increases as the key size increases in terms of the execution time and energy cost. Table 14.3 and Table 14.4 indicate that ECC is more appropriate than RSA for use in sensor networks.

The implementation of RSA and ECC cryptography on Mica2 motes further proved that a public key-based protocol is viable for WSNs. Two modules, TinyPK [43], based on RSA, and TinyECC [44], based on ECC, have been designed and implemented on Mica2 motes using the TinyOS development environment. Similar work was also conducted by Malan et al. on ECC cryptography using a Mica2 mote [45]. In their work, ECC was used to distribute a single symmetric key for the link layer encryption provided by the TinySec module [46].

While public key cryptography may be possible in sensor nodes, the private key operations are still expensive. The assumptions in [33,45] may not be satisfied in some applications. For example, the work in [33,45] concentrated on the public key operations only, assuming the private key operations will be performed by a base station or a third party. By selecting appropriate parameters, for example, using the small integer $e = 2^{16} + 1$ as the public key, the public key operation time can be extremely fast while the private key operation time does not change. The limitation of private key operation occurring only at a base station makes many security services using public key algorithms not available under these schemes. Such services include peer-to-peer authentication and secure data aggregation.

In contrast, Table 14.5 and Table 14.6 show the execution time and energy cost of two symmetric cryptography protocols on an Atmel ATmega128 processor. In Table 14.5, the execution time was measured on a 64-bit block using an 80-bit key. From the table we can see that symmetric key cryptography is faster and consumes less energy when compared to public key cryptography. In the remaining section, we focus on symmetric key cryptography.

Table 14.5 Symmetric Key Cryptography: Average RC5 and Skipjack Execution Times

Algorithm	Operation Time (ms)
Skipjack (C) [47]	0.38
RC5 (C, assembly) [48]	0.26

Source: Y. Wang, G. Attebury, and B. Ramamurthy, *IEEE Communications Surveys and Tutorials*, Vol. 8, no. 2, pp. 2–23, 2006. With permission.

Table 14.6   Symmetric Key Cryptography: Average Energy Numbers for AES and SHA-1

Algorithm	Energy
SHA-1 (C) [49]	5.9 μJ/byte
AES-128 Enc/Dec (assembly) [50]	1.62/2.49 μJ/byte

Source: Y. Wang, G. Attebury, and B. Ramamurthy, *IEEE Communications Surveys and Tutorials,* Vol. 8, no. 2, pp. 2–23, 2006. With permission.

14.4.2   Symmetric Key Cryptography in WSNs

The constraints on computation and power consumption in sensor nodes limit the application of public key cryptography in WSNs. Thus, most research studies focus on symmetric key cryptography in sensor networks.

Five popular encryption schemes, RC4 [51], RC5 [48], IDEA [51], SHA-1 [49], and MD5 [51,52], were evaluated on six different microprocessors ranging in word size from 8-bit (Atmel AVR) to 16-bit (Mitsubishi M16C) to 32-bit widths (StrongARM, XScale) in [53]. The execution time and code memory size were measured for each algorithm and platform. The experiments indicated uniform cryptographic cost for each encryption class and each architecture class. The impact of caches was negligible while Instruction Set Architecture (ISA) support is limited to specific effects on certain algorithms. Moreover, hashing algorithms (MD5, SHA-11) incur almost an order of magnitude higher overhead than encryption algorithms (RC4, RC5, and IDEA).

In [54], Law et al. evaluated two symmetric key algorithms: RC5 and TEA [55]. They further evaluated six block ciphers including RC5, RC6 [56], Rijndael [50], MISTY1 [57], KASUMI [58], and Camellia [59] on IAR Systems' MSP430F149 in [60]. The benchmark parameters were code, data memory, and CPU cycles. The evaluation results showed that Rijndael is suitable for high security and energy efficiency requirements and MISTY1 is suitable for good storage and energy efficiency. The evaluation results in [60] disagreed with the work in [8] in which RC5 was selected as the encryption/decryption scheme, and with the work in [22] in which RC6 was selected. The work in [60] provides a good resource for deciding which symmetric algorithm should be adopted in sensor networks.

The performance of symmetric key cryptography is mainly decided by the following factors:

■ Embedded data bus width: Many encryption algorithms prefer 32-bit word arithmetic, but most embedded processors usually use an 8- or 16-bit wide data bus.

■ Instruction set: The ISA has specific effects on certain algorithms. For example, most embedded processors do not support the variable-bit rotation instruction like ROL (rotate bits left) of the Intel architecture which greatly improves the performance of RC5.

Due to the constraints in sensor nodes, symmetric key cryptography is preferred in a WSN.

14.4.3 Open Research Issues

Selecting the appropriate cryptography method for sensor nodes is fundamental to provide security services in WSNs. However, the decision depends on the computation and communication capability of the sensor nodes. Open research issues range from cryptographic algorithms to hardware design as described below:

■ Recent studies on public key cryptography have demonstrated that public key operations may be practical in sensor networks. However, private key operations are still too expensive in terms of computation and energy cost to accomplish in a sensor node. The application of private key operations to sensor nodes needs to be studied further.

■ Symmetric key cryptography is superior to public key cryptography in terms of speed and low energy cost. However, the key distribution schemes based on symmetric key cryptography are not perfect. Efficient and flexible key distribution schemes need to be designed.

■ It is also likely that more powerful motes will need to be designed to support the increasing requirements on computation and communication in sensor nodes.

14.5 Key Management Protocols

Key management is a core mechanism to ensure the security of network services and applications in WSNs. The goal of key management is to establish required keys between sensor nodes which must exchange data. Further, a key management scheme should also support node addition and revocation while working in undefined deployment environments. Due to the constraints on sensor nodes, key management schemes in WSNs have many differences with the schemes in ad hoc networks.

As shown in Section 14.4, public key cryptography suffers from limitations in WSNs. Thus, most proposed key management schemes are based

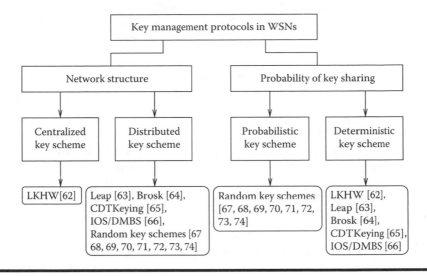

Figure 14.2 Key management protocols in WSNs: A taxonomy. (From Y. Wang, G. Attebury, and B. Ramamurthy, *IEEE Communications Surveys and Tutorials*, Vol. 8, no. 2, pp. 2–23, 2006. With permission.)

on symmetric key cryptography. Further, a straight pairwise private key sharing scheme between every pair of nodes is also impractical in WSNs. A pairwise private key sharing scheme requires pre-distribution and storage of $n - 1$ keys in each node, where n is the number of nodes in a sensor network. Due to the large amount of memory required, pairwise schemes are not viable when the network size is large. Moreover, most key pairs would be unusable because direct communication is possible only among neighboring nodes. This scheme is also not flexible for node addition and revocation. In this section, we discuss key management protocols in WSNs. Another investigation of key management mechanisms for WSNs can be found in [61].

Figure 14.2 shows a taxonomy of key management protocols in WSNs. According to the network structure, the protocols can be divided into centralized key schemes and distributed key schemes. According to the probability of key sharing between a pair of sensor nodes, the protocols can be divided into probabilistic key schemes and deterministic key schemes. In this section, we present a detailed overview of the main key management protocols in WSNs. We start with key management protocols based on network structure.

14.5.1 Network Structure-Based Key Management Protocols

The underlying network structure plays a significant role in the operation of key management protocols. According to the structure, the protocols can be divided into two categories: centralized key schemes and distributed key schemes.

14.5.1.1 Centralized Key Management Schemes

In a centralized key scheme, there is only one entity, which is often called a key distribution center (KDC), controlling the generation, regeneration, and distribution of keys. The only proposed centralized key management scheme for WSNs in the current literature is the LKHW scheme, which is based on Logical Key Hierarchy (LKH) [62]. In this scheme, the base station is treated as a KDC and all keys are logically distributed in a tree rooted at the base station.

The central controller does not have to rely on any auxiliary entity to perform access control and key distribution. However, with only one managing entity, the central server is a single point of failure. The entire network and its security will be affected if there is a problem with the controller. During the time when the controller is not working, the network becomes vulnerable as keys are not generated, regenerated, and distributed. Furthermore, the network may become too large to be managed by a single entity, thus affecting scalability.

14.5.1.2 Distributed Key Management Schemes

In the distributed key management approaches, different controllers are used to manage key generation, regeneration, and distribution, minimizing the risk of failure and allowing for better scalability. In this approach, more entities are allowed to fail before the whole network is affected.

Most proposed key management schemes are distributed schemes. These schemes also fall into deterministic and probabilistic categories, which are discussed in detail in the following sub-section.

14.5.2 Key Management Protocols Based on the Probability of Key Sharing

In the remainder of this section, we present the key management protocols based on the probability of key sharing between a pair of sensor nodes. We first discuss deterministic approaches and then discuss probabilistic approaches.

14.5.2.1 Deterministic Approaches

Zhu et al. proposed a key management protocol, Localized Encryption and Authentication Protocol (LEAP), for sensor networks in [63]. LEAP supports the establishment of four types of keys for each sensor node:

- An individual key shared with the base station (pre-distributed)
- A group key shared by all the nodes in the network (pre-distributed)
- Pairwise keys shared with immediate neighboring nodes
- A cluster key shared with multiple neighboring nodes

The pairwise keys shared with immediate neighboring nodes are used to protect peer-to-peer communication and the cluster key is used for local broadcast. The pairwise keys can be set up as follows: in the key pre-distribution stage, each sensor node is loaded with an initial key K_I and each node A generates a master key $K_A = f_{K_I}(A)$, where f is a pseudo-random function. Then, in the neighbor discovery stage, A broadcasts a Hello message and expects an acknowledgment from neighboring nodes, e.g., node B:

$$A \longrightarrow * : A$$

$$B \longrightarrow A : B, MAC(K_B, A|B)$$

Node A computes its pairwise key with B, $K_{AB} = f_{K_B}(A)$ and node B knows A, K_B and can also compute K_{AB} in the same way. Then, K_{AB} serves as their pairwise key.

Cluster key establishment follows the pairwise key establishment phase. Suppose node A wants to establish a cluster key with all its immediate neighbors $B_1, B_2, \ldots, B_m$. Node A first generates a random key K_A^c, then encrypts this key with the pairwise key shared with each neighbor, and finally transmits the encrypted key to each neighbor B_i, where $1 \leq i \leq m$.

$$A \longrightarrow B_i : \left(K_A^c\right)_{K_{AB_i}}$$

LEAP uses unicast for key exchange. Notice that most of the proposed security protocols were based on point-to-point handshaking procedures to negotiate session keys. Lai et al. proposed a BROadcast Session Key (BROSK) negotiation protocol [64]. BROSK assumes a master key is shared by all nodes in the network. To establish a session key K with its neighbors, such as node B, a sensor node A broadcasts a key negotiation message:

$$A \longrightarrow * : ID_A|N_A, MAC(K, ID_A|N_A)$$

$$B \longrightarrow * : ID_B|N_B, MAC(K, ID_B|N_B)$$

A and B will receive the broadcast message. They can verify the message using the master key K and both A and B can calculate the shared session key:

$$K_{AB} = MAC(K, N_A|N_B)$$

BROSK therefore establishes pairwise session keys between every two neighboring nodes. It is both scalable and energy efficient.

Camtepe and Yener proposed a deterministic key distribution scheme for WSNs using Combinatorial Design Theory [65]. The Combinatorial Design Theory based pairwise key pre-distribution (CDTKeying) scheme is based on block design techniques in combinatorial design theory. It employs symmetric and generalized quadrangle design techniques. The scheme uses a finite projective plane of order n (for prime power n) to generate a symmetric design with parameters $n^2 + n + 1, n + 1, 1$. The design supports $n^2 + n + 1$ nodes and uses a key pool of size $n^2 + n + 1$. It generates $n^2 + n + 1$ key chains of size $n + 1$ where every pair of key chains has exactly one key in common, and every key appears in exactly $n + 1$ key-chains. After the deployment, every pair of nodes finds exactly one common key. Thus, the probability of key sharing among a pair of sensor nodes is 1. The disadvantage of this solution is that the parameter n has to be a prime power, thus indicating that not all network sizes can be supported for a fixed key chain size.

Lee and Stinson proposed two combinatorial design theory based deterministic schemes: ID-based one-way function scheme (IOS) and deterministic multiple space Blom's scheme (DMBS) [66]. They further discussed the use of combinatorial set systems in the design of deterministic key pre-distribution schemes for WSNs in [67].

14.5.2.2 Probabilistic Approaches

Most proposed key management schemes in WSNs are probabilistic and distributed schemes.

Eschenauer and Gligor introduced a key pre-distribution scheme for sensor networks which relies on probabilistic key sharing among the nodes of a random graph [68]. The scheme consists of three phases: key pre-distribution, shared key discovery, and path key establishment. In the key pre-distribution phase, each sensor is equipped with a key ring held in the memory. The key ring consists of k keys which are randomly drawn from a large pool of P keys. The association information of the key identifiers in the key ring and sensor identifier is also stored at the base station. Further, the authors assumed that each sensor shares a pairwise key with the base station. In the shared key discovery phase, each sensor discovers its neighbors within wireless communication range with which it shares keys. Two methods to accomplish this are suggested in [68]. The simplest method is for each node to broadcast a list of identifiers of the keys in their key ring in plaintext allowing neighboring nodes to check whether they share a key. However, the adversary may observe the key-sharing patterns among sensors in this way. The second method uses the challenge–response technique to hide key-sharing patterns among nodes from an adversary. For every K_i

on a key ring, each node could broadcast a list $\alpha, E_{K_i}(\alpha), i = 1, \ldots, k$ where α is a challenge. The decryption of $E_{K_i}(\alpha)$ with the proper key by a recipient would reveal the challenge and establish a shared key with the broadcasting node. This method requires the challenge α be well known in the sensor network, allowing the recipient with the proper key to discover the challenge.

Finally, in the path key establishment phase, a path key is assigned for those sensor nodes within wireless communication range and not sharing a key, but connected by two or more links at the end of the second phase. If a node is compromised, the base station can send a message to all other sensors to revoke the compromised node's key ring. Re-keying follows the same procedure as revocation. The messages from the base station are signed by the pairwise key shared by the base station and sensor nodes, thus ensuring that no adversary can forge a base station. If a node is compromised, the attacker has a probability of approximately k/P to attack any link successfully. Because $k \ll P$, it only affects a small number of sensor nodes.

Inspired by the work in [68], which we call the basic random key scheme in the following section, additional random key pre-distribution schemes have been proposed in [69–74].

In the basic random key scheme, any two neighboring nodes need to find a single common key from their key rings to establish a secure link in the key setup phase. However, Chan et al. observed that increasing the amount of key overlap in the key ring can increase the resilience of the network against node capture [69]. Thus, they proposed a q-composite keying scheme. It is required to share at least q common keys in the key setup phase to build a secure link between any two neighboring nodes. Further, they introduced a key update phase to enhance the basic random key scheme. Suppose A has a secure link to B after the key setup phase and the secure key is k from the key pool P. Because k may be residing in the key ring memory of some other nodes in the network, the security of the link between A and B is jeopardized if any of those nodes are captured. Thus, it is better to update the communication key between A and B instead of using a key in the key pool. To address the problem, they presented a multipath key reinforcement for the key update. Assume there are j disjoint paths between A and B. A generates j random values $v_1, v_2, \ldots, v_j$ and then routes each random value along a different path to B. When B has received all j keys, the new link key can be computed by both A and B as:

$$k' = k \oplus v_1 \oplus v_2 \oplus \ldots \oplus v_j$$

The adversary has to eavesdrop on all j paths if he wants to reconstruct the communication key. This security enhancement comes at the cost of more communication overhead needed to find multiple disjoint paths. Further, Chan et al. [69] also developed a random-pairwise key scheme for node-to-node authentication.

Blundo et al. presented a polynomial-based key pre-distribution protocol for group key pre-distribution in [75], which can also be adapted to sensor networks. The key setup server randomly generates a bivariate t-degree polynomial $f(x, y) = \sum_{i,j=0}^{t} a_{ij} x^i y^j$ over a finite field F_q where q is a prime number that is large enough to accommodate a cryptographic key such that it has the property of $f(x, y) = f(y, x)$. For each sensor i, the setup server computes a polynomial share of $f(x, y)$, that is, $f(i, y)$. For any two sensor nodes i and j, node i can compute the common key $f(i, j)$ by evaluating $f(i, y)$ at point j, and node j can compute the same key $f(j, i) = f(i, j)$ by evaluating $f(j, y)$ at point i. In this approach, each sensor node i needs to store a t-degree polynomial $f(i, x)$, which occupies $(t + 1) \log q$ storage space. This scheme is unconditionally secure and t-collusion resistant. However, the storage cost for a polynomial share is exponential in terms of the group size, making it prohibitive in sensor networks.

Inspired by the work of [68,69,75], Liu and Ning proposed a polynomial pool-based key pre-distribution scheme in [70], which also includes three phases: setup, direct key establishment, and path key establishment. In the setup phase, the setup server randomly generates a set F of bivariate t-degree polynomials over the finite field F_q. For each sensor node, the setup server picks a subset of polynomials $F_i \subseteq F$ and assigns the polynomial shares of these polynomials to node i. In the direct key establishment stage, the sensor nodes find a shared polynomial with other sensor nodes and then establish a pairwise key using the polynomial-based key pre-distribution scheme discussed in [75]. The path key establishment phase is similar to that in the basic random key scheme. Further, the proposed framework allows for the study of multiple instantiations of possible pairwise key establishment schemes. Two of the possible instantiations, the key pre-distribution scheme based on random subset assignment and the grid-based key pre-distribution scheme, are also presented and analyzed in the paper.

Similar to [70], Du et al. presented another pairwise key pre-distribution scheme in [72] which uses Blom's method [76]. The key difference between [70] and [72] is that the scheme in [70] is based on a set of bivariate t-degree polynomials and Du's scheme is based on Blom's method. The proposed scheme allows any pair of nodes in a network to be able to find a pairwise secret key. As long as no more than λ nodes are compromised, the network is perfectly secure (which is called the λ-secure property). To use Blom's method, during the pre-deployment phase, the base station first constructs a $(\lambda + 1) \times N$ matrix G over a finite field $GF(q)$, where N is the size of the network and G is considered to be public information. Then the base station creates a random $(\lambda + 1) \times (\lambda + 1)$ symmetric matrix D over $GF(q)$, and computes an $N \times (\lambda + 1)$ matrix $A = (D \cdot G)^T$, where $(D \cdot G)^T$ is the transpose of $D \cdot G$. Matrix D needs to be kept secret, and should not be disclosed to adversaries. It is easy to verify that $A \cdot G$ is a symmetric matrix.

$$A \cdot G = (D \cdot G)^T \cdot G = G^T \cdot D^T \cdot G = G^T \cdot D \cdot G$$

$$= (A \cdot G)^T$$

Thus, we know $K_{ij} = K_{ji}$. The idea is to use K_{ij} (or K_{ji}) as the pairwise key between node i and node j. To carry out the above computation, in the pre distribution phase, for any sensor $k = 1, \ldots, N$:

■ Store the kth row of matrix A at node k.
■ Store the kth column of matrix G at node k.

Therefore, when nodes i and j need to find the pairwise key between them, they first exchange their columns of G, and then compute K_{ij} and K_{ji}, respectively, using their private rows of A.

In the proposed scheme in [72], each sensor node is loaded with G and τ distinct D matrices drawn from a large pool of ω symmetric matrices $D_1, \ldots, D_\omega$ of size $(\lambda + 1) \times (\lambda + 1)$. For each D_i, calculate the matrix $A_i = (D_i \cdot G)^T$ and store the jth row of A_i at this node. After deployment, each node needs to discover whether it shares any space with neighbors. If they find out that they have a common space, the nodes can follow Blom's method to build a pairwise key. The scheme in [72] is scalable and flexible. Moreover, it is substantially more resilient against node capture as compared to [70].

Hwang et al. extended the basic random key scheme and proposed a cluster key grouping scheme [74]. They further analyzed the trade-offs involved between energy, memory, and security robustness.

Notice that location information helps to avoid unnecessary key assignments and thus improve the performance of sensor networks, such as connectivity, memory usage, and network resilience against node capture. Taking this into account, two random key pre-distribution schemes were proposed in [73] and [77]. Although the presented schemes show improved performance, the deployment information, such as location, is required when sensors are deployed.

The above-mentioned schemes are classified and compared in Table 14.7.

14.5.3 Open Research Issues

Although some key management protocols have been proposed for sensor networks, the design of key management protocols is still largely open to research. Open research issues include the following:

Table 14.7 Classification and Comparison of Key Management Protocols in WSNs

	Protocol	Ref.	Master Key	Pairwise Key	Path Key	Cluster Key	Scalability	Resiliency	Processing Load	Comm. Load	Storage Load
I	All pairwise	—	n/a	Yes	No	No	Low	Low	Low	Low	High
	LEAP	[63]	Yes	Yes	Yes	Yes	Good	Low	Low	Low	Low
	BROSK	[64]	Yes	Yes	No	No	Good	Low	Low	Low	Low
	LKHW	[62]	Yes	Yes	No	Yes	Limited	Low	Low	Low	Low
	CDTKeying	[65]	n/a	Yes	No	No	Good	Good	Medium	Medium	High
	IOS & DMBS	[66]	n/a	Yes	No	No	Good	Good	Medium	Medium	High
II	q-composite	[68]	n/a	Yes	Yes	No	Good	Good	Medium	Medium	High
	Polynomial based	[69]	n/a	Yes	Yes	No	Good	Good	Medium	Medium	High
	Blom based	[70]	n/a	Yes	Yes	No	Good	Good	Medium	Medium	High
	Blom based	[72]	n/a	Yes	Yes	No	Good	Good	Medium	Medium	High
	Deployment knowledge based	[73]	n/a	Yes	Yes	No	Good	Good	Medium	Medium	Medium
	Cluster key grouping	[74]	n/a	Yes	Yes	No	Good	Good	Medium	Medium	High
	Location based	[77]	n/a	Yes	Yes	No	Good	Good	Medium	Medium	Medium

Note: Category I denotes deterministic approaches and category II denotes probabilistic approaches. Master key is the key shared by all the nodes in the network. Pairwise key is the key shared between two neighboring nodes. Path key denotes the key shared between any two nodes which need exchange data, but do not share a pairwise key. Cluster key denotes the common key shared by all cluster members.

■ The proposed key management protocols discussed in this section employ different strategies on the trade-off between memory, processing, and communication overhead. These protocols could be improved and new key management protocols need to be designed.

■ All key management protocols discussed in the literature so far are based on symmetric key cryptography. Recent progress in public key cryptography has shown that public key cryptography may be suitable for sensor networks. Key management schemes based on public key cryptography need to be designed.

■ Current proposed key management schemes assume that the base station is trustworthy. However, there may be situations, such as in the battlefield, where the security of a base station needs to be considered. New schemes need to be designed considering the security of base stations.

14.6 Secure Routing Protocols

Many routing protocols have been specifically designed for WSNs. These routing protocols can be divided into three categories according to the network structure: flat-based routing, hierarchical-based routing, and location-based routing [78]. In flat-based routing, all nodes are typically assigned equal roles or functionality. In hierarchical-based routing, nodes play different roles in the network. In location-based routing, sensor node positions are used to route data in the network. Although many sensor network routing protocols have been proposed in the literature, few of them have been designed with security as a goal. Lacking security services in the routing protocols, WSNs are vulnerable to many kinds of attacks.

Most network layer attacks against sensor networks fall into one of the categories described in Section 14.3.3, namely:

■ Spoofed, altered, or replayed routing information
■ Selective forwarding
■ Sinkhole
■ Sybil
■ Wormholes
■ Hello flood attacks
■ Acknowledgment spoofing

These attacks may be applied to compromise the routing protocols in a sensor network. For example, directed diffusion is a flat-based routing algorithm for drawing information from a sensor network [79]. In directed diffusion, sensors measure events and create gradients of information in their respective neighboring nodes. The base station requests data by broadcasting interest which describes a task to be conducted by the network. The interest is diffused through the network hop by hop, and broadcasted by each node to its neighbors. As the interest is propagated throughout the network, gradients are set up to draw data satisfying the query toward the requesting node. Each sensor that receives the interest sets up a gradient toward the sensor nodes from which it received the interest. This process continues until gradients are set up from the sources back to the base station. Interests initially specify a low rate of data flow, but once a base station starts receiving events, it will reinforce one or more neighboring nodes to request higher data rate events. This process proceeds recursively until it reaches the nodes generating events, causing them to generate events at a higher data rate. Paths may also be negatively reinforced. Directed diffusion is vulnerable to many kinds of attacks if authentication is not included in the protocol [26]. For example, it is easy for an adversary to add himself onto the path taken by a flow of events, as described in the following:

■ The adversary can influence the path by spoofing positive reinforcements. After receiving and rebroadcasting an interest, an adversary could strongly reinforce the nodes to which the interest was sent while spoofing high rate, low latency events to the nodes from which the interest was received.
■ The adversary can replay the interests intercepted from a legitimate base station and list himself as a base station. All events satisfying the interest will then be sent to both the adversary and the legitimate base station.

By using the attacks above, the adversary can add himself onto the path and thus gain full control of the flow. The adversary can eavesdrop, modify, and selectively forward packets of his choosing. He can drop all forwarded packets and act as a sinkhole. Further, a laptop-class adversary can exert great influence on the topology by using a wormhole attack. The adversary creates a tunnel between a node located near a base station and a node located close to where events are likely to be generated. By spoofing positive or negative reinforcements, the adversary can push data flows away from the base station and toward the nodes selected by the adversary.

Hierarchical and location-based routing protocols not incorporating security services are also vulnerable to many attacks [26]. For example, location-based routing protocols such as Geographic and Energy Aware Routing (GEAR) [80] require location information to be exchanged between neighbors. However, location information can be misrepresented. Regardless of the adversary's actual location, he may advertise false position data to place himself on the path of a known flow. Once on that path, the adversary can mount selective forwarding and Sybil attacks in the data flows. Simulations in [81] found that such attacks have great influence on the overall ratio of successfully delivered messages in the network.

Secure routing in ad hoc networks is similar to that in sensor networks and has been well studied in literature [14]. However, the defense mechanisms developed for ad hoc networks cannot be directly applied to sensor networks because of the differences between sensor and ad hoc networks discussed in Section 14.1.

Ideally, a secure routing protocol should guarantee the integrity, authentication, and availability of messages in the presence of adversaries of arbitrary power. In the presence of only outsider adversaries, it is conceivable to achieve these idealized goals. However, in the presence of compromised nodes or insider adversaries, especially those with laptop-class capabilities, it is most likely that some if not all of these goals are not fully attainable. In this situation, the best we can hope for is graceful degradation instead of a complete compromise of the network. To achieve the above goal requires that a routing protocol degrades no faster than a rate approximately proportional to the ratio of compromised nodes to total nodes in the network [26].

A secure routing protocol depends on an appropriate key management scheme in a WSN, which has been discussed in Section 14.5. Before a routing protocol starts, sensor nodes should have been loaded with proper keys, e.g., the key for confidentiality, authentication, etc. One of the fundamental security services in sensor networks is broadcast authentication, which enables the base station to broadcast authenticated data to the entire sensor network. In this section, we first discuss the broadcast authentication problem and then review several secure routing schemes.

14.6.1 Broadcast Authentication

Previous proposals for authenticated broadcast are impractical in WSNs for the following reasons:

- Most proposals rely on public key cryptography for the authentication. However, public key cryptography is impractical for WSNs.
- Even one-time signature schemes that are based on symmetric key cryptography have too much overhead.

μ*TESLA* (the "micro" version of the Timed, Efficient, Streaming, Loss-tolerant Authentication protocol) [10] and its extensions [82,83] have been proposed to provide broadcast authentication for sensor networks.

μ*TESLA* is an authenticated broadcast protocol which was proposed by Perrig et al. for the SPINS protocol [8]. μ*TESLA* introduces asymmetry through a delayed disclosure of symmetric keys resulting in an efficient broadcast authentication scheme. $\mu TESLA$ requires that the base station and nodes be loosely time synchronized, and that each node knows an upper bound on the maximum synchronization error.

To send an authenticated packet, the base station simply computes a MAC on the packet with a key that is secret at that point in time. When a node gets a packet, it can verify that the corresponding MAC key was not yet disclosed by the base station. Because a receiving node is assured that the MAC key is known only by the base station, the receiving node is assured that no adversary could have altered the packet in transit. The node stores the packet in a buffer. At the time of key disclosure, the base station broadcasts the verification key to all receivers. When a node receives the disclosed key, it can easily verify the correctness of the key. If the key is correct, the node can now use it to authenticate the packet stored in its buffer.

Each MAC key is a key from the key chain, generated by a public one-way function F. To generate the one-way key chain, the sender chooses the last key K_n from the chain, and repeatedly applies F to compute all other keys: $K_i = F(K_{i+1})$.

Figure 14.3 shows an example of $\mu TESLA$. The receiver node is loosely time synchronized and knows K_0 in an authenticated way. Packets P_1 and P_2 sent in interval 1 contain a MAC with key K_1. Packet P_3 has a MAC using key K_2. If P_4, P_5, and P_6 are all lost, as well as the packet that disclosed key K_1, the receiver cannot authenticate P_1, P_2, and P_3. In interval 4 the base station broadcasts key K_2, which the nodes authenticate by verifying $K_0 = F(F(K_2))$, and hence know also $K_1 = F(K_2)$, so they can authenticate packets P_1, P_2 with K_1, and P_3 with K_2.

SPINS limits the broadcasting capability to only the base station. If a node wants to broadcast authenticated data, the node has to broadcast the data through the base station. The data is first sent to the base station in an authenticated way. It is then broadcasted by the base station.

To bootstrap a new receiver, $\mu TESLA$ depends on a point-to-point authentication mechanism in which a receiver sends a request message to the base station and the base station replies with a message containing all the necessary parameters. Notice that $\mu TESLA$ requires the base station to

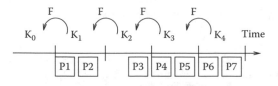

Figure 14.3 Using a time-released key chain for source authentication. (From Y. Wang, G. Attebury, and B. Ramamurthy, *IEEE Communications Surveys and Tutorials*, Vol. 8, no. 2, pp. 2–23, 2006. With permission.)

unicast initial parameters to individual sensor nodes, and thus incurs a long delay to boot up a large scale sensor network. Liu and Ning proposed a multi-level key chain scheme for broadcast authentication to overcome this deficiency in [82,83].

The basic idea in [82,83] is to predetermine and broadcast the initial parameters required by $\mu TESLA$ instead of using unicast-based message transmission. The simplest way is to pre-distribute the $\mu TESLA$ parameters with a master key during the initialization of the sensor nodes. As a result, all sensor nodes have the key chain commitments and other necessary parameters once they are initialized, and are ready to use $\mu TESLA$ as long as the starting time has passed. Furthermore, Liu and Ning introduced a multi-level key chain scheme, in which the higher key chains are used to authenticate the commitments of lower-level ones. However, the multi-level key chain scheme suffers from possible DoS attacks during the commitment distribution stage. Further, none of the $\mu TESLA$ or multi-level key chain schemes is scalable in terms of the number of senders. In [84], a practical broadcast authentication protocol was proposed to support a potentially large number of broadcast senders using $\mu TESLA$ as a building block.

$\mu TESLA$ provides broadcast authentication for base stations, but is not suitable for local broadcast authentication. This is because $\mu TESLA$ does not provide immediate authentication. For every received packet, a node has to wait for one $\mu TESLA$ interval to receive the MAC key used in computing the MAC for the packet. As a result, if $\mu TESLA$ is used for local broadcast authentication, a message traversing l hops will take at least l $\mu TESLA$ intervals to arrive at the destination. In addition, a sensor node has to buffer all the unverified packets. Both the latency and the storage requirements limit the scheme for authenticating infrequent messages broadcast by the base station. Zhu et al. proposed a one-way key chain scheme for one-hop broadcast authentication in LEAP [63]. In this scheme, every node generates a one-way key chain of certain length and then transmits the commitment (i.e., the first key) of the key chain to each neighbor, encrypted with their pairwise shared key. Whenever a node has a message to send, it attaches to the message to the next authenticated key in the key chain. The authenticated keys are disclosed in reverse order to their generation. A receiving neighbor can verify the message based on the commitment or an authenticated key it received from the sending node more recently.

14.6.2   Secure Routing

The goal of a secure routing protocol is to ensure the integrity, authentication, and availability of messages. The proposed secure routing protocols for WSNs in the literature were all based on symmetric key cryptography except the work in [85], which was based on public key cryptography.

SPINS is a suite of security protocols optimized for sensor networks [8]. SPINS includes two building blocks: SNEP (Secure Network Encryption Protocol) and μ*TESLA*. SNEP provides data confidentiality, two-party data authentication, and data freshness for peer-to-peer communication (node to base station). μ*TESLA* provides authenticated broadcast as discussed before. We discuss SNEP in this sub-section.

SPINS assumes that each node is pre-distributed with a master key K which is shared with the base station at creation time. All other keys, including a key K_{encr} for encryption, a key K_{mac} for MAC generation, and a key K_{rand} for random number generation, are derived from the master key using a strong one-way function. SPINS uses RC5 for confidentiality. If A wants to send a message to base station B, the complete message that A sends to B is

$$A \rightarrow B : D_{\langle K_{encr}C \rangle}, MAC(K_{mac}, C|D)_{\langle K_{encr}C \rangle}$$

where D is the transmitted data and C is a shared counter between the sender and the receiver for the block cipher in counter mode. The counter C is incremented after each message is sent and received in both the sender and the receiver side. SNEP also provides a counter exchange protocol to synchronize the counter value in both sides.

SNEP offers the following properties: semantic security, data authentication, replay protection, weak freshness, and low communication overhead. SPINS identifies two types of freshness: weak freshness and strong freshness. Weak freshness provides partial message ordering and carries no delay information; strong freshness provides a total order on a request-response pair and allows for delay estimation.

- ■ Semantic security: The counter value is incremented after each message and thus the same message is encrypted differently each time.
- ■ Data authentication: A receiver can be assured that the message originated from the claimed sender if the MAC verifies correctly.
- ■ Replay protection: The counter value in the MAC prevents replaying old messages.
- ■ Weak freshness: The counter also maintains a message ordering in the receiver side and yields weak freshness. SNEP provides weak data freshness only because there is no absolute assurance to node A that a message was created by node B in response to an event in node A.
- ■ Low communication overhead: The counter state is kept at each endpoint and does not need to be sent in each message.

Directed diffusion routing protocol was proposed by Intanagonwiwat et al. without considering security issues [79]. Pietro et al. proposed an extension of directed diffusion protocol which provides secure multicasting in [62]. The extended scheme, Logical Key Hierarchy for WSNs (LKHW), provides robustness in routing and security and supports both backward and forward secrecy for sensor join and leave operations. However, it does not provide data authentication.

Inspired by the work on public key cryptography [6,7,33,43], Du et al. investigated the public key authentication problem [85]. The use of public key cryptography eases many problems in secure routing, for example, authentication and integrity. However, before a node A uses the public key from another node B, A must verify that the public key is actually B's, i.e., A must authenticate B's public key; otherwise, man-in-the-middle attacks are possible. In general networks, public key authentication involves a signature verification on a certificate signed by a trusted third party Certificate Authority (CA) [86]. However, the signature verification operations are still too expensive for sensor nodes, as depicted in Table 14.3 and Table 14.4. Du et al. proposed an efficient alternative that uses only a one-way hash function for the public key authentication. The proposed scheme can be divided into two stages. In the pre-distribution stage, a Merkle tree R is constructed with each leaf L_i corresponding to a sensor node (more information on Merkle trees is given in Section 14.7). Let pk_i represent node i's public key, V be an internal tree node, and V_{left} and V_{right} be V's two children. The value of an internal tree node is denoted by ϕ. The Merkle tree can then be constructed as follows:

$$\phi(L_i) = h(id_i, pk_i), \quad for \ i = 1, \ldots, N$$

$$\phi(V) = h(\phi(V_{left}) \parallel \phi(V_{right}))$$

where "$\parallel$" represents the concatenation of two strings and h is a one-way hash function such as MD5 or SHA-1. Let R be the root of the tree. Each sensor node v needs to store the root value $\phi(R)$ and the sibling node values $\lambda_1, \ldots, \lambda_H$ along the path from v to R. If node A wants to authenticate B's public key, B sends its public key pk along with the value of $\lambda_1, \ldots, \lambda_H$ to node A. Then, A can use the same procedure to reconstruct the Merkle tree R' and calculate the root value $\phi(R')$. A will trust B to be authentic if $\phi(R') = \phi(R)$. A sensor node only needs $H + 1$ storage units for the extra hash values. Based on this scheme, Du et al. further extended the idea to reduce the height of the Merkle tree to improve the communication overhead of the scheme. The proposed scheme is more efficient than signature verification on certificates. However, the scheme requires that some hash values be distributed in a pre-distribution stage. This results in some scalability issues when new sensors are added to an existing WSN.

The discussion above is summarized in Table 14.8.

Table 14.8 Comparison of Secure Routing Protocols

	Ref.	Routing	Confidentiality	P2P Authentication	Broadcast Authentication	Integrity	Scalability
SNEP	[8]	Flat	Yes	Yes	No	Yes	Good
LKHW	[62]	Flat	Yes	No	No	No	Limited
μTESLA	[8]	Flat/hierarchy	No	No	Yes	Yes	Medium
Multi-level key chains	[82]	Flat/hierarchy	No	No	Yes	Yes	Good
LEAP	[63]	Hierarchy	Yes	Yes	Yes	Yes	Medium

14.6.3 Open Research Issues

The development of secure routing protocols is challenging because sensor nodes are prone to failures, and the topology of a sensor network changes frequently due to node failures and possible mobility. Key open research issues include the following:

- The proposed secure routing protocols for WSNs focus on static sensor networks only, ignoring mobility. Secure routing protocols for mobile sensor networks need to be investigated.
- Current broadcast authentication schemes such as $\mu TESLA$ and its extensions require the sensor network to be loosely time synchronized. This requirement is often hard to meet and new techniques that do not require time synchronization are desirable.
- New schemes with higher scalability and efficiency need to be developed for the authenticated broadcast protocols. The recent progress on public key cryptography may facilitate the design of authenticated broadcast protocols.
- Quality of service in WSNs needs to be evaluated with the addition of secure routing services.

14.7 Secure Data Aggregation

Data communication constitutes an important share of the total energy consumption of the sensor network. The simulation in [8] shows that data transmission accounts for 71 percent of the energy cost of computation and communication for the SNEP protocol. Thus, data aggregation can greatly help conserve the scarce energy resources by eliminating redundant data.

Data aggregation (fusion) protocols aim at eliminating redundant data transmitted across the network and are essential for energy-constrained WSNs. Traditional data aggregation techniques include simple types of queries such as SUM, COUNT, AVERAGE, and MIN/MAX. Some researchers also extend data aggregation to median, the most frequent (consensus) data values, a histogram of the data distribution, and range queries [87]. Data aggregation can be divided into two stages: detection and data fusion.

In a WSN, there are usually certain nodes, called aggregators, helping aggregate information requested by queries. When an aggregator node is compromised, it is easy for the adversary to inject false data into sensor networks. Thus, the aggregators are vulnerable to be attacked. Another possible attack is to compromise a sensor node and inject forged data through it. Without authentication, the attackers can fool the aggregators into reporting false data to the base station. Secure data aggregation requires authentication, confidentiality, and integrity. Moreover, secure data

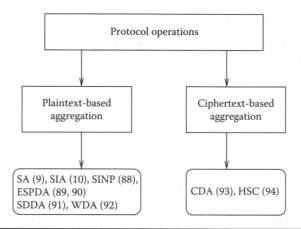

Figure 14.4   Secure data aggregation in WSNs: A taxonomy. (From Y. Wang, G. Atte-bury, and B. Ramamurthy, *IEEE Communications Surveys and Tutorials*, Vol. 8, no. 2, pp. 2–23, 2006. With permission.)

aggregation also requires the cooperation of sensor nodes to identify the compromised sensors.

However, requirements for confidentiality and data aggregation are at odds with each other. Confidentiality requires the data to be transmitted in ciphertext, data aggregation is usually based on plaintext. A straight-forward method is to invoke end-to-end encryption before evoking data aggregation. However, the trade-off is that the end-to-end encryption and decryption operations consume more energy, which is of great concern in WSNs. An alternative way is to provide data aggregation on concealed data, which requires a particular class of encryption transformation. How-ever, this method usually lowers the security level.

Figure 14.4 shows a taxonomy of secure data aggregation protocols in WSNs. According to the protocol operation, secure data aggregation can be classified into two categories: plaintext-based and ciphertext-based. This section reviews the techniques for secure data aggregation.

14.7.1   Plaintext-Based Secure Data Aggregation

Hu and Evans proposed a secure aggregation (SA) protocol for WSNs that is resilient to both intruder devices and single device key compromises [9]. However, the protocol may be vulnerable if a parent and a child node in the hierarchy are compromised.

Przydatek et al. proposed a secure information aggregation (SIA) frame-work for sensor networks [10]. The framework consists of three node cate-gories: a home server, base station(s), and sensor nodes. A base station is a resources-enhanced node which is used as intermediary between the home server and the sensor nodes, and it is also the candidate to perform the aggregation task. SIA assumes that each sensor has a unique identifier and shares a separate secret cryptographic key with both the home server and the aggregator. The keys enable message authentication and encryption if data confidentiality is required. Moreover, it further assumes that the home server and base station can use a mechanism, such as $\mu TESLA$, to broadcast authenticated messages. The proposed solution consists of three parts: computation of the result, committing to the collected data, and reporting the aggregation result while proving the correctness of the result.

In the first part, the aggregator collects the data from sensors and locally computes the aggregation result. The aggregator can verify the authenticity of each sensor reading.

In the second part, the aggregator commits to the collected data. The commitment to the input data ensures that the aggregator uses the data provided by the sensors, and that the statement to be verified by the home server about the correctness of computed results is meaningful. One efficient way of committing to the data is a Merkle hash-tree construction. In this construction, all the data collected from the sensors is placed at the leaves of the tree. The aggregator then computes a binary hash tree starting from the leaf nodes. Each internal node in the hash tree is computed as the hash value of the concatenation of its two child nodes. The root of the tree is called the commitment of the collected data. As the hash function in use is collision resistant, once the aggregator commits to the collected values, it cannot change any of the collected values. Figure 14.5 shows an example of a Merkle hash tree.

In the third part, the aggregator and the home server engage in a protocol in which the aggregator communicates the aggregation result and the commitment to the server while proving to the server that the reported results are correct using interactive proof protocols. Moreover, the authors also presented efficient protocols for secure computation of the median and the average of the measurements, for the estimation of the network size, and for finding the minimum and maximum sensor reading.

Deng et al. proposed a collection of mechanisms for securing in-network processing (SINP) for WSNs [88]. Security mechanisms were proposed to address the downstream requirement that sensor nodes authenticate commands disseminated from parent aggregators and the upstream requirement that aggregators authenticate data produced by sensors before aggregating that data. In the downstream stage, two techniques are involved: one way functions and $\mu TESLA$. The upstream stage requires that a pairwise key be shared between an aggregator and its sensor nodes.

Çam et al. proposed an energy-efficient secure pattern-based data aggregation (ESPDA) protocol for wireless sensor networks in [89,90]. ESPDA is applicable for hierarchy-based sensor networks. In ESPDA, a clusterhead first requests sensor nodes to send the corresponding pattern code for the

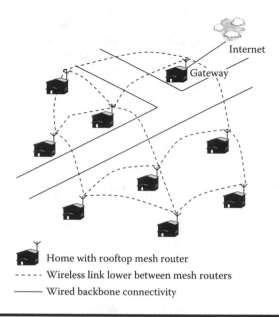

Home with rooftop mesh router
- - - - Wireless link lower between mesh routers
——— Wired backbone connectivity

Figure 14.5 Merkle hash tree used to commit to a set of values. The aggregator constructs the Merkle hash tree over the sensor measurement $m_0, \cdots, m_7$. To construct the Merkle hash tree, the aggregator first hashes the measurements with a cryptographic hash function, e.g., $v_{3,0} = H(m_0)$, assuming that the size of the hash is smaller than the size of the data. Then, each internal value of the Merkle hash tree is derived from its two child nodes: $v_{i,j} = H(v_{i+1,2j} \| v_{i+1,2j+1})$. The Merkle hash tree is a commitment to all the leaf nodes. Once the aggregator commits to the collected values, it cannot change any of the collected data. A verifier can authenticate any value by verifying that the leaf value is used to derive the root node given the authentic root node $v_{0,0}$. For example, to authenticate the measurement m_5, the aggregator sends m_5 along with $v_{3,4}, v_{2,3}, v_{1,0}$, and m_5 is authentic if the following equality holds: $v_{0,0} = H(v_{1,0} \| H(H(v_{3,4} \| H(m_5)) \| v_{2,3}))$. (From Y. Wang, G. Attebury, and B. Ramamurthy, *IEEE Communications Surveys and Tutorials*, Vol. 8, no. 2, pp. 2–23, 2006. With permission.)

sensed data. If multiple sensor nodes send the same pattern code to the clusterhead, only one of them is permitted to send the data to the clusterhead. ESPDA is secure because it does not require encrypted data to be decrypted by clusterheads to perform data aggregation.

Further, Çam et al. introduced another secure differential data aggregation (SDDA) scheme based on pattern codes [91]. SDDA prevents redundant data transmission from sensor nodes by implementing the following schemes: (1) SDDA transmits differential data rather than raw data, (2) SDDA performs data aggregation on pattern codes representing the main characteristics of sensed data, and (3) SDDA employs a sleep protocol to coordinate the activation of sensing units in such a way that only one of the sensor nodes capable of sensing the data is activated at a given time. In the SDDA data transmission scheme, the raw data from sensor nodes is compared to reference data with the difference data being transmitted. The reference data is obtained by taking the average of previously transmitted data.

Du et al. proposed a witness-based data aggregation (WDA) scheme for WSNs to assure the validation of the data sent from data fusion nodes to the base station [92]. To prove the validity of the fusion result, the fusion node has to provide proofs from several witnesses. A witness is one who also conducts data fusion like a data fusion node, but does not forward its result to the base station. Instead, each witness computes the MAC of the result and then provides it to the data fusion node, which must forward the proofs to the base station.

Wagner studied secure data aggregation in sensor networks and proposed a mathematical framework for formally evaluating their security [93]. In [11] and [94], the authors proposed two data fusion schemes for the filtering of injected false data in sensor networks, which will be introduced in Section 14.8.

14.7.2 Ciphertext-Based Secure Data Aggregation

Two ciphertext-based secure data aggregation schemes were proposed in [95] and [96]. The works in [95] and [96] are based on a particular encryption transformation: a privacy homomorphism (PH). A privacy homomorphism is an encryption transformation that allows direct computation on encrypted data. Let Q and R denote two rings, and let $+$ denote addition and $\times$ denote multiplication on both. Let K be the key space. We denote an encryption transformation $E : K \times Q \longrightarrow R$ and the corresponding decryption transformation $D : K \times R \longrightarrow Q$. Given $a, b \in Q$ and $k \in K$, we term

$$a + b = D_k(E_k(a) + E_k(b))$$

additively homomorphic and

$$a \times b = D_k(E_k(a) \times E_k(b))$$

multiplicatively homomorphic [12].

The proposed scheme, Concealed Data Aggregation (CDA), in [95] is based on the PH proposed in [97]. Although the study in [98] has shown that the proposed PH in [97] is unsecure against chosen plaintext attacks for some parameter settings, the authors in [95] claimed that for the WSN data aggregation scenario, the security level is still adequate and the proposed

PH method in [97] can be employed for encryption. CDA can be used to calculate SUM and AVERAGE in a hierarchical WSN. To calculate AVERAGE, an aggregator needs to know the number of sensor nodes n.

Castelluccia et al. proposed a simple and provable secure additively homomorphic stream cipher (HSC) that allows for the efficient aggregation of encrypted data [96]. The new cipher uses modular addition and is therefore very well suited for CPU-constrained devices such as those in WSNs. The aggregation based on this cipher can be used to efficiently compute statistical values such as the mean, variance, and standard deviation of sensed data while achieving significant bandwidth gain.

14.7.3 Open Research Issues

Data aggregation is essential for WSNs, and security is absolutely necessary to defend against compromised sensor nodes. Open research issues include the following:

■ Several secure data aggregation protocols have been proposed. However, no comparisons have been conducted on these protocols. Further evaluations and comparisons are desirable to learn the performance of these protocols. The performance matrices might include security, processing overhead, communication overhead, energy consumption, and data compression rate.
■ New data aggregation protocols need to be developed to address higher scalability and higher reliability against aggregator and sensor node cheating.

14.8 Intrusion Detection

The security mechanisms implemented in secure routing protocols and secure data aggregation protocols are configured ahead of time to inhibit an attacker from breaking the security of the network. These security mechanisms alone cannot ensure perfect security of a WSN. Because sensor nodes can be compromised, it is easy for an adversary to inject false data into a WSN through the compromised nodes. Authentication and data encryption are not enough for ensuring data security. Another approach to protect WSNs involves mechanisms for detecting and reacting to intrusions.

An intrusion detection system (IDS) monitors a host or network for suspicious activity patterns outside normal and expected behavior [5]. It is based on the assumption that there exists a noticeable difference in the behavior of an intruder and legitimate user in the network such that an

IDS can match those pre-programmed or possibly learned rules. Based on the analysis model used for analyzing the audit data to detect intrusions, intrusion detection systems in ad hoc networks are classified into rule-based and anomaly-based systems. The rule-based intrusion detection systems are used to detect known patterns of intrusions (e.g., [99] and [100]) while anomaly-based systems are used to detect new or unknown intrusions (e.g., [101] and [102]). A rule-based IDS has a low false-alarm rate when compared to an anomaly-based system, and an anomaly-based IDS has a high intrusion detection rate in comparison to a rule-based system.

However, WSNs are generally application-specific and lack basic information on topology, normal usage, expected communication patterns, etc. It is impractical to pre-install some fixed patterns in sensors before they are deployed. Moreover, due to constraints in sensors, to learn and detect these parameters after deployment is both time and energy consuming. Thus, existing intrusion detection schemes in ad hoc networks may not be adapted to WSNs.

The research on intrusion detection in WSNs is still preliminary. Current research focuses on how to detect and eliminate injected false information. Note that compromised nodes can always inject false information into a sensor network. Thus, cooperation among sensors, especially neighboring nodes, is necessary to decide the validity of a report. In this section, we discuss the intrusion detection techniques in WSNs.

14.8.1 Intrusion Detection in WSNs

Zhu et al. proposed an interleaved hop-by-hop authentication (IHOP) scheme in [11]. IHOP guarantees that the base station will detect any injected false data packets when no more than a certain number t of nodes are compromised. The sensor network is organized in a cluster-based hierarchy. Each clusterhead builds a route to the base station and each intermediate node has an upper associate node and a lower associate node that is $t + 1$ hops away.

IHOP uses a number of shared keys:

- Every node shares a master secret key with the base station.
- Each node knows its one-hop neighbors and has established a pairwise key with each of them.
- A node can establish a pairwise key with another node that is multiple hops away if needed.

Further, IHOP also assumes that the base station has a mechanism to authenticate broadcast messages, e.g., $\mu TESLA$.

A clusterhead collects information from its members and sends a report to the base station only when at least $t + 1$ sensors observe the same result.

Meanwhile, a clusterhead also collects the MACs from detecting nodes. Each detecting node sends two MACs to the clusterhead: a MAC using the key shared with the base station, referred to as the individual MAC, and a MAC using the key shared with its upper associate nodes, referred to as the pairwise MAC. The clusterhead then compresses the $t + 1$ individual MACs by XORing them to reduce the size of a report. However, the pairwise MACs are not compressed for transmission. If they were, a node relaying the message would not be able to extract the pairwise MACs of interest to it. Thus, a legitimate report includes $t + 1$ pairwise MACs and a compressed MAC for the base station. When an intermediate node receives a report, it verifies the MAC of its lower associate node. If it fails, the report is eliminated. Otherwise, it removes the MAC, generates a new MAC using its upper associate node pairwise key, and appends it to the report.

IHOP ensures that the base station can detect false data packets when no more than t nodes are compromised. However, the paper does not show how to select the parameter t for a sensor network.

Ye et al. proposed a statistical en-route filtering (SEF) mechanism that can detect and drop false data in [94]. SEF uses a similar key assignment scheme as the basic random key scheme presented in [68]. There is a global key pool and each sensor is pre-installed in a partition selected from the pool. When a stimulus occurs in the fields, the sensors detecting this event elect one of the nodes as the center-of-stimulus (CoS), a node which collects and summarizes the detection results from all detecting nodes and produces a synthesized report on behalf of the group. The CoS generates the report and broadcasts it to all detecting nodes. If a detecting node agrees with the report, it generates a MAC using a key in its partition and sends the MAC to the CoS. The CoS reports the stimulus to the base station only if it receives adequate MACs. A legitimate report carries multiple MACs and a single compromised node cannot fake all MACs. When an en-route node receives the report, it verifies the correctness of the MACs probabilistically and drops those with invalid MACs immediately. Finally, if a report reaches the base station, the base station checks all the MACs and filters out any remaining false reports that escaped the en-route filtering. When a stimulus appears, multiple nodes that detect it collaborate to process the signal and elect the CoS based on the sensing signal strength. The node with the strongest signal stands out as the CoS. To reduce the communication overhead, SEF further uses a Bloom filter [103] to reduce MAC sizes. SEF is designed to protect against injected false information and cannot defend against selective forwarding attacks.

Deng et al. proposed an intrusion-tolerant routing in wireless sensor networks (INSENS) in [104] and further evaluated its performance in [105]. INSENS is a proactive routing protocol. The sensors collect local topology information and send this information back to the base station. The base station generates a forwarding table based on the collected information and sends the routing table to the corresponding sensors. The base station is the central control point for calculating the routing table which relieves the computation load of individual sensors. Protecting against intrusions focuses on three attacks: DoS-type attacks, routing attacks, and select forwarding attacks. To protect against DoS-type attacks, only the base station is allowed to broadcast to the entire network and individual sensors can only send unicast messages. INSENS requires some broadcast authentication scheme such as $\mu TESLA$. Although a compromised node may still alter a valid message and broadcast that message to its neighbors, the damage is restricted to only nearby nodes and the downstream nodes. To protect against routing attacks which propagate erroneous control packets, a symmetric key is chosen for confidentiality and authentication. Further, to protect against select forwarding attacks, data is sent to base stations along two separate paths which are calculated by the base stations in the route discovery step. However, INSENS is built on a table-based routing protocol, and as such depends on the base stations to collect all needed topology information to calculate the forwarding table for each individual sensor. Thus, INSENS is not scalable in large sensor networks.

Wang et al. proposed a scheme to detect whether a node is faulty or malicious with the collaboration of neighbor nodes [106]. In the proposed scheme, when a node suspects that one of its neighbors is faulty, it sends out messages to request the opinions on the behavior of this suspected node from other neighbors of the suspect. After collecting the results, the node analyzes the results to diagnose whether the suspect has a fault. The authors formalized the problem as how to construct a dominating tree to cover all the neighbors of the suspect and further proposed two tree-based propagation collection protocols to construct a dominating tree and collect information via the tree structure.

14.8.2 Open Research Issues

Intrusion detection in WSNs is still largely open to research. Key research issues include the following:

■ Due to the constraints in WSNs, intrusion detection has many aspects not of concern in other network types. The problem of intrusion detection needs to be well defined in WSNs.
■ The proposed IDS protocols in the literature focus on filtering injected false information only [11,94,104]. These protocols need to be improved to address scalability issues.

14.9   Security in WSNs: Future Directions

WSNs are promising solutions for many applications, and security is often a key concern. Although research efforts have been made on cryptography, key management, secure routing, secure data aggregation, and intrusion detection in WSNs, there are still some challenges to be addressed. First, the selection of the appropriate cryptographic methods depends on the processing capability of sensor nodes, indicating that there is no unified solution for all sensor networks. Instead, the security mechanisms are highly application-specific. Second, sensors are characterized by the constraints on energy, computation capability, memory, and communication bandwidth. The design of security services in WSNs must satisfy these constraints. Third, most of the current protocols assume that the sensor nodes and the base station are stationary. However, there may be situations, such as battlefield environments, where the base station and possibly the sensors need to be mobile. The mobility of sensor nodes has a great influence on sensor network topology and thus raises many issues in secure routing protocols. In particular, we identify some of the future directions in the study of security issues in WSNs as follows:

- Exploit the availability of private key operations on sensor nodes: Recent studies on public key cryptography show that public key operations may be practical in sensor nodes. However, private key operations are still too expensive to accomplish in a sensor node. As public key cryptography can greatly ease the design of security in WSNs, improving the efficiency of private key operations on sensor nodes is highly desirable.
- Secure routing protocols for mobile sensor networks: Mobility of sensor nodes has a great influence on sensor network topology and thus on the routing protocols. Mobility can be at the base station, sensor nodes, or both. Current protocols assume the sensor network is stationary. New secure routing protocols for mobile sensor networks need to be developed.
- Continuous stream security in WSNs: Current work on security in sensor networks focuses on discrete events such as temperature and humidity. Continuous stream events such as video and images are not discussed. Video and image sensors for WSNs might not be widely available now, but will likely be in the future. Substantial differences in authentication and encryption exist between discrete events and continuous events, indicating that there will be distinctions between continuous stream security and the current protocols in WSNs.
- QoS and security: Performance is generally degraded with the addition of security services in WSNs. Current studies on security in

WSNs focus on individual topics such as key management, secure routing, secure data aggregation, and intrusion detection. QoS and security services need to be evaluated together in WSNs.

14.10 Summary

As WSNs grow in capability and are used more frequently, the need for security in them becomes more apparent. However, the nature of nodes in WSNs gives rise to constraints such as limited energy, processing capability, and storage capacity. These constraints make WSNs very different from traditional ad hoc wireless networks. As such, special protocols and techniques have been developed for use in WSNs.

While existing surveys in [12–15] discuss security in wireless networks, none focus specifically on security in WSNs and the constraints unique to them. In this chapter, we have surveyed the security issues in WSNs starting with the attacks and countermeasures in each network layer followed by the issues and solutions in cryptography, key management, secure routing, secure data aggregation, and finally intrusion detection. Although the discussed security services certainly add more computation, communication, and storage overhead in WSNs consuming more energy, they are highly desirable and often required in real-world applications.

Acknowledgments

This work is partially supported by NSF grant no. CCR-0311577.

References

[1] D. Estrin, L. Girod, G. Pottie, and M. Srivastava, Instrumenting the world with wireless sensor networks, in *Proceedings of the International Conference on Acoustics, Speech and Signal Processing*, Salt Lake City, Utah, May 2001.

[2] H. Chan and A. Perrig, Security and privacy in sensor networks, *IEEE Computer Magazine*, pp. 103–105, October 2003.

[3] E. Shi and A. Perrig, Designing secure sensor networks, *Wireless Communication Magazine*, vol. 11, no. 6, pp. 38–43, December 2004.

[4] I. F. Akyildiz, W. Su, Y. Sankarasubramaniam, and E. Cayirci, A survey on sensor networks, *IEEE Communications Magazine*, vol. 40, no. 8, pp. 102–114, August 2002.

[5] A. D. Wood and J. A. Stankovic, Denial of service in sensor networks, *IEEE Computer*, vol. 35, no. 10, pp. 54–62, 2002.

[6] N. Gura, A. Patel, A. Wander, H. Eberle, and S. C. Shantz, Comparing elliptic curve cryptography and RSA on 8-bit CPUs, in *CHES '04: Proceedings of Workshop on Cryptographic Hardware and Embedded Systems*, August 2004.

[7] A. S. Wander, N. Gura, H. Eberle, V. Gupta, and S. C. Shantz, Energy analysis of public-key cryptography for wireless sensor networks, in *PerCom '05: Proceedings of the 3rd IEEE International Conference on Pervasive Computing and Communication*, March 2005.

[8] A. Perrig, R. Szewczyk, V. Wen, D. Culler, and J. D. Tygar, SPINS: Security protocols for sensor networks, *Wireless Networks*, vol. 8, no. 5, pp. 521–534, September 2002.

[9] L. Hu and D. Evans, Secure aggregation for wireless networks, in *Workshop on Security and Assurance in Ad hoc Networks*, 2003.

[10] B. Przydatek, D. Song, and A. Perrig, SIA: secure information aggregation in sensor networks, in *SenSys '03: Proceedings of the 1st International Conference on Embedded Networked Sensor Systems*. New York, ACM Press, 2003, pp. 255–265.

[11] S. Zhu, S. Setia, S. Jajodia, and P. Ning, An interleaved hop-by-hop authentication scheme for filtering of injected false data in sensor networks, in *Proceedings of IEEE Symposium on Security and Privacy*, Oakland, CA, May 2004, pp. 259–271.

[12] R. L. Rivest, L. Adleman, and M. L. Dertouzos, On data banks and privacy homomorphisms, in *Foundations of Secure Computation (Workshop, Georgia Institute of Technology, Atlanta, 1977)*, Academic, New York, 1978, pp. 169–179.

[13] F. Stajano and R. J. Anderson, The resurrecting duckling: Security issues for ad-hoc wireless networks, in *Proceedings of the 7th International Workshop on Security Protocols*, London, Springer-Verlag, 2000, pp. 172–194.

[14] Y.-C. Hu and A. Perrig, A survey of secure wireless ad hoc routing, *IEEE Security & Privacy, special issue on Making Wireless Work*, vol. 2, no. 3, pp. 28–39, May/June 2004.

[15] D. Djenouri, L. Khelladi, and N. Badache, A survey on security issues in mobile ad hoc and sensor networks, *IEEE Communications and Surveys and Tutorials*, vol. 7, no. 4, 2005.

[16] P. Levis and D. Culler, Mate: A tiny virtual machine for sensor networks, in *ASPLOS-X: Proceedings of the 10th International Conference on Architectural Support for Programming Languages and Operating Systems*, New York, ACM Press, 2002, pp. 85–95.

[17] EYES project, March 2002–February 2005 [online], available at http://www.eyes.eu.org.

[18] A. Savvides and M. B. Srivastava, A distributed computation platform for wireless embedded sensing, in *ICCD '02: Proceedings of the 2002 IEEE International Conference on Computer Design: VLSI in Computers and Processors*, Washington, DC, IEEE Computer Society, 2002, p. 220.

[19] R. Kling, Intel research mote, in *Network Embedded Systems Technology, Winter 2003 Retreat*, January 15–17 2003.

[20] J. Hill, R. Szewczyk, A. Woo, S. Hollar, D. Culler, and K. Pister, System architecture directions for networked sensors, in *ASPLOS-IX: Proceedings of the 9th International Conference on Architectural Support for Programming Languages and Operating Systems*, New York, ACM Press, 2000, pp. 93–104.

[21] J. Hill, R. Szewczyk, A. Woo, S. Hollar, D. Culler, and K. Pister, System architecture directions for networked sensors, *SIGOPS Operating Systems Review*, vol. 34, no. 5, pp. 93–104, 2000.

[22] S. Slijepcevic, M. Potkonjak, V. Tsiatsis, S. Zimbeck, and M. B. Srivastava, On communication security in wireless ad-hoc sensor networks, in *Proceedings of 11th IEEE International Workshop on Enabling Technologies: Infrastructure for Collaborative Enterprises (WETICE '02)*, 2002, pp. 139–144.

[23] L. Yuan and G. Qu, Design space exploration for energy-efficient secure sensor networks, in *ASAP '02: IEEE International Conference on Application-Specific Systems, Architectures, and Processors*, July 2002, pp. 88–100.

[24] J. Deng, R. Han, and S. Mishra, Enhancing Base Station Security in Wireless Sensor Networks, Department of Computer Science, University of Colorado, Technical report CU-CS-951-03, 2003.

[25] B. Deb, S. Bhatnagar, and B. Nath, Information assurance in sensor networks, in *WSNA '03: Proceedings of the 2nd ACM International Conference on Wireless Sensor Networks and Applications*, New York, ACM Press, 2003, pp. 160–168.

[26] C. Karlof and D. Wagner, Secure routing in wireless sensor networks: Attacks and countermeasures, in *Proceedings of the 1st IEEE International Workshop on Sensor Network Protocols and Applications*, May 2003, pp. 113–127.

[27] J. Newsome, R. Shi, D. Song, and A. Perrig, The sybil attack in sensor networks: Analysis and defenses, in *IPSN '04: Proceedings of IEEE International Conference on Information Processing in Sensor Networks*, April 2004.

[28] Y.-C. Hu, A. Perrig, and D. B. Johnson, Packet leashes: A defense against wormhole attacks in wireless networks, in *Proceedings of IEEE Infocomm 2003*, April 2003.

[29] W. Diffie and M. E. Hellman, New directions in cryptography, *IEEE Transactions on Information Theory*, vol. 22, no. 6, pp. 644–654, November 1976.

[30] R. L. Rivest, A. Shamir, and L. Adleman, A method for obtaining digital signatures and public-key cryptosystems, *Communications of the ACM*, vol. 26, no. 1, pp. 96–99, 1983.

[31] D. W. Carman, P. S. Kruus, and B. J. Matt, Constraints and Approaches for Distributed Sensor Network Security, NAI Labs, Technical report 00-010, 2000.

[32] M. Brown, D. Cheung, D. Hankerson, J. L. Hernandez, M. Kirkup, and A. Menezes, PGP in constrained wireless devices, in *Proceedings of 9th USENIX Security Symposium*, August 2000.

[33] G. Gaubatz, J.-P. Kaps, and B. Sunar, Public key cryptography in sensor networks—Revisited, in *ESAS '04: 1st European Workshop on Security in Ad-hoc and Sensor Networks*, 2004.

[34] M. O. Rabin, Digitalized Signatures and Public-Key Functions as Intractable as Factorization, Cambridge, MA, Technical report, 1979.

[35] J. Hoffstein, J. Pipher, and J. H. Silverman, Ntru: A ring-based public key cryptosystem, in *ANTS-III: Proceedings of the 3rd International Symposium on Algorithmic Number Theory*, London, Springer-Verlag, 1998, pp. 267–288.

[36] V. S. Miller, Use of elliptic curves in cryptography, in *Lecture Notes in Computer Sciences; 218 on Advances in Cryptology—CRYPTO 85*, New York, Springer-Verlag, 1986, pp. 417–426.

[37] N. Koblitz, Elliptic curve cryptosystems, *Mathematics of Computation*, vol. 48, pp. 203–209, 1987.

[38] Elliptic Curve Cryptography, SECG Std. SEC1, 2000, available at www.secg.org/collateral/sec1.pdf.

[39] B. Kaliski, TWIRL and RSA Key Size, RSA Laboratories, Technical note, May 2003.

[40] Recommended Elliptic Curve Domain Parameters, SECG Std. SEC2, 2000, available at www.secg.org/collateral/sec2.pdf.

[41] D. Hankerson, A. Menezes, and S. Vanstone, *Guide to Elliptic Curve Cryptography*, New York, Springer-Verlag, 2004.

[42] A. Freier, P. Karlton, and P. Kocher, The SSL Protocol, Version 3.0., available at http://home.netscape.com/eng/ssl3/.

[43] R. Watro, D. Kong, S. fen Cuti, C. Gardiner, C. Lynn, and P. Kruus, TinyPK: Securing sensor networks with public key technology, in *SASN '04: Proceedings of the 2nd ACM Workshop on Security of Ad hoc and Sensor Networks*, New York, ACM Press, 2004, pp. 59–64.

[44] A. Liu and P. Ning, TinyECC: Elliptic Curve Cryptography for Sensor Networks (version 0.1), September 2005, [online], available at http://discovery.csc.ncsu.edu/software/TinyECC/.

[45] D. J. Malan, M. Welsh, and M. D. Smith, A public-key infrastructure for key distribution in TinyOS based on elliptic curve cryptography, in *Proceedings of the 1st IEEE International Conference on Sensor and Ad Hoc Communications and Networks*, Santa Clara, California, October 2004.

[46] C. Karlof, N. Sastry, and D. Wagner, Tinysec: A link layer security architecture for wireless sensor networks, in *SenSys '04: Proceedings of the 2nd International Conference on Embedded Networked Sensor Systems*, New York, ACM Press, 2004, pp. 162–175.

[47] U.S. National Institute of Standards and Technology (NIST), SKIPJACK and KEA algorithm specifications, *Federal Information Processing Standards Publication 185 (FIPS PUB 185)*, June 1998.

[48] R. L. Rivest, The RC5 encryption algorithm, in *Fast Software Encryption*, B. Preneel, Ed., Springer, 1995, pp. 86–96.

[49] D. Eastlake III and P. Jones, U.S. Secure Hash Algorithm 1 (SHA1), RFC 3174 (Informational), September 2001.

[50] J. Daemen and V. Rijmen, AES proposal: Rijndael, in *Proceedings of 1st AES Conference*, August 1998.

[51] A. J. Menezes, S. A. Vanstone, and P. C. V. Oorschot, *Handbook of Applied Cryptography*, CRC Press, Boca Raton, FL, 1996.

[52] R. L. Rivest, The MD5 Message-Digest Algorithm, RFC 1321, April 1992.

[53] P. Ganesan, R. Venugopalan, P. Peddabachagari, A. Dean, F. Mueller, and M. Sichitiu, Analyzing and modeling encryption overhead for sensor network nodes, in *WSNA '03: Proceedings of the 2nd ACM International Conference on Wireless Sensor Networks and Applications*, New York, ACM Press, 2003, pp. 151–159.

[54] Y. W. Law, S. Dulman, S. Etalle, and P. H. Hartel, Assessing Security-Critical Energy-Efficient Sensor Networks, in *Proceedings of 18th IFIP TC11 International Conference on Information Security, Security and Privacy in the Age of Uncertainty (SEC)*, Athens, Greece: Kluwer Academic Publishers, Boston, May 2003, pp. 459–463.

[55] D. J. Wheeler and R. M. Needham, TEA, a tiny encryption algorithm, in *Proceedings of Fast Software Encryption: Second International Workshop, Lecture Notes in Computer Science*, B. Preneel, Ed., vol. 1008, December 14–16 1994.

[56] R. L. Rivest, M. J. B. Robshaw, R. Sidney, and Y. L. Yin, The RC6 Block Cipher, submitted to NIST as a candidate for the AES.

[57] M. Matsui, New block encryption algorithm misty, in *Proceedings of 4th International Workshop of Fast Software Encryption*, E. Biham, Ed., LNCS 1267, Springer-Verlag, 1997, pp. 54–68.

[58] ETSI/SAGE, Specification of the 3GPP Confidentiality and Integrity Algorithms Document 2: KASUMI Specification, December 1999.

[59] K. Aoki, T. Ichikawa, M. Matsui, S. Moriai, J. Nakajima, and T. Tokita, Specification of Camellia—A 128-Bit Block Cipher, Specification Version 2.0, Nippon Telegraph and Telephone Corporation and Mitsubishi Electric Corporation, 2001.

[60] Y. W. Law, J. M. Doumen, and P. H. Hartel, Benchmarking block ciphers for wireless sensor networks (extended abstract), in *1st IEEE International Conference on Mobile Ad-hoc and Sensor Systems*, IEEE Computer Society Press, October 2004.

[61] S. A. Camtepe and B. Yener, Key distribution mechanisms for wireless sensor networks: A survey, Computer Science Department at RPI, Technical report TR-05-07, 2005.

[62] R. D. Pietro, L. V. Mancini, Y. W. Law, S. Etalle, and P. J. M. Havinga, LKHW: A directed diffusion-based secure multicast scheme for wireless sensor networks, in *ICPPW '03: Proceedings of the 32nd International Conference on Parallel Processing Workshops*, IEEE Computer Society Press, 2003, pp. 397–406.

[63] S. Zhu, S. Setia, and S. Jajodia, LEAP: efficient security mechanisms for large-scale distributed sensor networks, in *CCS '03: Proceedings of the 10th ACM Conference on Computer and Communications Security*, New York, ACM Press, 2003, pp. 62–72.

[64] B. Lai, S. Kim, and I. Verbauwhede, Scalable session key construction protocols for wireless sensor networks, in *IEEE Workshop on Large Scale Real Time and Embedded Systems*, 2002.

[65] S. A. Cametepe and B. Yener, Combinatorial design of key distribution mechanisms for wireless sensor networks, in *Proceedings of 9th European Symposium on Research Computer Security*, 2004.

[66] J. Lee and D. R. Stinson, Deterministic key predistribution schemes for distributed sensor networks, in *Proceedings of Selected Areas in Cryptography*, 2004, pp. 294–307.

[67] J. Lee and D. R. Stinson, A combinatorial approach to key predistribution for distributed sensor networks, in *Proceedings of the IEEE Wireless Communication and Networking Conference*, 2005.

[68] L. Eschenauer and V. D. Gligor, A key-management scheme for distributed sensor networks, in *CCS '02: Proceedings of the 9th ACM Conference on Computer and Communications Security*, New York, ACM Press, 2002, pp. 41–47.

[69] H. Chan, A. Perrig, and D. Song, Random key predistribution schemes for sensor networks, in *Proceedings of IEEE Symposium on Security and Privacy*, May 2003.

[70] D. Liu and P. Ning, Establishing pairwise keys in distributed sensor networks, in *CCS '03: Proceedings of the 10th ACM Conference on Computer and Communications Security*, New York, ACM Press, 2003, pp. 52–61.

[71] R. D. Pietro, L. V. Mancini, and A. Mei, Random key-assignment for secure wireless sensor networks, in *SASN '03: Proceedings of the 1st ACM Workshop on Security of Ad hoc and Sensor Networks*, New York, ACM Press, 2003, pp. 62–71.

[72] W. Du, J. Deng, Y. S. Han, and P. K. Varshney, A pairwise key predistribution scheme for wireless sensor networks, in *CCS '03: Proceedings of the 10th ACM Conference on Computer and Communications Security*, New York, ACM Press, 2003, pp. 42–51.

[73] W. Du, J. Deng, Y. S. Han, S. Chen, and P. K. Varshney, A key management scheme for wireless sensor networks using deployment knowledge, in *Proceedings of IEEE INFOCOM*, Hong Kong, 2004, pp. 586–597.

[74] D. D. Hwang, B. Lai, and I. Verbauwhede, Energy-memory-security tradeoffs in distributed sensor networks, in *Proceedings of 3rd International Conference on Ad-hoc Networks and Wireless*, July 2004, pp. 70–81.

[75] C. Blundo, A. D. Santis, A. Herzberg, S. Kutten, U. Vaccaro, and M. Yung, Perfectly-secure key distribution for dynamic conferences, in *CRYPTO '92: Proceedings of the 12th Annual International Cryptology Conference on Advances in Cryptology*, London, Springer-Verlag, 1993, pp. 471–486.

[76] R. Blom, An optimal class of symmetric key generation systems, in *Proceedings of the EUROCRYPT '84 Workshop on Advances in Cryptology: Theory and Application of Cryptographic Techniques*, New York, Springer-Verlag, 1985, pp. 335–338.

[77] D. Liu and P. Ning, Location-based pairwise key establishments for static sensor networks, in *Proceedings of the ACM Workshop on Security in Ad hoc and Sensor Networks*, October 2003.

[78] J. N. Al-Karaki and A. E. Kamal, Routing techniques in wireless sensor networks: A survey, *IEEE Wireless Communications*, vol. 11, no. 6, pp. 6–28, December 2004.

[79] C. Intanagonwiwat, R. Govindan, and D. Estrin, Directed diffusion: A scalable and robust communication paradigm for sensor networks, in *Mobi-Com '00: Proceedings of the 6th Annual International Conference on Mobile Computing and Networking*, New York, ACM Press, 2000, pp. 56–67.

[80] Y. Yu, R. Govindan, and D. Estrin, Geographical and Energy Aware Routing: A Recursive Data Dissemination Protocol for Wireless Sensor Networks, UCLA Computer Science Department, Technical report UCLA/CSD-TR-01-0023, May 2001.

[81] T. Leinmüller, E. Schoch, F. Kargl, and C. Maihöfer, Influence of falsified position data on geographic ad-hoc routing, in *2nd European Workshop on Security and Privacy in Ad hoc and Sensor Networks (ESAS 2005)*, LNCS, July 2005.

[82] D. Liu and P. Ning, Efficient distribution of key chain commitments for broadcast authentication in distributed sensor networks, in *Proceedings of the 10th Annual Network and Distributed System Security Symposium*, San Diego, CA, February 2003, pp. 263–276.

[83] D. Liu and P. Ning, Multi-level μTESLA: Broadcast authentication for distributed sensor networks, *Transactions on Embedded Computing Systems*, vol. 3, no. 4, pp. 800–836, 2004.

[84] D. Liu, P. Ning, S. Zhu, and S. Jajodia, Practical broadcast authentication in sensor networks, in *MobiQuitous '05: Proceedings of The 2nd Annual International Conference on Mobile and Ubiquitous Systems: Networking and Services*, July 2005, pp. 118–129.

[85] W. Du, R. Wang, and P. Ning, An efficient scheme for authenticating public keys in sensor networks, in *MobiHoc '05: Proceedings of the 6th ACM International Symposium on Mobile Ad hoc Networking and Computing*, New York, ACM Press, 2005, pp. 58–67.

[86] Public-Key Infrastructure (X.509) (pkix) [online], available at http://www.ietf.org/html.charters/pkix-charter.html.

[87] N. Shrivastava, C. Buragohain, D. Agrawal, and S. Suri, Medians and beyond: new aggregation techniques for sensor networks, in *SenSys '04: Proceedings of the 2nd International Conference on Embedded Networked Sensor Systems*, New York, ACM Press, 2004, pp. 239–249.

[88] J. Deng, R. Han, and S. Mishra, Security support for in-network processing in wireless sensor networks, in *SASN '03: Proceedings of the 1st ACM Workshop on Security of Ad hoc and Sensor Networks*, New York, ACM Press, 2003, pp. 83–93.

[89] H. Çam, D. Muthuavinashiappan, and P. Nair, ESPDA: Energy-efficient and secure pattern-based data aggregation for wireless sensor networks, in *Proceedings of IEEE Sensors*, Toronto, Canada, October 2003, pp. 732–736.

[90] H. Çam, D. Muthuavinashiappan, and P. Nair, Energy-efficient security protocol for wireless sensor networks, in *Proceedings of IEEE VTC Conference*, Orlando, Florida, October 2003, pp. 2981–2984.

[91] H. Çam, S. Ozdemir, H. O. Sanli, and P. Nair, *Sensor Network Operations*, Wiley, 2004, ch. Secure Differential Data Aggregation for Wireless Sensor Networks.

[92] W. Du, J. Deng, Y. S. Han, and P. K. Varshney, A witness-based approach for data fusion assurance in wireless sensor networks, in *GLOBECOM '03: Proceedings of IEEE Global Telecommunications Conference*, San Francisco, December 2003, pp. 1435–1439.

[93] D. Wagner, Resilient aggregation in sensor networks, in *SASN '04: Proceedings of the 2nd ACM workshop on Security of Ad hoc and Sensor Networks*, New York, ACM Press, 2004, pp. 78–87.

[94] F. Ye, L. Z. Haiyun Luo, and Songwu Lu, Statistical en-route filtering of injected false data in sensor networks, in *Proceedings of IEEE INFOCOM*, Hong Kong, 2004.

[95] J. Girao, D. Westhoff, and M. Schneider, CDA: Concealed data aggregation for reverse multicast traffic in wireless sensor networks, in *ICC '05: Proceedings of IEEE International Conference on Communications*, Seoul, Korea, May 2005.

[96] C. Castelluccia, E. Mykletun, and G. Tsudik, Efficient aggregation of encrypted data in wireless sensor network, in *Proceedings of ACM/IEEE Mobiquitous*, San Diego, July 2005.

[97] J. Domingo-Ferrer, A provably secure additive and multiplicative privacy homomorphism, *Lecture Notes in Computer Science*, vol. 2433, pp. 471–483, 2002.

[98] D. Wagner, Cryptanalysis of an algebraic privacy homomorphism, in *ISC '03: Proceedings of the 6th Information Security Conference*, Bristol, UK, October 2003.

[99] S. Marti, T. J. Giuli, K. Lai, and M. Baker, Mitigating routing misbehavior in mobile ad hoc networks, in *MobiCom '00: Proceedings of the 6th Annual International Conference on Mobile Computing and Networking*, New York, ACM Press, 2000, pp. 255–265.

[100] Y. Zhang, W. Lee, and Y.-A. Huang, Intrusion detection techniques for mobile wireless networks, *Wireless Networks*, vol. 9, no. 5, pp. 545–556, 2003.

[101] Y. Huang, W. Fan, W. Lee, and P. S. Yu, Cross-feature analysis for detecting ad-hoc routing anomalies, in *ICDCS '03: Proceedings of the 23rd International Conference on Distributed Computing Systems*, Providence, RI, May 2003.

[102] Y. Huang and W. Lee, Attack analysis and detection for ad hoc routing protocols, in *RAIS '04: Proceedings of the 7th International Symposium on Recent Advances in Intrusion Detection*, Sophia Antipolis, France, September 2004.

[103] B. H. Bloom, Space/time trade-offs in hash coding with allowable errors, *Communications of the ACM*, vol. 13, no. 7, pp. 422–426, 1970.

[104] J. Deng, R. Han, and S. Mishra, INSENS: Intrusion-Tolerant Routing in Wireless Sensor Networks, Department of Computer Science, University of Colorado, Technical report CU CS-939-02, November 2002.

[105] J. Deng, R. Han, and S. Mishra, A performance evaluation of intrusion-tolerant routing in wireless sensor networks, in *IPSN '03: Proceedings of IEEE 2nd International Workshop on Information Processing in Sensor Networks*, Palo Alto, California, 2003, pp. 349–364.

[106] G. Wang, W. Zhang, C. Cao, and T. L. Porta, On supporting distributed collaboration in sensor networks, in *Proceedings of MILCOM*, 2003.

[107] Y. Wang, G. Attebury, and B. Ramamurthy, A survey of security issues in wireless sensor networks, in *IEEE Communications Surveys and Tutorials*, Vol. 8, no. 2, pp. 2–23, 2006.

Chapter 15

Key Management in Wireless Sensor Networks

Falko Dressler

Contents

Wireless sensor networks and corresponding applications greatly benefit from the proliferation of energy-aware embedded systems. Various application scenarios have successfully shown that the usage of sensor network technology is applicable in different domains. At the same time, the need for security solutions is rising. This includes mechanisms for secure management and control, e.g., routing and software management, as well as for data communication. Similarly, the demand for higher availability including the protection against attacks and misbehaving nodes emerged. Security architectures have been proposed to address these requirements. All these solutions are based on cryptographic algorithms and appropriate key management and key distribution solutions. The objective of this chapter is to provide an overview to state-of-the-art key management and key distribution techniques. Additionally, a classification of key management and key distribution solutions is provided, followed by an in-depth study of selected key distribution approaches. The chapter also includes an outlook to application scenarios and outlines the open issues for further research on key management and key exchange.

15.1 Introduction

Wireless sensor networks (WSN) have become a major research domain in the communications community [1]. Besides other issues that have been studied so far [2], energy consumption and security were identified to be the most challenging problem spaces. These properties are influenced by the massively distributed operating principle based on self-organization mechanisms [3]. Similarly, the lifetime of sensor networks [4] depends strongly on the operation mode, i.e., the used routing algorithms, the application behavior, and, finally, the employed security methods.

A survey of security issues in ad hoc and sensor networks can be found in [5]. Additional related work in the security area, focused on WSN, is summarized in [6].

The primary requirements on a successful security architecture are availability, authentication, data confidentiality, integrity, and non-repudiation. Most of these objectives can be addressed using cryptographic hash functions and appropriate encryption schemes. In ad hoc and sensor networks, many proposals were published concerning the use of security measures for particular applications [5]. Security protocols such as SPINS [6] define complex architectures to be used in a sensor network environment.

Most of these proposals defer the problem of key management — one of the most sophisticated problems — to be solved elsewhere. Fortunately, several approaches seem to be adequate in this domain as already studied in ad hoc networks [7,8]. In this chapter, we discuss various key management solutions for sensor networks and provide an overview to general key pre-distribution and proactive key exchange solutions. This survey also provides a classification of key management solutions for wireless sensor networks and an outline of open research issues including efficient public-key encryption in sensor networks [9]. Further discussion on key management solutions can be found in [10].

Besides security architectures and special solutions for routing or key management, the aggregation of encrypted data in WSN was discussed [11] as well as the integration of particular security layers for reliable and secured communication [12]. Finally, secure overlays were proposed to address the security concerns in WSN [13].

In summary, it can be said that many promising proposals can be found in the literature that address the security objectives in sensor networks. Nevertheless, most of these papers only outline the principles or use simulation environments for verification. Experimentation on real sensor nodes is necessary to analyze the behavior of proposed security architectures and to contribute to the sensor network security domain.

All approaches for enabling security in WSN are very scenario dependent. There are different requirements, for example, in an agriculture scenario [14] compared to a habitat monitoring scenario [15]. Other requirements appear in the operation and control domain. Sensor nodes must be reconfigured, calibrated, and reprogrammed [16]. Such operations are very sensible for possible attacks. Finally, it must be mentioned that we ignore the problem of key management. Several solutions have been proposed that address this issue, e.g., [17].

The rest of this chapter is organized as follows. Section 15.2 outlines the major security objectives in sensor networks. Then, Section 15.3 discusses application scenarios that strongly depend on security mechanisms, and therefore profit from efficient and secure key management. This is followed by an overview to key management solutions and mechanisms in Section 15.4. Selected key management schemes are presented in detail in Section 15.5. Research challenges and open issues in key management are outlined in Section 15.6. Finally, Section 15.7 concludes the chapter.

15.2 Sensor Network Security Objectives

In this section, we summarize the security properties required by communication networks focusing on the specific capabilities of sensor networks. The necessary security services in sensor networks are not altogether different from those of other networks [5]. The goal of these services is to protect information and resources from attacks and misbehavior. In the context of sensor network security, the following requirements must be ensured for an effective security architecture.

■ Data confidentiality: Ensures that the transmitted data cannot be understood by anyone other than the desired recipient. Concentrating on sensor networks, it is commonly agreed that the level of necessary confidentiality grows with the concentration or aggregation of multiple sensor measures. Confidentiality is typically enabled by applying either symmetric or asymmetric data encryption techniques. Therefore, keys must be exchanged before a transmission can occur.

■ Message authentication: Data or message authentication is of paramount importance for many applications in sensor networks. Technically, message authentication ensures the genuineness of received messages. Also covered is data integrity (see below). Usually, cryptographic hash functions using appropriate key material are used to fulfill this objective. In summary, data authentication ensures that received messages were sent by the expected source and not modified during the transmission.

■ Data integrity: Ensures that the received data was not modified during the transmission. In contrast to message authentication, there is no key material involved in processes to ensure data integrity. Similar cryptographic hash functions can be applied in this context. Looking at the properties of sensor networks, data integrity alone is not sufficient due to the inherent property of multi-hop sensor networks that any node can intercept messages, modify them (including the computation of a new hash value), and transmit the modified messages to the final destination.

A detailed analysis of security solutions for WSN is out of the scope of this discussion. More information on this topic can be found in [5,6,18]. In summary, it can be said that cryptographic hash functions and encryption schemes can be employed to ensure the most prominent security objectives in sensor networks. A prerequisite for this is the exchange of key material. This step must occur before any sensor data can be exchanged.

15.3 Application Scenarios

The security objectives as outlined in the previous section must be considered in various application scenarios for wireless sensor networks. In this section, we summarize selected applications that need to be secured by means of network security solutions. Additionally, we discuss the need for inherently integrating key management solutions into the security approaches to validate the efficiency and performance.

One of the first applications of network security mechanisms was secure routing in ad hoc and sensor networks [18,19]. In most routing protocols, routers exchange information on the topology of the network to establish routes between nodes. Such information could become a target for malicious adversaries who intend to bring the network down. There are two sources of threats to routing protocols. The first comes from external attackers [20]. By injecting erroneous routing information, replaying old routing information, or distorting routing information, an attacker could successfully partition a network or introduce excessive traffic load into the network by causing retransmission and inefficient routing. The second and also the more severe kind of threat comes from compromised nodes, which advertise incorrect routing information to other nodes. Detection of such incorrect information is difficult: merely requiring routing information to be signed by each node would not work, because compromised nodes are able to generate valid signatures using their private keys. Several solutions have been proposed [18,21] that all rely on an efficient key management, including the detection of compromised or malicious nodes, and appropriate revocation mechanisms are strongly demanded.

Similarly, the data dissemination and data forwarding needs to be secured. Proposals such as SPINS [6] address this issue. Key management techniques become even more critical if data must be aggregated, modified, or pre-processed within the network [22,23]. This case was, for example, discussed by Castelluccia and co-workers in their study on efficient aggregation of encrypted data in wireless sensor networks [11]. In this case, every node that receives a packet needs to share a key with the sender to process the message. Key management can easily become unserviceable if too many keys need to be stored in each device or if too many nodes become involved in a single-hop message exchange. We discuss this issue later in Section 15.5. Higher-layer solutions also rely on efficient key management that is assumed to support end-to-end communication as well in a reliable and secure fashion [12].

If software modules are distributed in a sensor network, it must be verified that no attacker will be able to compromise a single node and distribute modified, i.e., infected, software modules. Software management solutions for sensor nodes were discussed in several proposals [16,24,25]. Key management solutions must provide the basis for secured incremental network programming for wireless sensors [25].

Service discovery is a more generalized form of knowledge distribution. If specific services should be announced and used in a dynamic way, it must be ensured that the identity of the service provider is unambiguous and it has not been compromised so far [26]. A case study for secure distributed service directory for wireless sensor networks outlined the needs of key management solutions [27]. In this context, a secure overlay for service-centric sensor networks was proposed [13].

Looking at middleware applications such as service discovery, coordination issues must be considered. Some of the most interesting solutions in the context of ad hoc and sensor networks address security issues, including key management objectives as well as particular challenges that emerge in such massively distributed systems. For example, a distributed coordination framework for wireless sensor and actor networks was proposed [28] as well as a cooperation technique for self-organizing mobile ad hoc networks [29].

15.4   Key Management in Sensor Networks

15.4.1   Overview to Key Management

The organization of key management techniques strongly depends on the selected cryptographic scheme. As mentioned above, we only consider cryptographic hash and encryption mechanisms. In this section, we focus on symmetric schemes that rely on appropriate key exchange and key distribution instead of key verification. In Section 15.6, Open Research Challenges, we give an outlook to issues for key management and verification for asymmetric operations.

Key management includes several functionalities. The most prominent, and in several solutions the only one, is key distribution. Nevertheless, key management is also responsible for issues such as key revocation and re-keying. Additionally, it must ensure resiliency to sensor-node capture. All these issues are outlined in Section 15.4.2. In this sub-section, we present a general classification of key distribution and key exchange solutions.

In theory, key management can be addressed in three ways:

1. Key pre-distribution
2. Proactive key distribution
3. On-demand key exchange

To date, the only practical option for the distribution of keys to sensor nodes in a large-scale sensor network would have to rely on key pre-distribution [30]. Keys would have to be installed in sensor nodes to accommodate secure connectivity between nodes. However, traditional key pre-distribution offers two inadequate solutions: either a single mission key or a set of separate $n-1$ keys, each being pairwise privately shared with another node, must be installed in every sensor node. These and more recent solutions that rely on probabilistic schemes [31] or on deployment information [32] are discussed in Section 15.4.3.

Proactive key distribution stands for key exchange after the deployment of the sensor network, but before any data communication occurs. Proactive solutions usually rely on central base stations that provide the necessary key material. On the other hand and to provide more reliability, probabilistic solutions have been proposed that reduce the necessary keys to a minimum, but still cover secure communication paths between all nodes [33]. Some of the proactive key distribution mechanisms also require some pre-deployment actions such as the computation and selection of key rings to be stored in all nodes [30]. Finally, tree-based key distribution algorithms belong to this domain such as [10,34]. More detailed information on proactive solutions is provided in Section 15.4.4.

Finally, on-demand key exchange mechanisms address the needs of typical applications not to focus on previously exchanged key material, but to set up security relations on demand [35]. Public key solutions can be seen to be on-demand solutions as the verification step takes place after the communication was initiated [36]. In general, there are only a few approaches available that make use of public-key cryptography. The primary reason is the strong resource limitations in sensor networks, e.g., the computational power or the available memory. Novel approaches that counteract these limitations are still works in progress such as [9].

15.4.2 Key Management Issues

In this sub-section, we present the basic features of key management solutions. All solutions for key management basically concentrate on key distribution or key pre-distribution. Nevertheless, issues such as revocation and re-keying must be considered as well.

■ Key distribution: Key distribution is the basis of all key management schemes [30]. It can be solved either by key pre-distribution prior to deployment or proactive in a sensor network prior to any data communication. Key distribution is the main topic of this chapter and is outlined in the following sub-sections.

■ Revocation: When a sensor node is compromised, it is essential to be able to revoke keys associated with this sensor node. This may involve a complete new key distribution in case of a single mission key. Usually, only the according key rings need to be discarded and re-built. Revocation procedures rely on an agreement that defines which keys need to be discarded. In most schemes, a controller node coordinates such a process. If there is no central controller available, election algorithms are used to select a node that performs the necessary tasks.

■ Re-keying: The lifetime of (particular) keys can be limited using expiration times. Although such mechanisms are rarely used in sensor networks, the expiration of keys and the necessary re-keying is a fundamental function in key management solutions. Basically, re-keying is equivalent to a self-revocation of a key by a node. It involves all nodes that share the specific key. Re-keying schemes were categorized into two classes: stateful and stateless [17].

■ Resiliency to sensor-node capture: The unattended operation of sensor nodes in hostile areas raises the possibility of sensor-node capture. Although node capture is a general threat that affects all security mechanisms, key management solutions must be aware of such situations and provide adequate mechanisms to counteract such captures. Basically, similar mechanisms as for general key revocation can be used in this case.

15.4.3 Key Pre-Distribution

Traditional Internet-based key exchange and key distribution protocols require an infrastructure providing trusted third parties. Such approaches are not feasible for large-scale sensor networks because the network topology is not known prior to deployment, the communication range is very limited, and the networks are dynamic in terms of sleep cycles or even node failures. Therefore, most key management approaches are based on key pre-distribution. Keys would have to be installed in sensor nodes to accommodate secure connectivity between nodes. Figure 15.1 depicts well-known key pre-distribution schemes. The intention of key pre-distribution is to make key material available during or before the deployment to minimize subsequent cryptographic overhead for key generation. In the following sub-section, the schemes are explained and discussed.

■ Single mission key: This approach deals with a pre-installed key on all sensor nodes. Usually, this key cannot be changed, and lasts for the whole lifetime of the network. Depending on the scenario, a single mission key might be a feasible approach considering a small network that needs to perform an application with a limited runtime. In any other case, such a solution is inadequate because the capture of any single node may compromise the complete network. Additionally, attacks can be initiated to recover the key using eavesdropped packets. Because all nodes use the same key, an attacker

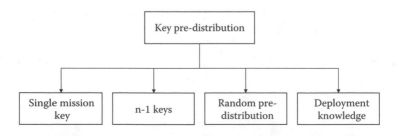

Figure 15.1 Overview of key pre-distribution techniques.

will be able to collect enough data for such an attack in quite a short time. The selective revocation is not possible in this scenario.

■ Set of $n-1$ keys: In contrast to the single mission key approach, the pairwise private sharing of keys between every two sensor nodes avoids the compromising of the entire sensor network upon node capture because selective key revocation becomes possible. However, this solution requires pre-distribution and storage of $n-1$ keys in each sensor node and $n(n-1)/2$ per sensor network. It was shown in [30] that this approach is impractical for sensor networks consisting of more than 10,000 nodes, for both intrinsic and technological reasons. First, pairwise private key sharing between any two sensor nodes would be unusable because direct node-to-node communication is achievable only in small node neighborhoods delimited by communication range and sensor density. Second, incremental addition and deletion as well as re-keying of sensor nodes would become both expensive and complex as they would require multiple keying messages to be broadcast networkwide to all nodes during their non-sleep periods (i.e., one broadcast message for every added/deleted node or re-key operation). Third, a dedicated RAM memory for storing $n-1$ keys would push the on-chip, sensor-memory limits for the foreseeable future, even if only short, 64-bit keys are used and would complicate fast key erasure upon detection of physical sensor tampering. More scalable approaches in this context were proposed in [30,37].

■ Random pre-distribution: The overhead due to the storage requirements for $n(n-1)/2$ keys can, for example, be reduced using randomized techniques. Instead of storing the whole key ring for all $n \times n$ communication relationships, only samples of the complete key ring are stored in each sensor node. To simplify the deployment of the sensor network as well as to allow the adding of nodes at any time without the necessity of key exchange procedures, probabilistic methods can be used to choose part of the key ring for each sensor. Such scenarios were investigated by several groups [30,31,38]. The complexity of such approaches does not lie in the key management, but in the identification of paths through the network that represent trusted chains. In such a chain, two neighboring nodes must share identical keys out of their key ring samples. So the problem of key distribution can be reduced to the problem of path finding or routing. Specific solutions using random subset assignment and grid assignment techniques were studied in [39].

■ Pre-distribution using deployment knowledge: Finally, another approach can be used to reduce the storage requirements known from the set of $n-1$ key solutions, the use of state information. Such solutions exploit the deployment knowledge, i.e., the state of the sensors, to avoid unnecessary key assignments and to reduce the number of required keys that each sensor node should carry. At the same time, it is possible to support higher connectivity and better resilience against node failures. In this context, state information means the classification of sensor node states into active and sleep [32,40]. Using this information, the efficiency of pure probabilistic schemes can be noticeably improved.

15.4.4 Proactive Key Distribution

In contrast to key pre-deployment strategies, proactive key distribution schemes are based on dynamic key generation or key exchange algorithms, respectively. Most of these approaches need to be initialized by a key predeployment mechanism as described above. Afterward, keys can be generated and replaced dynamically. It must be mentioned that the dynamics in proactive solutions are limited. Compared to on-demand algorithms that can create new keys just in time with a forthcoming communication [35], proactive mechanisms need to be executed prior to any data communication, i.e., before the key material might be needed. Figure 15.2 depicts an overview of typical proactive key distribution methodologies. In the following, possible solutions for such schemes are discussed.

■ Base station approach: Bootstrapping any further secured communication can be initiated by selected base stations. Considering typical sensor network architectures, base stations are used to provide connectivity between the sensor network and a fixed communication infrastructure. Therefore, compromising the base station could render the entire sensor network useless. Thus, the base stations are a

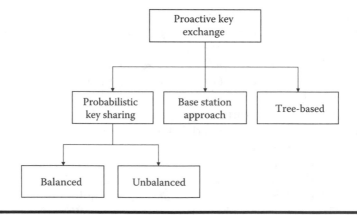

Figure 15.2 Proactive key management techniques.

necessary part of the trusted computing base [6]. A trust setup mimics this, and so all sensor nodes intimately trust the base station: at creation time, each node is given a master key, which is shared with the base station. All other keys are derived from this key.

■ Probabilistic key sharing: Another solution space is again based on probabilistic schemes. Initially, trust is created by the use of subsets of key rings. The subsets can be either balanced, i.e., each node is required to store the same amount of keys [30]. This procedure results in a homogeneous distribution of both, keys and subsequent processing requirements, due to key management actions. Depending on the topology of the sensor network and the communication relationships, e.g., arbitrary communication vs. base station solutions, this approach can lead to unfair exhaustion of resources of single sensor nodes. Additionally, heterogeneity of sensor nodes cannot be exploited, e.g., if the network consists of small nodes with very limited resources and larger ones that are able to store huge amounts of keys. Unbalanced approaches have been discussed that promise to solve this problem [33].

■ Tree-based key management: In many sensor network scenarios, either the communication can be compared to a tree with a single base station or gateway at the root [9] or the deployment follows a hierarchical structure [10]. In both cases, the key management can be adapted to the tree structure to reduce the number of keys that need to be pre-distributed or proactively computed.

15.5   Selected Key Management Schemes

In this section, we provide more details on selected key management schemes. Again, we follow the classification presented in the previous section. Many proposed solutions are constructed on top of each other. Therefore, we try to follow the chronological order as well. The first three methods, i.e., balanced random pre-distribution, unbalanced random pre-distribution, and state-based pre-distribution, can directly be compared in terms of $p(\lambda)$, the probability that two sensors share at least one key after the pre-distribution phase. This parameter is outlined in each sub-section. Afterward, tree-based key distribution is discussed.

15.5.1   *Balanced Random Pre-Distribution in Homogeneous Networks*

Eschenauer and Gligor presented a scheme for key management in distributed sensor networks using probabilistic key sharing and a simple protocol for shared-key discovery and path-key establishment, and for key revocation, re-keying, and incremental addition of nodes [30]. Here, we discuss the three phases key pre-distribution, shared-key discovery, and path-key establishment.

The key pre-distribution phase consists of five offline steps:

1. Generation of a large pool of P keys (e.g., 2^{17}–2^{20} keys) and of their key identifiers
2. Random drawing of k keys out of P without replacement to establish a key ring of a sensor
3. Loading the key ring into the memory of each sensor node
4. Saving key identifiers of a key ring and associated sensor identifier on a trusted controller node
5. For each node, loading the ith controller node with the key shared with that node

This procedure ensures that only a small number of keys need to be placed on each sensor node's key ring to ensure that any two sensor nodes share at least a key with a chosen probability.

The shared-key discovery phase takes place during the sensor network initialization. where every node discovers its neighbors in the wireless communication range with which it shares keys. The simplest way to discover neighboring nodes that share a key with a specific node is to broadcast, in cleartext, the list of identifiers of the keys on the local key ring. Therefore, this phase establishes the topology of the sensor network as seen by the network layer. A link between any two neighboring nodes exists if they share a key. The other way around, if a link exists between two nodes, all communication between these nodes can be secured using appropriate cryptographic algorithms.

The path-key establishment phase finally assigns a path-key to selected pairs of nodes that do not share a key, but are connected by two or more links at the end of the shared-key discovery phase.

Using random graph theory, Eschenauer and Gligor have shown that, given a pool of P keys and randomly choosing k keys for the key ring, the probability p of sharing a key between any two nodes in a neighborhood can be calculated as follows:

$$p = 1 - Pr[\text{two nodes do not share any key}]$$

$$= 1 - \frac{((P-k)!)^2}{(P-2k)!P!} \tag{15.1}$$

In [30], the following numerical example was depicted. Let us assume a sensor network consisting of $n = 10{,}000$ nodes and a desired probability of $P_c = 0.99999$ for obtaining an "almost certainly" connected network, and a wireless communication range that allows the neighborhood connectivity of 40 nodes. Then $k = 250$ out of $P = 100,000$ keys must be stored in each node. If the connectivity increases to 60, only 200 keys are needed.

15.5.2 Unbalanced Random Pre-Distribution in Heterogeneous Networks

Traynor and co-workers demonstrated that a probabilistic unbalanced distribution of keys throughout the network that leverages the existence of a small percentage of more capable sensor nodes can not only provide an equal level of security, but also reduce the consequences of node compromise. They demonstrated the effectiveness of this approach on small networks using a variety of trust models and then demonstrated the application of this method to very large systems [33].

As shown in the previous sub-section, random key pre-deployment in sensor networks has assumed very large random-graph arrangement such that all neighbors within the transmission radius of a given node are reachable. Communication between adjacent nodes is therefore limited only by key matching. This model is not always realistic for a number of reasons. In the unbalanced case, the network now consists of a mix of nodes with different capabilities and missions. The sensing or Level 1 (L1) nodes are assumed to be very limited in terms of memory and processing capability, and perform the task of data collection. Level 2 (L2) nodes have more memory and processing ability. These nodes are equipped with additional keys, and take on the role of routers and gateways between networks.

Again, the connectivity must be analyzed. In the following, n is the number of L1 nodes in a neighborhood, and g is the number of L2 nodes in a neighborhood, where applicable. The scheme for the unbalanced distribution of keys throughout a wireless sensor network builds upon the previously described balanced approach of Eschenauer and Gligor. Given the same generated key pool of size P, we store a key ring of size k keys in each sensor (L1) node, and a key ring of size m keys in each L2 node, where $m \gg k$. Then, the probability of an L2 and L1 having at least one key in common can be calculated as follows:

$$p = 1 - Pr[\text{two nodes do not share any key}]$$

$$= 1 - \frac{(P-k)!(P-m)!}{(P-m-k)!P!} \tag{15.2}$$

Traynor and co-workers demonstrated that their unbalanced approach has similar security capabilities as the balanced case. In a simulation, they have proven that a key ring of 328 keys (considering 40 neighboring nodes) is comparable to 5 L2-nodes with 711 keys and 35 L1-nodes with 30 keys, respectively. Therefore, they achieved a noticeable reduction of the load of typical sensor nodes by exploiting heterogeneous sensor network environments. Additionally, the unbalanced scheme not only reduces the number of transmissions necessary to establish session-keys, but also reduces the effects of both single and multiple node captures. Lastly, the unbalanced scheme allows for even the most memory constrained platforms, from sensor nodes to RFID tags, to hold enough keys to establish secure connections for communication.

15.5.3 State-Based Key Pre-Distribution Supporting Busy–Sleep Cycles

Location information can be facilitated as deployment knowledge for improvement of the previously discussed key pre-distribution schemes. If two sensor nodes are closely located to each other, they have very low probability to be in active-state at the same time. Therefore, unnecessary key assignments can be eliminated because keys shared only between such closely located nodes may be hardly used. In [32,40], Park and co-workers propose a random key pre-distribution scheme that exploits new deployment knowledge, the state of the sensors, to avoid unnecessary key assignments and to reduce the number of required keys that each sensor node must carry while supporting higher connectivity and better resilience against node captures.

In Figure 15.3, an example is shown for key assignments in a sensor network. s_i and k_j (with $i = 1, 2, \ldots$ and $j = 1, 2, \ldots$) denote the sensor nodes and their pre-distributed keys, respectively. Let T_i denote the time-interval when sensor s_i is supposed to be in active-state with high probability. Two sensors, s_1 and s_2, are deployed closely, so they may share more keys as

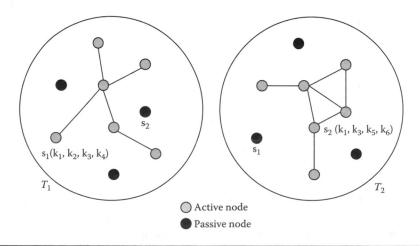

○ Active node
● Passive node

Figure 15.3 Typical key assignments in sensor networks monitored at time T_1 and T_2.

proposed in [32]. Suppose that s_1 and s_2 have key set $\{k_1, k_2, k_3, k_4\}$ and $\{k_1, k_3, k_5, k_6\}$, respectively. During T_1, s_1 and s_2 are in active-state and sleep-state, respectively. Then, as time goes by, s_1 and s_2 transit their states to sleep and active, respectively. If s_1 and s_2 are in active-state at the same time with very low probability, the shared key only between them, $\{k_1, k_3\}$, may be hardly used. Therefore, the key assignments of these keys to s_1 and s_2 are unnecessary.

Park and co-workers used this idea to develop a state-based key management scheme [40]. They assumed that sensor nodes are implemented to be in active-state at specific time-intervals with high probability and in other time-intervals the probability is relatively low. Then, sensor nodes can be grouped by the time-intervals when they have high probabilities to be in active-state. For instance, if sensor s_1 has high probability to be in active-state at time-interval T_1, it may be grouped within the first group. Using these assumptions, the active-state group (ASG) can be defined as the group of sensor nodes with high probability to be in active-state at the same time interval. The calculation of the active-probability is depicted in [40].

For key distribution, Park et al. use two key pools:

1. Global key pool (GlP): A GlP S is a pool of random symmetric keys, from which a group key pool is generated. The cardinality of S is equal to $|S|$.
2. Group key pool (GrP): A GrP S_i is a subset of GlP S for ith group, from which a key ring is generated. The cardinality of S_i is equal to $|S_G|$.

These pools are used for the key pre-distribution phase. Assuming L groups defined during the modeling of the ASG, the key server generates a large GlP S and divides it into L GrPs S_i for each ASG G_i. The purpose of setting up the GrP is to allow the time-neighbor ASGs to share more keys. After completing the GrP setup, for each sensor node j in ASG G_i, a randomly selected key ring $R_{j,i}$ from its corresponding GrP S_i is loaded into the memory of the sensors. For the assignment, an overlapping factor a is used that determines a certain number of common keys between two nearby time-interval groups. Because keys selected from the other groups are all distinct, the sum of all the number of keys should be equal to $|S|$. Therefore, $|S_G|$ can be calculated as follows:

$$|S_G| = \frac{|S|}{L - aL + a} \tag{15.3}$$

The probability that two sensors share at least one common key can be expressed as $1 - Pr$ [two nodes do not share any key]. Because the size of GrP is $|S_G|$, the number of keys shared between two GrPs is $\lambda |S_G|$, where is λ is 1, a, or 0. According to the value of λ, we should consider three cases for finding the required probability: two sensors come from same group ($\lambda = 1$), the neighbor two groups ($\lambda = a$), and the different groups which are not neighbors of each other ($\lambda = 0$). The same overlapping key pool method used in [32] can be adopted. The first node selects i keys from the $\lambda|S_G|$ shared keys; it then selects the remaining $R - i$ keys from the non-shared keys. The second node selects R keys from the remaining $|S_G| - i$ keys from its GrP. Therefore, $p(\lambda)$, the probability that two sensors share at least one key when their GrPs have $\lambda|S_G|$ keys in common, can be calculated as:

$$p(\lambda) = 1 - Pr[\text{two nodes do not share any key}]$$

$$= 1 - \frac{\displaystyle\sum_{i=0}^{min(R,\lambda|S_G|)} \binom{\lambda|S_G|}{i}\binom{(1-\lambda)|S_G|}{R-i}\binom{|S_G|-i}{R}}{\binom{|S_G|}{R}^2} \quad (15.4)$$

A detailed performance analysis of this approach is presented in [40]. In many scenarios, this scheme offers a better performance compared to the approaches from Eschenauer and Gligor [30] and Du et al. [32].

15.5.4 Tree-Based Key Distribution

Chen and Drissi contributed to the proactive key management by arranging the sensor nodes in a hierarchical form [10]. They express the communication in a sensor network in a well-structured way and provide several application examples that support and confirm this approach. Given such a hierarchical design of a sensor network as depicted in Figure 15.4, two forms of communication are necessary: between neighboring nodes at the

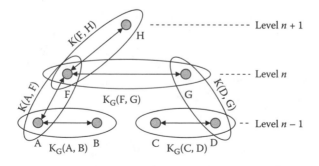

Figure 15.4 Hierarchical or tree-based organization of sensors and the according keys.

same level n (and the same group) and between sensors and their direct leaders in the next higher level $n+1$.

Appropriate keys must be distributed according to the communication paths in the network. Chen et al. propose the following scheme in which all nodes (except leaves and the root) are given four types of keys, namely, the group key (only one), the uplevel pairwise key (only one), the downlevel group key (only one), and the downlevel pairwise key (can be many). These keys and their usage are described in the following. Hereby, we follow the notation as used in Figure 15.4.

- Group key: The group key must be known by each group member to communication in the direct neighborhood, i.e., in the local group. Examples are nodes A and B, C and D, and F and G, respectively. A and B belong to the same group. Therefore, they must share the key $K_G\{A, B\}$ for secure communication. This group key must also be known by the direct group leader, i.e., node F in our example. This knowledge is used for key management and command issues instead of data communication.
- Downlevel group key: The downlevel group key is the same key as the group key described above. This key is only used for command purposes, e.g., key management issues for sensor node addition, replacement, and deletion.
- Uplevel pairwise key: Communication between disjunctive groups must occur via the network-inherent hierarchy, e.g., communication between A and C must use node F as a gateway. Therefore, each sensor node must share a private key with its uplevel group leader. Examples are pairwise keys $K\{A, F\}$ between nodes A and F and $K\{F, H\}$ between F and H.
- Downlevel pairwise key: This key was is the same as the uplevel pairwise key, but seen from the different angle.

As already mentioned, the communication paths follow the hierarchy as do the key sharings. If node A wants to send a message to D, the following transmissions will occur: A→F using $K\{A, F\}$, F→G using $K_G\{F, G\}$, and G→D using $K\{D, G\}$.

Considering the performance of this approach, we examine the amount of keys necessary for communication and key management in such a hierarchical design. As described in [10], a network of n sensor nodes with a depth of the tree of d (assuming a complete tree) results in $\log_d n$ sensor nodes per group. Each leaf sensor only needs to store two keys; the root sensor needs to store approximately $\log_d n + 1$ keys. All the other nodes need to store about $\log_d n + 3$ keys. Therefore, the key storage requirement is $O(\log_d n)$.

A similar tree-based approach for secure key distribution is described by Bla et al. [34]. In this work, the primary objective is on securely integrating new nodes in an existing tree. Additionally, the hierarchical structure is not based on a pre-defined setup, but on the real communication paths that can be observed in the network.

15.6 Open Research Challenges

The typical hardware and software constraints make it impractical to use the majority of the current secure algorithms, which were designed for powerful workstations. For example, the working memory of a sensor node is insufficient even to hold the variables (of sufficient length to ensure security) that are required in asymmetric cryptographic algorithms (e.g., RSA and Diffie–Hellman), let alone perform operations with them [6]. A particular challenge is broadcasting authenticated data to the entire sensor network. Current proposals for authenticated broadcast are impractical for sensor networks. First, most proposals rely on asymmetric digital signatures for the authentication, which are impractical for multiple reasons (e.g., long signatures with high communication overhead of 50 to 1000 bytes per packet, very high overhead to create and verify the signature). The main problem of any public key-based security system is to make each user's public key available to others in such a way that its authenticity is verifiable. In mobile ad hoc networks, this problem becomes even more difficult to solve because of the absence of centralized services and possible network partitions. More precisely, two users willing to authenticate each other are likely to have access only to a subset of nodes of the network (possibly those in their geographic neighborhood). Self-organized public key management is a first approach to address the security requirements in a scalable way [36]. On the other hand, cryptographic primitives are the fundamental building blocks of every secure protocol and the knowledge of algorithm usability is crucial for the design of new protocols for sensor networks. More acceptable encryption schemes using elliptic curve cryptography are proposed in [9].

Broadcast authentication is another problem. Even previously proposed purely symmetric solutions for broadcast authentication are impractical: Gennaro and Rohatgi's initial work required over 1 KB of authentication information per packet [41], and Rohatgi's improved k-time signature scheme requires over 300 bytes per packet [42]. Perrig et al. implemented the necessary primitives [6]. The available computational resources are usually very limited and often not concerned security solutions. A typical performance evaluation must employ adequately calibrated simulation models [43]. In this reference, measurements of typical sensor nodes are depicted that show that even symmetrical cryptography has practical limitations in real sensor networks.

A common characteristic of sensor networks is their severely limited energy supply. Ultimately, the available energy determines that, for example, base stations differ from nodes in having longer-lived energy supplies and having additional communications connections to outside networks. To minimize the energy usage, a security sub-system should place minimal requirements on the processor, and add minimal information to each message transmitted. On the other hand, the limited lifespan of each node limits the lifetime of usable keys. Given the severe hardware and energy constraints, we must be careful in the choice of cryptographic primitives and the security protocols in the sensor networks.

Key agreement is necessary based on scalable and efficient solutions. In [44], three approaches to the problem of user-friendly key agreement (and mutual authentication) in settings where the users do not share any authenticated information in advance were proposed. The first approach belongs to the family of solutions requiring the users to compare strings of words, whereas the other two approaches are based on radio channel specific techniques, namely, distance-bounding and integrity-codes (I-codes). Scalable key management with inherent self-configuration will allow the deployment of even larger networks [45].

Last but not least, group key management including group re-keying mechanisms for sensor networks are needed. Most existing group re-keying schemes are not suitable for sensor networks because they have large overhead and are not scalable. This problem was addressed by a family of pre-distribution and local collaboration-based group re-keying (PCGR) schemes [17]. These schemes are designed based on the ideas that future group keys can be preloaded to the sensor nodes before deployment, and neighbors can collaborate to protect and appropriately use the preloaded keys.

In summary, the following research aspects and challenges for key management solutions can be formulated:

■ Energy-aware key management
■ Public key management (key infrastructure)
■ Feasible public key cryptography
■ Key agreement mechanisms
■ Group key management

15.7 Conclusion

Security issues in wireless sensor networks have been studied by various groups to fulfill the raising demands of applications in this domain. In these works, special requirements on security solutions have been identified that are correlated to the specific characteristics of sensor networks (strongly limited resources in terms of processing and storage capacity, communication bandwidth, and energy). Based on the results, many proposals for security in WSNs are available that focus on routing, data aggregation, and cooperation issues. All of them rely on appropriate key management solutions that must be made available for sensor network installations.

In this chapter, we presented an overview to key management and key distribution approaches for application in wireless sensor networks. We started with a first categorization of key management solutions in the area of WSN. Basically all proposals are based on efficient key pre-distribution or proactive key exchange supporting symmetric cryptographic techniques. The different classes can be distinguished by the presumed knowledge about network topology and routing mechanisms.

Based on this classification, we described selected examples in detail to demonstrate the basic principles of the available solutions. We added a brief discussion on the performance to each of these mechanism.

Besides a few academic proposals and testbeds, asymmetric solutions cannot be found in sensor networks. There are two reasons for this observation: first, asymmetric cryptographic operations cannot be efficiently used in small embedded systems and, second, to date there is no public key infrastructure available for use in wireless sensor networks.

Finally, we also provided a section outlining open issues and challenges in the domain of security in WSN focusing on key management. This roundup is intended to motivate further research work in this domain.

References

[1] I. F. Akyildiz, W. Su, Y. Sankarasubramaniam, and E. Cayirci, A survey on sensor networks, *IEEE Communications Magazine*, vol. 40, no. 8, pp. 102–116, August 2002.

[2] C.-Y. Chong and S. P. Kumar, Sensor networks: Evolution, opportunities, and challenges, *Proceedings of the IEEE*, vol. 91, no. 8, pp. 1247–1256, August 2003.

[3] F. Dressler, Self-Organization in Ad Hoc Networks: Overview and Classification, University of Erlangen, Dept. of Computer Science 7, Technical report 02/06, March 2006.

[4] F. Dressler and I. Dietrich, Lifetime analysis in heterogeneous sensor networks, in *9th EUROMICRO Conference on Digital System Design — Architectures, Methods and Tools (DSD 2006)*, Dubrovnik, Croatia, August 2006, pp. 606–613.

[5] D. Djenouri and L. Khelladi, A survey of security issues in mobile ad hoc and sensor networks, *IEEE Communication Surveys and Tutorials*, vol. 7, no. 4, pp. 2–28, December 2005.

[6] A. Perrig, R. Szewczyk, V. Wen, D. Culler, and J. D. Tygar, SPINS: Security protocols for sensor networks, *Wireless Networks*, vol. 8, no. 5, pp. 521–534, September 2002.

[7] B. Wu, J. Wu, E. B. Fernandez, and S. Magliveras, Secure and efficient key management in mobile ad hoc networks, in *1st International Workshop on Systems and Network Security (SNS 2005)*, April 2005.

[8] J.-P. Hubaux, L. Buttyan, and S. Capkun, Security, testbeds and applications: The quest for security in mobile ad hoc networks, in *ACM International Symposium on Mobile and Ad Hoc Networks (ACM MobiHoc)*, October 2001.

[9] F.-O. Bla and M. Zitterbart, Towards acceptable public-key encryption in sensor networks, in *the 2nd International Workshop on Ubiquitous Computing (ACM SIGMIS)*, May 2005.

[10] X. Chen and J. Drissi, An efficient key management scheme in hierarchical sensor networks, in *2nd IEEE International Conference on Mobile Ad Hoc and Sensor Systems (IEEE MASS 2005): International Workshop on Wireless and Sensor Networks Security (WSNS '05)*, Washington, DC, November 2005.

[11] C. Castelluccia, E. Mykletun, and G. Tsudik, Efficient aggregation of encrypted data in wireless sensor networks, in *Mobile and Ubiquitous Systems: Networking and Services (MobiQuitous 2005)*, July 2005, pp. 109–117.

[12] F. Dressler, Reliable and semi-reliable communication with authentication in mobile ad hoc networks, in *2nd IEEE International Conference on Mobile Ad Hoc and Sensor Systems (IEEE MASS 2005): International Workshop on Wireless and Sensor Networks Security (WSNS '05)*, Washington, DC, November 2005, pp. 781–786.

[13] H.-J. Hof, E.-O. Bla, and M. Zitterbart, Secure overlay for service centric wireless sensor networks, in *1st European Workshop on Security in Ad-Hoc and Sensor Networks (ESAS 2004)*, August 2004.

[14] A. Baggio, Wireless sensor networks in precision agriculture, in *ACM Workshop on Real-World Wireless Sensor Networks (REALWSN 2005)*, Stockholm, Sweden, June 2005.

[15] A. Mainwaring, J. Polastre, R. Szewczyk, D. Culler, and J. Anderson, Wireless sensor networks for habitat monitoring, in *1st ACM Workshop on Wireless Sensor Networks and Applications*, Atlanta, September 2002.

[16] G. Fuchs, S. Truchat, and F. Dressler, Distributed software management in sensor networks using profiling techniques, in *1st IEEE/ACM International Conference on Communication System Software and Middleware (IEEE/ACM COMSWARE 2006): 1st International Workshop on Software for Sensor Networks (SensorWare 2006)*, New Dehli, India, January 2006, pp. 1–6.

[17] W. Zhang and G. Cao, Group rekeying for filtering false data in sensor networks: A predistribution and local collaboration-based approach, in *24th IEEE Annual Joint Conference of the IEEE Computer and Communications Societies (IEEE INFOCOM 2005)*, March 2005, pp. 503–514.

[18] L. Zhou and Z. J. Haas, Securing ad hoc networks, *IEEE Network*, vol. 13, no. 6, November/December 1999.

[19] B. R. Smith, S. Murphy, and J. J. Garcia-Luna-Aceves, Securing distance-vector routing protocols, in *Symposium on Network and Distributed Systems Security*, Los Alamitos, CA, February 1997, pp. 85–92.

[20] C. Karlof and D. Wagner, Secure routing in wireless sensor networks: Attacks and countermeasures, in *Workshop on Sensor Network Protocols and Applications*, 2003.

[21] K. Sanzgiri, B. Dahill, B. N. Levine, C. Shields, and E. Belding-Royer, A secure routing protocol for ad hoc networks, in *International Conference on Network Protocols (ICNP)*, November 2002.

[22] L. Hu and D. Evans, Secure aggregation for wireless sensor networks, in *Workshop on Security and Assurance in Ad hoc Networks*, 2003.

[23] B. Przydatek, D. Song, and A. Perrig, SIA: Secure information aggregation in sensor networks, in *ACM SenSys*, 2003, pp. 255–265.

[24] D. Culler, J. Hill, P. Buonadonna, R. Szewczyk, and A. Woo, A network-centric approach to embedded software for tiny devices, in *1st International Workshop on Embedded Software (EMSOFT 2001)*, Tahoe City, CA, October 2001.

[25] J. Jeong and D. Culler, Incremental network programming for wireless sensors, in *1st IEEE International Conference on Sensor and Ad hoc Communications and Networks (IEEE SECON)*, June 2004.

[26] F. Almenarez and C. Campo, SPDP: A secure service discovery protocol for ad hoc networks, in *9th Open European Summer School and IFIP Workshop on Next Generation Networks*, Budapest, Hungary, 2003.

[27] H.-J. Hof, E.-O. Bla, T. Fuhrmann, and M. Zitterbart, Design of a secure distributed service directory for wireless sensor networks, in *1st European Workshop on Wireless Sensor Networks*, January 2004.

[28] T. Melodia, D. Pompili, V. C. Gungor, and I. F. Akyildiz, A distributed coordination framework for wireless sensor and actor networks, in *6th ACM International Symposium on Mobile Ad Hoc Networking and Computing (ACM Mobihoc 2005)*, Urbana-Champaign, Il, May 2005, pp. 99–110.

[29] L. Buttyn and J.-P. Hubaux, Stimulating cooperation in self-organizing mobile ad hoc networks, *Mobile Networks and Applications*, vol. 8, no. 5, pp. 579–592, October 2003.

[30] L. Eschenauer and V. D. Gligor, A key-management scheme for distributed sensor networks, in *9th ACM Conference on Computer and Communication Security (ACM CCS)*, Washington, DC, November 2002.

[31] H. Chan, A. Perrig, and D. Song, Random key management predistribution schemes for sensor networks, in *IEEE Symposium on Research in Security and Privacy*, 2003.

[32] W. Du, J. Deng, Y. S. Han, S. Chen, and P. Varshney, A key management scheme for wireless sensor networks using deployment knowledge, in *IEEE Infocom 2004*, March 2004, pp. 586–597.

[33] P. Traynor, H. Choi, G. Cao, S. Zhu, and T. L. Porta, Establishing pairwise keys in heterogeneous sensor networks, in *25th IEEE Conference on Computer Communications (IEEE INFOCOM 2006)*, Barcelona, Spain, April 2006.

[34] E.-O. Bla, M. Conrad, and M. Zitterbart, A tree-based approach for secure key distribution in wireless sensor networks, in *The REALWSN*, June 2005.

[35] N. Asokan and P. Ginzboorg, Key agreement in ad hoc networks, *Computer Commmunications*, vol. 23, pp. 1627–1637, 2000.

[36] S. Capkun, L. Buttyn, and J.-P. Hubaux, Self-organized public-key management for mobile ad hoc networks, *IEEE Transactions on Mobile Computing*, vol. 2, no. 1, pp. 52–64, January 2003.

[37] W. Du, J. Deng, Y. S. Han, and P. Varshney, A pairwise key predistribution scheme for wireless sensor networks, in *10th ACM Conference on Computer and Communications Security (CCS)*, October 2003, pp. 42–51.

[38] S. Zhu, S. Xu, S. Setia, and S. Jajodia, Establishing pair-wise keys for secure communication in ad hoc networks: A probabilistic approach, in *IEEE International Conference on Network Protocols (ICNP)*, November 2003.

[39] D. Liu and P. Ning, Establishing pairwise keys in distributed sensor networks, in *10th ACM Conference on Computer and Communications Security*, Washington DC, October 2003, pp. 52–61.

[40] J. Park, Z. Kim, and K. Kim, State-based key management scheme for wireless sensor networks, in *2nd IEEE International Conference on Mobile Ad Hoc and Sensor Systems (IEEE MASS 2005): International Workshop on Wireless and Sensor Networks Security (WSNS '05)*, Washington, DC, November 2005.

[41] R. Gennaro and P. Rohatgi, How to sign digital streams, in *Advances in Cryptology — Crypto '97*, vol. LNCS 1294, Berlin, Germany, 1997, pp. 180–197.

[42] P. Rohatgi, A compact and fast hybrid signature scheme for multicast packet authentication, in *6th ACM Conference on Computer and Communication Security*, November 1999.

[43] M. Passing and F. Dressler, Experimental performance evaluation of cryptographic algorithms on sensor nodes, in *3rd IEEE International Conference on Mobile Ad Hoc and Sensor Systems (IEEE MASS 2006): 2nd IEEE International Workshop on Wireless and Sensor Networks Security (WSNS '06)*, Vancouver, Canada, October 2006, pp. 882–887.

[44] M. Cagalj, S. Capkun, and J.-P. Hubaux, Key agreement in peer-to-peer wireless networks, *Proceedings of the IEEE (Special Issue on Cryptography and Security)*, vol. 94, no. 2, pp. 467–478, February 2006.

[45] F. Liu and X. Cheng, A self-configured key establishment scheme for large-scale sensor networks, in *3rd IEEE International Conference on Mobile Ad Hoc and Sensor Systems (IEEE MASS 2006)*, Vancouver, Canada, October 2006, pp. 447–456.

Index

A

Access control, *See* Authorization; MAC address security; Medium access control (MAC) layer; *specific mechanisms*

Access Control Lists (ACLs), 367
  IEEE 802.15.4 networks, 417–418
  MAC spoofing vulnerability, 117
  ZigBee security vulnerabilities, 372–374

Access/One, 39, 77, 93–94

Access points (APs), 6, 48, 384
  auto-configurability, *See* Auto-configuration
  friend nodes, 274–275
  hand-off mechanisms, 29
  home networking, 10
  IEEE 802.11i authentication model, 272–275
  intrusion detection issues, 153
  secure routing approach, 178

Acknowledgment (ACK), 21, 101, 449

Adaptive Robust Tree (ART), 63–65

Ad hoc networks, 7–8, 47, *See also* Mobile ad hoc networks
  energy-aware protocols, 30
  energy constraints, 7–8
  IEEE 802.11 mode, 35
  transport layer protocols, 23–24

Ad hoc On-Demand Distance Vector (AODV)
  AODVSTAT, 160–161
  AOSR performance vs., 314–317
  hop count, 20–21
  IEEE 802.11s, 71, 175
  message formats and mutable fields, 188–189
  routing security issues, 179–182
  rushing attack vulnerability, 119
  SAODV, *See* Secure AODV
  secure extensions, 191–193
  security flaws of, 181–182
  subnet communications, 180
  trusted routing (TCAODV), 286–287

Ad-hoc On-demand Secure Routing (AOSR), 298, 306–310
  performance evaluation, 313–317
  security analysis, 310–313

Ad hoc QoS Routing (AQOR) protocol, 73

Administrative distances, 178, 336

Admission control, 74–75

Advanced Encryption Standard (AES), 354
  AES-CBC-MAC, 419
  AES-CCM, 368–369, 386, 419
  counter mode (AES-CTR), 418
  IEEE 802.11i standard, 288
  IEEE 802.15.4 standard, 354–355, 418–419
  ZigBee, 369–370, 424

Aggregator nodes, 473, 475

Alarms, 149, 159

Algebraic attacks, 364

Algorithmic key attacks, 389–390

Analog-to-digital converters (ADCs), 439

Anomaly detection, 151–152, 155–156, 165, 479

Anonymity, 287

Antenna technologies, 9, 16–17, 99

AODV, *See* Ad hoc On-Demand Distance Vector

AODVSTAT, 160–161